SYSTEMS ANALYSIS AND DESIGN

SYSTEMS ANALYSIS AND DESIGN

H. L. CAPRON

The Benjamin/Cummings Publishing Company, Inc.

Reading, Massachusetts • Menlo Park, California
Don Mills, Ontario • Wokingham, U.K. • Amsterdam • Sydney
Singapore • Tokyo • Mexico City • Bogota • Santiago • San Juan

Sponsoring Editor: Sally Elliott
Production Editor: Hal Lockwood, Bookman Productions
Production Supervisor: Mary Picklum
Copy Editor: Carol Dondrea
Interior and Cover Designer: Hal Lockwood
Illustration: Cyndie Clark-Huegel
Photo Research: Monica Suder
Composition: Graphic Typesetting Service

Benjamin/Cummings Publishing Company has made every attempt to supply trade-mark information about company names, products, and services mentioned in this book. The trademarks listed at the end of the book were derived from various sources. Benjamin/Cummings Publishing Company cannot attest to the accuracy of this information.

Library of Congress Cataloging-in-Publication Data

Capron, H. L.
  Systems analysis and design.

  Includes index.
  1. System analysis.   I. Title.
QA402.C357   1986   003   85-30797
ISBN 0-8053-2241-8

GHIJ-DO-943210

The Benjamin/Cummings Publishing Company, Inc.
2727 Sand Hill Road
Menlo Park, California 94025

To my mother,

EMER GILLICK CAPRON,

who raised her children to be independent

Preface

This text covers the systems analysis and design content recommended by the DPMA Model Curriculum, but it does not stop there. In addition to the core material, the text presents information on many necessary extra topics such as microcomputers in systems, the information center, and office automation.

◇ ORGANIZATION

The text is organized into six parts. Part 1, An Introduction to Systems Thinking, discusses the characteristics of a system and then introduces the key systems concept, the systems life cycle. Part 2, Initial Phases of the Systems Life Cycle, begins with preliminary investigation and systems analysis. Both of these phases involve background work that is necessary before any action can be taken. Part 3, Systems Design, thoroughly covers the design process, from preliminary design through the detailed design of output, input, processing, and files. Part 4, Final Phases of the Systems Life Cycle, completes the systems life cycle with the systems development and implementation phases. In Part 5, System Control, the critical elements of project management and system security are examined. Part 6, System Trends Today, provides detailed discussion on key topics such as the impact of microcomputers in systems, the information center, and office automation systems.

There is a strong case for covering the chapters in the book in the order presented. However, let's examine this order and then mention other possibilities:

- Chapter 1 is introductory and can be covered quickly.
- Chapter 2 provides an important overview of the systems life cycle and should be presented early in the quarter or semester—and with care.
- Chapter 3 gives the student a glimpse of the analyst life; this chapter can be absorbed without a lot of guidance if you are anxious to move to the meat of the book.

- Chapter 4 presents the subject of system documentation by explaining its rationale and techniques in some detail. Since documentation pervades every phase of the systems life cycle and, indeed, is a critical facet of the analyst life, we think that the students should understand its importance right away. This chapter also provides information on presentations.
- Chapters 5 through 13 cover the systems life cycle in great detail: Chapter 5 is on preliminary investigation, Chapter 6 on analysis, Chapters 7 through 11 on design, Chapter 12 on development, and Chapter 13 on implementation.
- The next two chapters relate to control: Chapter 14 is on project management and Chapter 15 on security.
- The remaining chapters present supplementary topics: Chapter 16 discusses microcomputers, Chapter 17 the information center, and Chapter 18 office automation.

Some variations are possible and perhaps desirable in individual classroom circumstances. For example, some instructor will want to cover the overview of the systems life cycle in Chapter 2, then proceed directly to the details of the cycle in Chapters 5 through 13; it is possible to do so without any loss in continuity. Instructors who plan to have their students working on a project related to microcomputers may wish to cover Chapter 16 at an earlier date. Instructors who want their students to experiment with project management software may want to have them read Chapter 14 early in the term; this could be done immediately after Chapter 2 with little loss of continuity.

◇ IMPORTANT FEATURES

The most fundamental reason for using *Systems Analysis and Design* is that it presents valuable material in a very readable style. The students will enjoy reading the book and will retain a high percentage of what they read. In addition, instructors will find all of the elements they have been asking for in this text and its supplements. Listed below are key pedagogical features that make this text enjoyable to read:

- Independent chapter minicases
- Ongoing King Books case study
- Boxes with relevant material
- Boldface key terms
- Photos
- Clear line drawings

- Chapter summaries
- Review questions
- Lists of key terms
- Glossary
- Index
- References list

◇ SUPPLEMENTS PACKAGE

The supplements package is the most complete of any systems analysis and design text on the market. This package includes a full, separate case study book and answers, free project management software from Microsoft, an instructor's guide, and a complete test bank. Let us now examine each of these items in detail.

Case Study Book

Many instructors want to give their students the opportunity to work in a real world environment within the classroom. The student case book, *The Pacific Health Club,* written by H. L. Capron and Patricia L. Clark, provides a simulated case situation. This case study is optional and may be purchased separately from the text. However, it does follow the material presented in the text and is offered as a true supplement to the text.

The case study presents a scenario of a business with some problems that can be solved with an automated solution. The student is given sufficient background to carry out assignments in the systems life cycle, especially preliminary investigation, analysis, and design. Answers for case study assignments are provided in the instructor's guide.

Software

We offer (free of charge to all schools adopting the text) a software tutorial called *Learning Microsoft Project*. This self-paced tutorial provides a hands-on introduction to a leading project management software package for the IBM PC. Students will learn how the software works and how they might use Microsoft *Project* in their work to manage schedules and resources.

Instructors may make copies of *Learning Microsoft Project* for student use. The tutorial consists of 30 independent lessons, which a student can follow with no documentation and a minimum of supervision. The program keeps track of each student's progress through the lessons. Suggestions on the use of the software are provided in the instructor's guide.

Instructor's Guide

To help instructors, particularly part-time instructors, the instructor's guide by H. L. Capron offers many insights on the subject matter as well as practical suggestions for getting the material across to students. The instructor's guide has three parts:

Part I: Chapter Notes. For each chapter in the book, the instructor's guide presents:

- Learning objectives

- Chapter overview
- Chapter outline
- Suggested answers to chapter overview questions
- Discussion of text case studies

Part II: Guide for Using Software. This section provides the instructor with information for use of the Microsoft project management software that is available for classroom use as a supplement to the text.

Part III: Answers for Student Case Study. Complete answers are provided for the assignments in the separate student case study.

Test Bank

There is a complete set of objective and essay test questions. Answers to objective questions are supplied. Each chapter asks the following types of questions:

- True/false
- Multiple choice
- Completion
- Short answer
- Essay

This test bank, together with individual and/or team exercises, should provide a sufficient basis for student evaluation.

◇ SPECIAL NOTE TO THE STUDENT

We have tried to speak directly to you, in ways that will prepare you for your role in the analysis and design of systems. We think this book will give you needed tools and expertise, broaden your background, and raise your confidence level. We wish you well in your career.

We welcome your reactions to this book. Any comments, favorable or otherwise, will be read with care. Write in care of the publisher, whose address is listed on the copyright page. All letters that supply a return address will be answered.

◇ ACKNOWLEDGMENTS

Many people contribute to make a successful book, and this one is no exception. I would like to thank them now.

The employees of Benjamin/Cummings have been competent and creative, each filling a niche in a special way. Sponsoring editor Sally Elliott has been the guiding hand on this project, and I thank her for her support and patience. Associate editor Jenny DeGroot managed the supplements with a finesse that made it look easy. Other leading players who made valuable and timely contributions are copy editor Carol Dondrea, photo researcher Monica Suder, and, especially, production editor Hal Lockwood.

Reviewers and consultants from both industry and academia have provided valuable contributions that improved the quality of the book. I choose to name them individually, and they are so listed at the end of this section. In particular I wish to thank:

- Marilyn Bohl, for outstanding and comprehensive review, which made a significant contribution to the overall quality of the book.
- M. Elizabeth Campbell, for early inspiration and substantial material contributions to the book.
- Ralph Duffy, for moral support and review contributions.

Reviewers and Consultants

Robert Barrett	Indiana University
James Blaisdell	Humboldt State University
Marilyn Bohl	IBM Corporation
M. Elizabeth Campbell	Safeco Insurance Corporation
Patricia Clark	Management Information Systems
John Couture	San Diego City College
Timothy Cramer	University of Puget Sound
Barbara Crim	Rainier National Bank
Ralph Duffy	North Seattle Community College
Paul Duval	Northeastern University
Kenneth Kendall	University of Nebraska
Ann Kinsinger	Programming Resource Organization
Paul Maxwell	Bentley College
Barbara McCormick	The Seattle Times
Fred McFadden	University of Colorado
Perry Sanders	Indiana University Northwest
Lee Ann Smith	Microsoft Corporation
Rod Southworth	Laramie Community College
William Spezeski	North Adams State College
Janice Strietelmeier	Indiana VOC Tech College
Julia Tinsely	Indiana Central University
Ken Thomason	Lane Community College
John Tomei	Diablo Valley College

Brief Contents

Detailed Contents

Chapter 8
Detail Design: Output

PART 1

AN INTRODUCTION TO SYSTEMS THINKING

Systems work is known to be both frustrating and fascinating. In the process of bringing order to systems, a systems analyst will never be bored. In this first section, we discuss the characteristics of a system and then introduce the key systems concept, the systems life cycle. This information forms the foundation on which subsequent chapters are built. Before moving on to those chapters, however, we pause to discuss the analyst life and then make a strong case for communication skills, which are so much a part of that life.

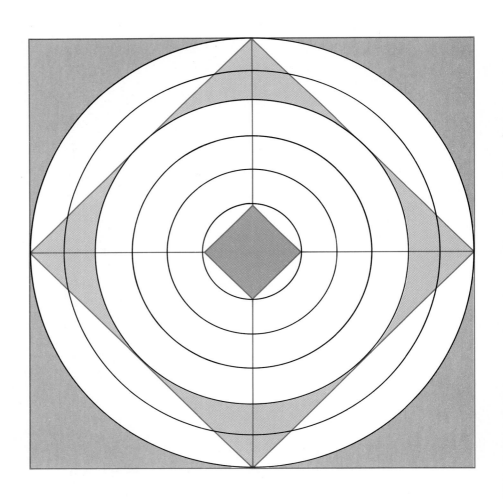

Chapter 1

Information Systems in Business

The last time you withdrew money from your bank account, it was probably pretty simple. You gave a withdrawal slip to the teller, who checked your account balance on the CRT screen and gave you the money. Routine transaction. You probably did not walk away all aglow, marveling at the computer system that made it all possible. But maybe you should have, because it is impressive.

A system that keeps bank accounts current and makes that information instantly available is difficult and costly to develop. To make everything work properly, the system components—computer, teller terminals, disk storage, communications lines, programs, operators, written procedures, and much more—must mesh perfectly. In this chapter we shall consider what a system is and what it takes to put it together.

◇ DATA: LOTS OF IT

We need systems to organize our data. A common media phrase is the "explosion of data," which implies that we really have more data than we know what to do with. The data is distributed through many media, including expanded educational opportunities, television, and, of course, the computer.

Just a few years ago only large organizations—the federal government, the aerospace industry, insurance companies—had computers. Then, while most of the computer industry was straining to make even more powerful machines, other companies began introducing minicomputers, which made computers affordable to a much larger segment of the market. Microcomputers appeared soon after and the rest is history. Computers are everywhere, and they all process data.

Processed data—information—is a company's greatest material asset. Information is a key corporate resource. Buildings, equipment, and materials are important, but they are replaceable. Where would a business be, however, if it lost its information on sales, markets, salaries, product design, taxes, inventory, and accounts receivable? Most of this information is now produced by computers. (See Figure 1-1.) Although the computer has become an indispensable tool, it is only one component of a larger concept, the system.

◇ SYSTEMS

"You can't beat the system" or "The system isn't working" are common phrases, and we know—more or less—what they mean. We might find it challenging, however, to explain *exactly* what they mean. We are surrounded by many kinds of systems: the economic system, the solar system, the ecological system. A **system** is a set of related components that work together to fulfill a purpose.

Business Computer Systems

In this book we are concerned with systems in business. The term *business* is not limited to privately owned factories or offices. In a broad sense business systems are found in government, hospitals, schools, agencies, transportation organizations, and a host of other places. We are concerned, in particular, with computer systems.

A **computer system** is a system whose components include one or more computers. Some people think of a computer system as a set of machines, but actually it is people who run the system, not machines. Novice programmers tend to focus their attention on one program at a time and to think that a computer system is just a set of programs. To show the total

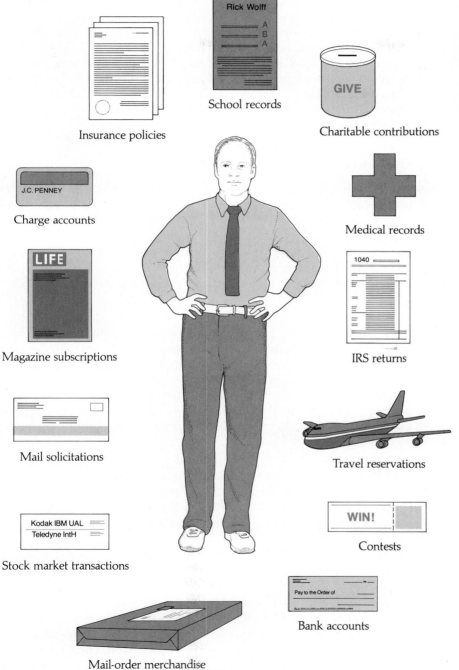

FIGURE 1-1

"Explosion of data" on a personal basis. Almost all our daily transactions are recorded on a computer file somewhere. Can you think of other transactions?

Insurance policies

School records

Charitable contributions

Charge accounts

Medical records

Magazine subscriptions

IRS returns

Mail solicitations

Travel reservations

Stock market transactions

Contests

Bank accounts

Mail-order merchandise

picture, we have to say that computer system components include hardware, software, data, procedures, and people.

We can think of a computer system as bridging the gap between technology and an organization, such as the personnel department or the shipping department or a small marketing firm. The systems analyst represents the technology, and the employee of the company needing a system represents the organization. Before we consider how they might work together to produce a computer system, let us examine the general nature of systems.

System Characteristics

Certain characteristics are common to many systems. A system must have a purpose, structure, and interdependence. A system may also be composed of subsystems. Let us consider each of these.

Purpose.   Systems do not just appear out of nowhere; they exist for a purpose. For example, an economic system tracks the economy, an accounts receivable system collects money owed, an airline system transports passengers.

Structure.   A system must be organized in a way that gives it structure. The components of the system must work together in some preplanned way so the system can operate in an orderly manner. A department store's financial system, for example, is structured to reflect the interaction of the pricing, purchasing, inventory, and billing activities. The store's management system probably has a hierarchical structure, with the president at the top, various levels of managers in the middle, and a large number of workers at the bottom.

Interdependence.   In order for a system to function, it must receive input from other systems. The system processes the input to produce output, which is used as required input for still another system. In other words, a system interacts with the environment in which it operates. (See Figure 1-2.) For example, a purchasing department might receive requests for materials from the factory and then generate purchase orders to be routed to vendors.

Subsystems.   A system may be made up of several subsystems—that is, components that are systems themselves. A company's finance system, for example, may comprise accounts payable, accounts receivable, general ledger, credit, and payroll subsystems. (See Figure 1-3.) A hospital may have separate subsystems for the pharmacy, the laboratory, and for patient medical records. In addition, each of these subsystems may have subsystems of its own.

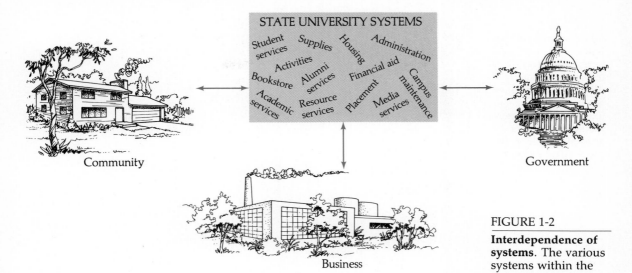

Community

Government

Business

FIGURE 1-2

Interdependence of systems. The various systems within the university are interdependent with service systems in business, government, and the community.

◇ HOW TO APPROACH SYSTEMS

Over the years, computer systems people have tried to bring order out of chaos by using various tools and techniques. A certain subset of early tools, such as system flowcharts, is known collectively as "classic" systems analysis. During the 1960s, however, the industry struggled to develop more sophisticated computer systems, such as those for airline reservations. Many times those systems were late, over budget, and failed to do what the users thought they should do.

As a partial answer to this problem, specific systems analysis and design methodologies—techniques based on some concept—were developed. The authors of these methodologies were saying, in effect, that if we did it their way everything would turn out all right. The originators wrote books to explain their techniques. Some people embraced the new approaches, while others waited. Meanwhile, back on the campus, instructors wondered what to teach their students.

The problem has been resolved in an interesting way. Unlike so many aspects of the computer industry, the new did not replace the old. Many parts of the new methodologies have been incorporated into the classic tradition. This book will do the same, discussing the principal methodologies by name in Chapter 10 on detail design processing.

The Systems Life Cycle

Breathing life into a new system requires a step-by-step process known as the **systems life cycle**. The participants in this process are the systems analyst and the employees of the organization needing the system. The

FIGURE 1-3

Systems and subsystems. The major systems in the top drawing represent a typical business. Each of the major systems is likely to have subsystems. The subsystems shown here for the finance system may have further subsystems of their own.

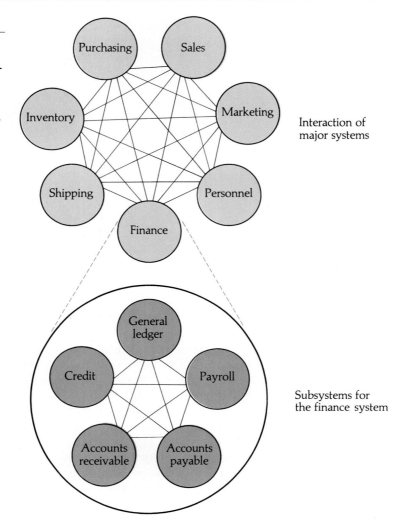

Interaction of major systems

Subsystems for the finance system

organization employees—often called **users**—understand the organization's needs and the existing ways of making it work. The **systems analyst** understands computer technology and how to proceed through the systems life cycle.

Chapter 2 introduces you to the systems life cycle. The initial step in a systems project is to study the nature of the problem and determine its scope. The analyst then works closely with the users to analyze the existing system and define the users' needs. After the analysis phase, a new system is designed and, eventually, developed and implemented. All of these activities must be carefully documented. The bulk of this book is concerned with the detailed activities involved in the various phases of the systems life cycle.

Recent Systems Trends

In Part 6 we examine the most important and prevalent trends in the development of business systems today: microcomputers in systems, the information center, and the automated office. All three of these demonstrate that the walls of the traditional data processing department, once so impregnable to users, are tumbling down.

Microcomputers.   Low prices and ease of use are bringing microcomputers into many areas of the business organization. In fact, it has been embarrassing for some computer professionals to discover that users know more than they do about microcomputers. As an analyst, you need to know how microcomputers fit into the overall systems picture.

Information Center.   A distinct trend in industry today is to give users the opportunity to take care of themselves. This is not always appropriate, obviously, but users are moving into areas once thought sacrosanct. The information center concept has been developed to provide users with the assistance they need. Many analysts will be involved in this venture, either directly or by recognizing when it is suitable to recommend the information center option to users.

Office Automation.   New products are flooding the office automation market. Most potential users are finding that the technology available surpasses their current needs. Aside from word processing, many users do not feel ready to find their way among the tangled choices. As an analyst, you need to have a basic understanding of how office automation techniques impact systems.

Human Factors in Systems

The wise analyst always keeps the user in mind. Analyst and user must work closely together throughout the systems life cycle. A common phrase to describe this is "user involvement": A primary goal is to have the user involved in the development of the system. This point will be stressed repeatedly as we discuss the systems life cycle. What really matters is that the users get the system they want and need.

Another human factor consideration is the way people will interact with the system once it is up and running. **Ergonomics**, the study of the human factors in computing, is concerned with every aspect of the people/machine interface. One appropriate concern, for example, is how a system presents

BOX 1-1	My Favorite Home System

We used to talk about the hi-fi system. Now it's the sound system. Some of us talk about our computer system. But all of these are pretty tame, compared to the—what else—home system. This is not an isolated, single-purpose system. No. It is grandiose, it is all-encompassing, it will run your home.

None of this is new. We all know that the technology exists today to take care of many household chores automatically: adjust the heat or air conditioning, turn on the coffee, or record a TV show. The point that may not have sunk in yet is that all these tasks can be assimilated into one computer-based system, a set of related components that will take care of every important aspect of running your home. Add these to the list: take phone messages, send phone messages, order groceries, place stock orders, dial the fire department. Perhaps we should think of this system, not as just running your home, but as helping you run your life.

output on a CRT screen. This type of output must be well-organized and easy to read. The analyst and user will select the method that is easy and efficient. Ergonomic considerations are an integral part of systems planning.

◇ ONWARD

You probably already know what it takes to develop a good program. You now need to understand how programs fit into the systems picture. It is time to see a new perspective, to broaden your horizons to include all phases of the systems life cycle. The challenge lies ahead.

FROM THE REAL WORLD

Computers Go to the Olympics

What kinds of things would you have to worry about if you were in charge of planning for the Olympic Games? Housing the athletes? Reporting the race results? Personnel and payroll? The answer, of course, is: all these items and many more. The Olympics, although temporary, are a major and multifaceted undertaking and must be run like a business. The size and complexity of the operation make the Olympics a prime candidate for automation.

Computers have been humming in the background of the Olympic Games since 1960, when an IBM RAMAC® was used to tabulate race results at the Winter Games in Squaw Valley, California. But that was many computer generations ago. The 1984 Summer Games in Los Angeles used 3 mainframes, 3 minicomputers, 100 microcomputers, 100 word processors, and 3,000 electronic mail stations. And all of these were linked by a telecommunications system. The computer system served the needs of 10,000 athletes, 8,000 reporters, 45,000 employees, and 7 million spectators!

Software for the Olympics must handle a variety of tasks: event schedules, ticket sales, team rosters, equipment budgets, race reports, hotel accommodations, accounting and inventory control, transportation routes, personnel, officials accreditation, communications, and many lesser tasks. Some highlights:

Ticketing. Ticketing for the Olympics is a complicated matter, due to the political overtones of the games and the fact that demand exceeds supply by a wide margin. Software must ensure that ticket sales comply with ticketing rules and constraints. To minimize scalping, tickets must be printed only a month in advance and distributed quickly.

Race Results. The results of each race are immediately keyed to nearby terminals for instant statistical analysis and final tabulation. Athletes, coaches, and reporters see the entire outcome on terminal screens in less than 60 seconds.

Communications links. Computer terminals, not telephones, provide the main communications link among participants. Sitting in front of a terminal, a player can scout the competition, page his or her coach, order a sandwich, dash off a telex home, or check the event schedule.

Transportation. Estimated traffic conditions must be incorporated into transportation software so athletes can be bussed between event sites and the Olympic villages over the easiest, most efficient route.

You see the magnitude of this undertaking. The computer system for the Olympics is not just set up every four years, but instead is an ongoing project. The system must be improved, fine-tuned, and adjusted to each new location. At whatever time you read this, the planning is already well under way for the next Olympics.

SUMMARY

- Data is a company's most important material resource. Most information is produced by computer systems.

- A system is a set of related components that work together to fulfill a purpose. A computer system is a system whose components include one or more computers.

- A system must have certain characteristics. The system must have a purpose, it must have structure, and it must have interdependence with other systems. A system may also be composed of subsystems.

- The systems life cycle is a step-by-step process to develop a new system. The new system is developed by the systems analyst, who understands computer technology and the systems life cycle, in conjunction with the users, employees of the organization needing the new system.

- Recent systems trends include the proliferation of microcomputers in the office, establishment of information centers to help users, and office automation.

- The study of human factors related to computers is called ergonomics.

KEY TERMS

Computer system	Systems analyst
Ergonomics	Systems life cycle
Subsystems	Users
System	

REVIEW QUESTIONS

1. The "explosion of data" is often related directly to individuals. What data is on file about you? Are you able to estimate how many files contain your name and address?

2. Can you name systems with which you interact in your daily life? How many of them are partially or fully automated?

3. This chapter has only touched on the systems life cycle and the systems analyst. Before you study these in more detail in the next two chapters, describe what you think the systems life cycle is and what a systems analyst does. It might be interesting to see how your perceptions change.

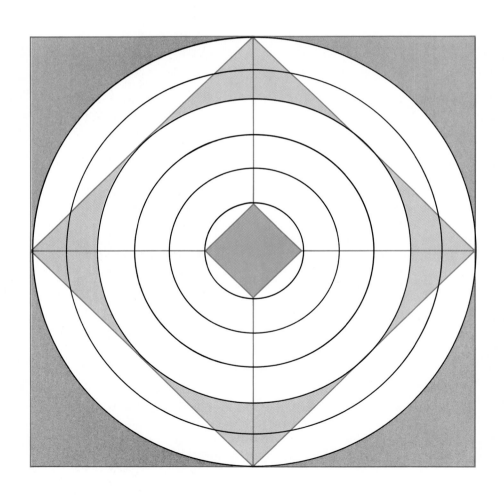

Chapter 2

Overview of the Systems Life Cycle

Mike Tripp knew a good thing when he saw it. As the analyst called in by the Pay and Go discount house, he figured out right away that their sales visibility problems could be solved by producing end-of-the-week net sales reports. This solution had worked nicely for a retail distributor he had worked with before. Mike had skipped the meetings he should have had with the owners of the discount house—it was clear to him that they knew little about computers.

Unfortunately for Mike, he forgot that the discount house owners knew a lot more than he did, not about computers, but about their own business. They soon realized that the reports foisted on them were useless. Mike had not hung around long enough to discover that one week is not necessarily the same as another in the discount business. A discount house has significant fluctuations the first week of the month, when social security checks are cashed; it also has wildly varying seasonal changes.

There is an important lesson here—one that should not be lost on an aspiring analyst. The most significant people the analyst deals with are those for whom the system is being developed: the users. This issue will be emphasized here and in subsequent chapters, as we consider what it takes to complete a successful systems project.

◇ THE SYSTEMS LIFE CYCLE

As the computer age unfolded, it became clear that some sort of method would be needed to organize, plan, and control systems projects. The problem could be stated this way: How do you find out what your customers want, and how do you deliver a product that meets their needs within a reasonable budget and schedule? Or, put another way: How do you do your job?

The **systems life cycle** evolved to answer this need. Today, most systems people would agree with the concept of the systems life cycle. But they do not always agree on the nature or sequence of the life cycle steps. In fact, they do not even agree on the number of steps in the cycle. You can pick up ten different books on systems analysis and design and find ten different sets of steps in the systems life cycle. No two books are alike.

In this book we propose a straightforward, five-part life cycle. The following list gives the names of the steps—often called "phases"—and simple explanations for what they mean:

1. Preliminary investigation—Determine the problem.
2. Analysis—Understand the existing system.
3. Design—Plan the new system.
4. Development—Do the work to bring the new system into being.
5. Implementation—Convert to the new system.

These descriptions are oversimplified, but the list does give you the flavor of the systems life cycle. (See Figure 2-1.)

In reality, each of the five phases is complicated, being composed of many subtasks. Furthermore, each phase is based on a certain philosophy—that is, has a certain rationale for being carried out the way it is.

In this chapter, the rationale for each phase will be emphasized as we give you the big picture, an overview of the entire systems life cycle. In Chapters 3 and 4 we examine the important topics of the analyst and documentation. Then, in Chapters 5 through 13, we study in detail the information needed to perform the work in the various phases of the systems life cycle.

◇ PHASE 1: PRELIMINARY INVESTIGATION

We begin at the beginning, with simple questions that anyone might ask when starting something new. Whose idea is this? What seems to be the problem? What do you expect us to accomplish if we try to do something new?

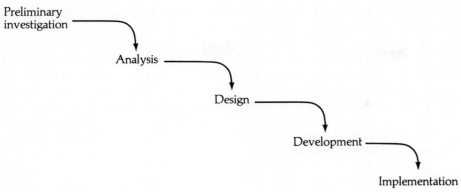

FIGURE 2-1

Systems life cycle.
The systems life cycle
is a step-by-step
process.

The Impetus for Change

There are many reasons why new systems are begun or old systems revised. The systems project does not spring out of thin air. The impetus—the cause—for the change can come from many sources. These sources are usually classified in one of two ways: an internal source or an external source. An *internal source* is one within the company or organization, such as a manager trying to improve inventory control or a user asking for an improved reporting system. An *external source* is outside the organization— for example, an agreement with union officials for payroll deduction of dues or customers complaining about incorrect bills. Whatever the reasons, some impetus will cause management to give the analyst the authority to pursue the problem.

The True Nature of the Problem

The analyst must differentiate between symptoms and actual problems. The impetus for a new system or system change may be a perceived problem, but its cause could turn out to be quite different from what was originally thought. The thoughtful analyst will investigate carefully and not jump to conclusions. In addition, the analyst must help the user resist jumping to conclusions. It is very easy to say "I need an online system," or "I need a microcomputer," but these statements do not define *needs*. Instead, they suggest premature solutions. The analyst, with the user, must identify and define the problem.

The Scope of the Problem

It is in everyone's best interests to define clearly what the system is supposed to do—and not do. It is easy to add to the wish list of things it would be "nice" to have the system do—both because users see the advantages and because systems personnel want to be helpful. It is much more difficult to define a specific area of study and then stick to it.

A contractor likes to tell stories about the financial perils of building a home. The buyers drop by daily to measure progress and, typically, find some "may-as-wells": "As long as we're going to pour a patio in the back, we *may as well* put a roof over it." And so it goes. Items small, medium, and large are added to the construction. Is it any wonder that the schedule and budget are soon ballooning out of sight? The buyers are dismayed; each item had seemed so manageable at the time. They have developed a new awareness level—the hard way.

Analysts must learn the lesson of containment, too, either by being forewarned or the hard way. Identifying the scope of the problem early, and describing it very specifically in writing, will help the analyst and user stay on course. Today, this defined project. Tomorrow, the world. (See Figure 2-2.)

Objectives

The most successful projects are those that truly do what their users expected. However, more projects fail because of inflated and unreasonable expectations than for any other reason. User expectations must be clearly defined. The definitions will be expressed in this phase as a set of objectives.

A key consideration at this point is **user involvement**. Users must be involved in the decision-making process from the beginning. Together, users and systems personnel can prepare a list of expectations that are satisfactory to all participants.

An Approach

The preliminary investigation is sometimes called by other names, such as the **feasibility study** or the **system survey**. Any of these three names is appropriate, since each conveys the idea that we are making an approach to the project, establishing a beachhead. The preliminary investigation phase may be brief—a few days for a small project—or quite lengthy, several weeks for a substantial project. In any case, it is concluded with appropriate documentation reports (and possibly a presentation) that define the problem and recommend an approach to a solution. The suggested approach

FIGURE 2-2

Phase 1 of the systems life cycle: Preliminary investigation.

will be very general at this point. If management elects to proceed, it is time for an in-depth study of the system under consideration, the systems analysis phase.

◇ PHASE 2: ANALYSIS

Suppose that someone asks your advice on what car to buy. You quickly size up the person and pop out an answer: "Well, you look like the sporty type to me. How about a red XKE?" You might wish you had been less hasty when you discover that the potential buyer is an active outdoor person with four children—perhaps a van would be more suitable. As simplistic as this example is, it illustrates the ridiculous position into which analysts sometimes put themselves: They design solutions when they do not understand the problems.

The systems life cycle is deliberately a step-by-step process. Before design can begin, the analyst must have a thorough understanding of the current system. If this process is short-circuited, as it is sometimes tempting to do, the analyst may court the wrong solution.

Understanding the System

Analysts must understand the system before they can do anything to improve it. The process of understanding includes becoming involved, sometimes deeply involved, in the system as it currently exists. During this time analysts learn a great deal about the subject matter; one of the benefits of being a systems analyst is this exposure to a variety of applications. Although they are computer professionals, analysts can apply their expertise, their analyst skills, to many areas—to stores, hospitals, farms, factories, government—the list is endless. This opportunity for enrichment, if we care to view it that way, comes mainly in this analysis phase. Analysts must find out what is going on. Toward that end, they must turn to the user.

User Involvement

Some analysts are so confused that they have it backwards: They see the users as a necessary burden on "their" system. That is akin to seeing customers as a burden on a retail store. We need only consider who pays the bills and who will be using the system to realize whose system it really is.

Look at it another way. The analyst and the user have a symbiotic relationship: Each needs the other to survive. The user needs the analyst's technical skills and the analyst needs the user's knowledge of the system. Although they work together during the preliminary investigation, their one-on-one relationship broadens during the analysis phase, as they delve into the system. An analyst, of course, can easily have such working relationships with more than one user.

The analyst/user relationship can develop through meetings, memos, phone calls, and/or interviews. Chapter 3 examines some of the problems inherent in this relationship and discusses how the analyst might approach them. For now, recognize that the user must be an early and permanent participant in the systems project. If the user is given only nominal attention, the analyst can anticipate problems throughout the entire systems life cycle. The most successful systems are those in which the user plays an active and vital role. Put as simply as possible: The analyst *needs* the user.

Data Gathering

The basic work of the analysis phase is gathering and analyzing data. Two common data gathering techniques are interviews and questionnaires. The analyst will also want to gather written materials related to the existing system and to observe the system in operation. In addition, the analyst may gather sample data from large volumes of data. There is no standard way to collect data: Both the type of data collected and the method by which this is done depend on the nature of the project. No two projects are ever the same.

The data gathering process may take weeks or even months and involve many people in the user organization. During this period, the analyst gradually develops an understanding of the existing system and user needs. He or she may collect a great deal of data in the form of interview notes, meeting minutes, questionnaire replies, documents, and forms. That data now needs to be analyzed.

Data Analysis

The collected data must be translated into a set of written products that will serve as a foundation for the documentation of the systems analysis phase. Although the results will vary somewhat according to the type of project, an analyst will typically produce charts and narratives to describe the system.

Through the process of analysis, the analyst increases his or her understanding of the system. The analyst spends a lot of time organizing the material and making written records. This written base serves as a communication channel between the analyst and the user. Now the gathered data can be verified. For example, an analyst might come away from a long user interview with a few pages of notes and a good understanding of the flow of the data in the organization. But words can be ambiguous. A smart analyst will probably put the data flow information into chart form and confirm its accuracy in a later meeting with the user. Methods of gathering and analyzing data will be discussed in detail in Chapter 6 on systems analysis. (See Figure 2-3.)

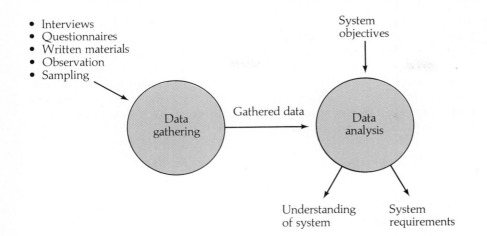

- Interviews
- Questionnaires
- Written materials
- Observation
- Sampling

FIGURE 2-3

Phase 2 of the systems life cycle: Analysis.

System Requirements

The purpose of gathering and analyzing data is twofold: to understand the system and, as a by-product of that understanding, to establish the system requirements. The description of the system was quite broad in the preliminary investigation phase, but now the analyst lists precise system requirements. Specific needs of the users are determined and documented. For example, a bank teller needs to be able to retrieve a customer record on the CRT screen within five seconds. The importance of accurate requirements cannot be overemphasized, since the design of the new system will be based on these system requirements.

Wrapping Up Analysis

Completion of the analysis and requirements document is usually followed by a presentation, giving users and user management an open forum for discussion. As the analyst, this is your opportunity to demonstrate your knowledge of the existing system. Even though you have established a certain rapport with the user organization, this formal presentation will enhance your credibility. It also will establish you as a person who can be trusted to design a system that will meet their needs. Assuming a decision to proceed, you enter the design phase.

◇ PHASE 3: DESIGN

You now could plunge headlong into design, producing comprehensive and detailed plans for all aspects of the system. There was a time when analysts did just this, but it is not a good idea. Details at this point are premature, and you need agreement on a design concept first. There is no way you can consider, for example, report formats, when you have not even decided whether the system will need batch reports or online screen queries. Look at it this way: Do you want to do all that detailed work and then discover that your fundamental approach is flawed or unsuitable? The systems design phase, therefore, is divided into two parts: preliminary design, to establish the new system concept, followed by detail design, to determine exact design specifications.

Preliminary Design

This will not be the first time you have considered a design for the new system. Recall that one of the results of the preliminary investigation phase was to a suggest a solution to the client. Also, in the analysis phase, as requirements were developed, the design solution became clearer. So, although design is not the announced purpose of the first two phases, some design possibilities have already been considered. Now it is time to study a design concept in earnest.

The first task of preliminary design is to review the system requirements. The next is to consider some of the major aspects of a system:

- Should the system be batch or online?
- Should the system be centralized or distributed?
- Should packaged software be purchased?
- Can the system be run on the user's microcomputers?
- Will new equipment be needed?
- How will input data be captured?
- What kinds of reports will be needed?

The questions can go on and on. Eventually, together with key personnel from the user organization, you determine an overall plan. The plan is expanded and described so that both the user and analyst can understand it. Consider these short illustrations of how plans are developed.

Example 1: Payroll for an Office Supplier. An office supply company has a payroll of 28. The company wishes to switch to an automated payroll system. It has very standard payroll requirements, and no desire to have any equipment of its own. The analyst locates a service bureau that has a suitable payroll program. The company will deliver payroll data to the service bureau, which will use a batch process to maintain payroll files, and will produce checks and a payroll register monthly.

Example 2: Audit System for a Department Store. The store already has department point-of-sale terminals tied to a minicomputer. These are used for sales transactions and inventory control. The store owners are happy with the existing system but want additional information. The analyst determines their general needs to be checking and reporting on sales transactions, correcting and adjusting sales transactions, and balancing the registers. The analyst and users agree that packaged software from the hardware vendor will fill their needs, and the cost of developing in-house software cannot be justified.

Example 3: Increased Efficiency for a Public Utility.   A city utility has an extensive batch system to process data for its electricity, gas, water supply, and waste disposal needs. As it faces record demands for services, the utility needs to find ways to make existing operations more efficient to meet existing revenue limitations. After months of study, the analyst and users conclude that a new online design is needed to provide more timely information. The design solution involves the purchase of a new minicomputer and several terminals. Software will be developed by the utility's own programmers.

These brief statements do not begin to convey the amount of work it takes to reach such conclusions. In fact, it is common practice to offer alternative plans: usually Volkswagen, Chevrolet, and Cadillac versions. That is, one version is simpler and less expensive, one is of medium complexity and cost, and one is more complex and more expensive. Users choose one of the alternatives, based on need and financial resources. As the basic design concept is established, it must be carefully documented. Taking a "high level"—as opposed to detail level—approach, the analyst shows the general processes for input, process, and output in the new system. Other preliminary design activities include cost analysis and scheduling, items of particular concern to users.

The documentation and presentation related to the preliminary design are critical. At this point the user becomes a buyer and you, the analyst,

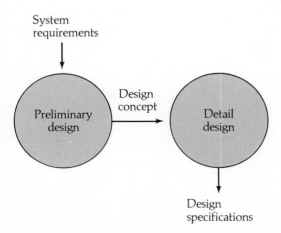

FIGURE 2-4

Phase 3 of the systems life cycle: Design.

become the seller. The system is either accepted or it is not. It is crucial, therefore, that your presentation be user-oriented and carefully planned. These activities will be discussed in Chapter 7 on preliminary systems design. (See Figure 2-4.)

Detail Design

The users have accepted your design proposal—you are on your way. You must now develop detail design specifications. This is a time-consuming part of the project but is relatively straightforward. The reward of a thorough job is that the development stage—software purchase/lease or actual coding and testing—will go smoothly.

A small warning is appropriate here. Although it is universally agreed that the systems life cycle is a step-by-step process, analysts are sometimes tempted to take a shortcut. They may jump in and do a little coding in advance, for example, before the design stage is complete. (In fact, sometimes they begin coding before the analysis phase ends, a usually fatal error.) Analysts may take this out-of-phase step for two reasons. First, they have been "burned" before by schedules and want to get a head start. Second, they want some relief from meetings; hence, they do a little coding instead. Don't do it. If you put the cart before the horse, you eventually will be faced with two unpleasant alternatives: (1) recode when you have the correct design, or (2) force the design to fit the already coded portions.

In this phase every facet of the system is considered in detail. Here is a partial list of design activities, to give you the flavor of what is to come:

- Design output forms and screens
- Design input data forms
- Design system flowcharts
- Plan file access methods and record formats
- Plan data dictionary entries
- Plan database interfaces
- Plan data communication interfaces
- Design system security controls
- Consider human factors

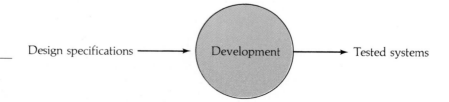

FIGURE 2-5

Phase 4 of the systems life cycle: Development.

Design specifications ⟶ Development ⟶ Tested systems

This list is not comprehensive, and not all activities on it will be used for all systems. These are just some of the possibilities. They will be discussed in Chapters 8 through 11. As before, the results of this phase will be documented. This large and detailed document, usually referred to as the detail design specifications, will be an outgrowth of the preliminary design document. A presentation often accompanies the completion of this stage. Unless something unexpected has happened, it is normal to proceed now with the development of the system.

◇ PHASE 4: DEVELOPMENT

In many organizations, the heaviest load is now shifted to programmers, who will proceed to develop the system based on the detail design specifications document. The analyst continues in a liaison role, coordinating between users and programmers, for understanding and clarification. Some organizations have the analyst do some or all of the programming, depending on departmental policies and the size and nature of the system.

The activities at this point are familiar to programmers. Programmers may refine program design tools that have been prepared by the analyst during the design phase. Or, if organization policy assigns this task to them, programmers may complete these tasks during the development phase. Program design tools may include flowcharts or structure charts or pseudocode or some other design tool. Programmers may participate in walk-throughs to receive comments on their programs. Programs are written in the language approved for the system. Programs are tested thoroughly, using test data, alone and then in conjunction with related programs. Finally, the entire system is tested with volume data. Program listings and test results are added to the accumulated documentation.

Of course, the development stage may actually bear little resemblance to these activities—that is, if the software for the system is purchased or leased. This possibility is especially likely in a microcomputer environment. (See Figure 2-5.)

◇ PHASE 5: IMPLEMENTATION

Moving from one phase to the next is not quite as clear-cut as it may seem in this discussion. In particular, the life cycle process does not call for all programs to be completed and tested before implementation activities are begun.

The key point in implementation is that users are now going to switch to the new system. This is not a simple change. For a smooth transition, groundwork must be laid carefully. Criteria must be evaluated to establish that the system is actually ready to be converted. These criteria are related

FIGURE 2-6

Phase 5 of the sys-
tems life cycle:

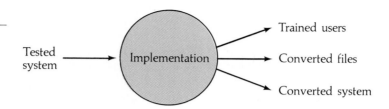

to what the system is ready to do, based on software testing, hardware availability, user readiness, and other factors.

Conversion to the new system must be planned carefully. Data files must be prepared to be used by the new system. Personnel who will use or interact with the system must be trained. All conversion-related activities must be scheduled and coordinated. Overall, the conversion process may be a trying one for those involved. Nerves become frayed and tempers sometimes grow short, as everyone tries to turn plans into reality.

When the new system has at last been converted, it is usually evaluated by some neutral party. The evaluator is probably another analyst, either from the same organization or from an independent consulting firm. And, of course, the system is evaluated by the users themselves, who are the key evaluators. The basic question is: How well does the system meet the specifications? Some fine-tuning usually follows this investigation. This final developmental activity is followed by maintenance, an ongoing process to improve the system. (See Figure 2-6.)

◇ SYSTEMS LIFE CYCLE: AN ITERATIVE PROCESS

There was a time when "frozen specs" really meant something. If your specifications were "frozen," it meant no changes could be made to them. Period. To the analyst, of course, the source of never-ending changes was the user, who was seen as a reed in the wind, blowing in all directions. The only way to get on with the work was to insist that the change deadline had passed; the system would be developed as the specifications stated.

In retrospect, this approach seems unreasonable. In addition to fostering an us-against-them mentality, it ignored the simple reality that humans are often unable to get it all—and get it all correctly—the first time around. And, of course, what good is it to "get on with the work" if the work specifications are incorrect? Mindlessly pushing forward is a waste of time and money. The workability of our approach increases as our knowledge increases over time.

Most systems people now acknowledge that each previous phase has to be reviewed and adjusted a little further down the line. This is not to say that every new frill should automatically be incorporated, but significant changes cannot be ignored.

◇ MOVING ON

You are now ready to tackle the phases of the systems life cycle in detail. Before we begin, however, let us consider two very important topics. The first is a discussion of the role, functions, qualities, and career paths of the systems analyst. Chapter 3 gives you an intimate look at the analyst life and at how the analyst does his or her job.

The second topic is communication, a subject that permeates every facet of the systems life cycle. As an analyst, you must appreciate its importance and know how to communicate effectively on paper and in person. Chapter 4 discusses this thoroughly.

FROM THE REAL WORLD

The Air Traffic Control System

The computer's speed is usually considered an advantage that produces quick results from high-volume data. Air traffic control, however, is one application where we need the computer's speed to monitor something else known for its speed: airplanes. The Air Traffic Control (ATC) computer system is an enormous undertaking. Some of its complexity stems from the fact that human life is directly at stake, so unchecked computer errors are intolerable. The computer-based ATC system has been implemented in all 48 contiguous states.

Each air traffic controller, or a team of controllers, is responsible for a certain geographic area. Each airplane within that geographic range is represented on the controller's large circular screen by a moving blip. The blip is accompanied by an electronically generated data tag that looks like this:

U A L 7 8 6
0 5 0 6 5 0

UAL786 is United Airlines flight number 786. The "050" in the second line of the data tag represents the altitude of 5,000 feet, and the "650" represents the speed of the plane.

As an airplane approaches the boundary of one controller's geographic area, the controller prepares to "hand off" control to the appropriate adjacent area. The controller makes a keyboard entry to the computer that causes a blinking data tag to appear on the screen of the new controller. That controller now makes a keyboard entry, indicating acceptance, that causes the data tag on the old screen to blink. Now the original controller knows that the plane is safely in someone else's hands, and the old blip can be removed from the screen. As a plane crosses the continent, it is moved from controller to controller in this way. The computer system also includes air traffic management functions such as warning systems that alert controllers when aircraft under their control are getting too close to the ground or to each other.

If you want to have a fascinating tour, visit your nearest Air Traffic Control Center. Tours are gladly given for any small group. For you, the highlight of the tour will be the computer hardware, which is a vast array of modern technology. Each system is duplicated in total. In the event of a system failure, there is an automatic switch to the other system in less than a second. Your tour will be complete as you observe the air traffic controllers at work in their semidark room, intently monitoring their screens. Computers, it would seem, have never been put to better use.

SUMMARY

- The systems life cycle has five phases. These phases are preliminary investigation, analysis, design, development, and implementation.

- Phase 1, preliminary investigation, is concerned with determining the problem. The reason for the problem, the impetus for change, may be internal or external. The analyst must differentiate between symptoms and the true nature of the problem. The analyst must also define the scope of the problem, the specific area of study. User expectations must be clearly defined as a set of objectives. User involvement in the system must begin in this first phase. Other names for the preliminary investigation phase are feasibility study and system survey.

- Phase 2, analysis, is concerned with understanding the existing system. The analyst must understand how the existing system works before he or she can take any action to improve it. Analysis begins with data gathering. This process uses techniques such as interviewing and questionnaires to produce data about the system. The next activity is data analysis, the process of translating the collected data into written products that will be the foundation for the documentation of the existing system. The purpose of gathering and analyzing data is to understand the existing system and to establish the system requirements, a precise list of user needs.

- Phase 3, design, is concerned with planning the new system. During preliminary design, the analyst determines an overall design plan that will meet system requirements. During detail design, every facet of the new system is planned in detail.

- Phase 4, development, is the actual work required to bring the system into being, based on the specifications prepared during the design phase. The primary activities of this phase are coding and testing.

- Phase 5, implementation, is the process of converting to the new system. Implementation phase activities include data file conversion, the actual change to the new system, and evaluation.

KEY TERMS

Analysis	Objectives
Data analysis	Preliminary design
Data gathering	Preliminary investigation
Design	Scope
Detail design	System requirements
Development	System survey
Feasibility study	Systems life cycle
Impetus for change	Nature of the problem
Implementation	User involvement

REVIEW QUESTIONS

1. Why do you suppose there is no universal agreement on the steps in the systems life cycle? Examine descriptions of the systems life cycle in two or three other books that you may get from your instructor or your library. Do the steps described in these seem to accomplish the same thing as the phases listed in this text?

2. Think of a specific example that might be an impetus for a new or revised system.

3. Can you think of an example in which the scope of a system is so loosely defined that the system project grows larger than originally intended?

4. In some ways, the preliminary investigation phase resembles the entire systems life cycle. Differentiate between the two.

5. In which of the five phases will user involvement be most extensive? Why? Do you think user involvement will diminish toward the end of the project?

6. Explain why it would be foolish to jump ahead into design before the existing system is thoroughly understood.

7. What is the purpose of splitting design into two parts, preliminary and detail?

8. Do you think users make the final switch to the new system easily? Are there factors that may influence the changeover?

9. Why is the systems life cycle considered an iterative process?

CASE 2-1
ORGANIZING THE
ORGANIZATION HEADQUARTERS

A national choral singing organization has 20,000 members in 450 chapters across the country. Local affiliates remain in touch with the organization through the national headquarters in Cleveland. The headquarters office has a paid staff of 27. These people answer to a board of directors made up of organization members. The staff manages member records and activities, including several local and national conventions each year.

Noting the excessive paper burden of the organization, all manually prepared, the board decided to consider a computer system. They checked with Susan Bradshaw, a member who was a computer professional. She recommended that they find a local consulting firm. President Teresa Dalton followed through on the assignment and eventually collaborated with Susan on reviewing the proposal from the consulting firm for a full analysis of the current system.

After the consultants had completed their investigation, they wrote a detailed analysis report and made a presentation to the board. In the process, they effectively described the existing system and outlined the requirements for a new system.

Although the consulting team had been quite thorough, they got a cool reception from the board. Teresa Dalton slipped out of the meeting and placed a call to Susan Bradshaw, to whom she indignantly noted that the consulting firm had just told them a lot of things they already knew! Were they to pay money for that?

1. This (mostly) true story had a happy ending after Susan explained what the consulting firm was trying to do. How would you have explained it to Teresa?
2. How could the consulting firm have made its position clearer?

CASE 2-2
THE SMALL PROFIT MARGIN

Metro Foods is a food wholesaler, or broker, that buys foods from manufacturers and commodity suppliers, and then sells them to retail markets. Food wholesaling is a tough business. All financial transactions must be monitored carefully to maintain even a small profit margin.

Metro Foods was concerned, in particular, about changing costs (usually increasing) on items they purchased. Finance manager Chris Parsons was familiar with existing computer systems at the brokerage and thought it would be easy to get a daily report on current item costs. As he explained to Kirk Williams, the analyst assigned to the project, an item cost list with changes marked would help the brokerage make corresponding changes in the prices charged to retailers.

Kirk was impressed by Chris's knowledge of both the business and the impact of the firm's computer systems. It was reasonable, he felt, to respond quickly to the request by moving directly to the design phase and then to program development. All went smoothly, as expected, and the customer seemed satisfied with the new report.

A few months later Chris called Kirk for further discussion. Metro was still being caught in the squeeze between rising costs and stable prices. Despite the new cost report, the brokerage could not react to the cost changes fast enough to bill customers properly.

Many months later Metro had a new system that worked. Cost changes are now submitted, as they occur, to an online system that causes the changes to be reflected automatically in the prices on customer billings.

1. All's well that ends well? Not really. What went wrong here?
2. What lessons will Kirk take with him for future assignments?

CASE 2-3
THE WRONG FORMULA

As a fledgling analyst, Dave Macklin had participated in the system process with senior analysts. His first solo assignment was appropriately straightforward: to work with the marketing department on a new sales analysis package. Dave did his homework thoroughly, meeting with users and progressing through the phases of the systems project.

Dave felt pretty confident about the system, particularly since parts of it were similar to a customer preference system he had worked on as a programmer. In fact, as he was beginning the preliminary design phase, Dave recognized that certain statistical analysis packages needed by the new system were almost identical to programs he had written before. He decided to code six small programs himself to get these routines out of the way early. Then the programmers could concentrate on the more general parts of the system.

However, as Dave became enmeshed in the details of the system design, it gradually became clear to him that the types of statistics needed would not, in fact, be produced by the programs already written. He began to feel an acute pang: writing the programs had not only put him behind schedule but had also impacted the overall design, which he now saw as incorrect.

1. How will Dave extricate himself from this mess? Will this impact the system schedule further?
2. We have a naive analyst here. As his supervisor, what advice/admonitions would you have for him?

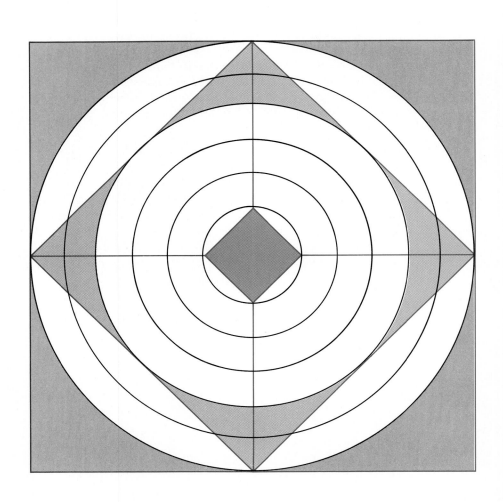

Chapter 3

The Systems Analyst

Denise Wharton had always wanted to be a systems analyst. And after a few years as a programmer, she had an opportunity, within her current company, to become an analyst. She was happy to take it. She looked forward, in particular, to designing new systems and envisioned that the users of the system would be pleased with her creative approaches.

After a few months Denise began to feel uncomfortable in her new job. Some of her planned changes were not working out well, and users were heard to grumble that they had not really been consulted. For her part, Denise felt that the users were deliberately not cooperating. Since meetings were becoming progressively more unpleasant, she gradually stopped having them. It would be many more months before Denise would come to understand the nature of the change process and how to work effectively with users. Let us now look at the systems analyst, the person who brings change to an organization by studying old systems and planning new ones.

◇ HUMAN RESPONSE TO CHANGE

Sometimes we talk about people being "set in their ways." Or we mention a friend who seems to be stuck "in a rut." Statements of this kind have a negative connotation. In contrast, we usually see ourselves in flattering terms, such as *flexible* or *easygoing*. Unfortunately, this is rarely true.

Most of us are, in fact, rather rigid. We do not like change. We like change even less if it is someone else's idea. Picture yourself in an office environment, comfortably established at the very heart of an important department such as accounts receivable. There have been an unusual number of problems lately—customer complaints, confusion over account numbers, frequent overtime, and a certain tension in the air. Still, nothing you cannot handle. But someone else feels otherwise, for here is a memo from your manager—two levels up!—requesting that employees cooperate in a study to consider a new computer system. Communiques such as this one suggest change. Enter the systems analyst, agent of change.

Change Agent.   Will the analyst be warmly welcomed, greeted, in fact, as a savior? Hardly. The analyst will be viewed as an interloper, as the person bringing the burden of change, as the change agent. Reactions depend somewhat on employees' previous experience with computers—or lack of same. Employees in the organization considering a new system—the users—sometimes feel threatened and may assume defensive postures. Some users just "clam up" and others resist in less subtle ways.

Change Theory.   An experienced analyst, of course, knows all this. He or she is familiar with the subject known as change theory. The analyst recognizes that people find change difficult. A key element of change theory is that too much change can cause people to become ineffective or even ill. A common phenomenon in offices experiencing layoffs, for example, is that activity grinds to a halt. Just when you would expect people to be working feverishly to appear indispensable, they are paralyzed by the fear of losing their jobs.

Change Agent Roles.   Even so, people do eventually adapt to change. But they adapt at varying rates. Young people, although not immune, are more inclined to adjust quickly. The analyst's initial task is to overcome the natural resistance to change. This involves establishing some level of trust. As the communication process with the user—such as the accounts receivable employee mentioned earlier—begins, the analyst may assume some role to make the user more comfortable. One role is to act as a *persuader*, gradually convincing the user that the road traveled is in his or her best interests. A more successful approach is that of *catalyst*, spurring the user to make discoveries and reach conclusions. The catalyst role encourages user involvement in the system, which is of critical importance for success. Other roles for the analyst assume noncooperation and are more heavy-handed: The analyst acts as a *confronter* by forcing the problem issues or as an *imposer* by invoking higher authority. The latter roles must be used with care and may cause dissension.

It is said that the only constant is change. As a systems analyst, it is important to understand that people's reaction to you may actually be their reaction to the change you represent. Is it possible to learn to handle change effectively? Once you are aware of the dynamics of change, it is useful to practice making small changes in your own daily life. Surely the systems analyst should not get "in a rut"!

◇ PROFILE OF THE SYSTEMS ANALYST

The analyst is viewed in different ways by different people. We have already seen how the analyst takes on the role of change agent. There are other roles to consider. He or she may assume different roles, depending on the people and the situation. When working on a project, the analyst performs a variety of functions. To do all of this well, the analyst should have certain personal assets and suitable training. Let us consider each of these in turn.

The Role of the Analyst

The basic role of the systems analyst is to find solutions for systems problems. Systems problems are as varied as the systems themselves: incorrect billing, slow reaction to market changes, poor management visibility, insuf-

BOX 3-1	The Analyst Advisor

You have heard all the old jokes about doctors and lawyers being asked for free professional advice at parties. Now, as a computer professional, you may find even the most remote acquaintance approaching you with the line, "I'm thinking about buying a computer."

If subtle, the free-advice-seeker may ask you to lunch, and for a $7.50 breaded cutlet you cheerfully give two hours of advice. You may even succumb to a hand-holding trip to the computer store. Your only reward may be a pat on the back but, more likely, you will be marked for maintenance phone calls. Your best defense may be to tell the buyer you don't make house calls, so he or she must unplug the machine and bring it to you. That should cause hesitation. The true solution, of course, is to make sure *beforehand* that the seller gives good support, and refer the person back to the store.

ficient inventory tracking—the list is endless. As a problem solver, the analyst is the main contact with the user. The analyst helps the user evaluate the problem and then begin to define expectations for the future. This process often calls for crossing organizational lines. Suppose, for example, that you are asked to check into complaints that customers are receiving certain products three or four weeks late. You may begin your investigation in the shipping department but eventually discover that the problem lies in the way those products were coded in the inventory control system.

In the process of seeking answers to puzzles such as this, the analyst spends time with a variety of people. The effective analyst soon discovers that he or she also takes on a variety of unofficial people-oriented roles, including detective, advocate, go-between, and—sometimes—amateur psychologist.

A seldom-acknowledged role of analysts is that of power broker. A respected analyst wields the power to recommend solutions that impact budgets and the way work is done. And, of course, he or she can send tremors of change throughout the organization.

Functions of the Analyst

So . . . what does a systems analyst do all day? There is an amazing lack of knowledge about systems analysts by the general public—they know the functions of brokers and bankers much better. Here is a somewhat unrefined list of what you, as an analyst, might do on a given day: review the specification changes with the new programmer, haggle with the facilities czar to get the training room changed to one that is in the same building with the users, add an overlooked file to a diagram in the requirements document and make 14 copies, stop by the post office on your lunch hour to pick up mass mailing regulations, contact users to confirm output layouts, prepare a list of invitees for a walkthrough, produce drafts for two new input forms, and review the revised schedule with your boss. Although it is not exactly a typical day, all these tasks are certainly things you may find yourself doing.

Perhaps the most noteworthy aspect of a day's activity list is its variety. An analyst is rarely bored. Analyst activities can be classified into three functional categories: coordination, communication, and planning and design.

Coordination. An analyst must coordinate schedules and activities with a variety of people: management, the system's users, vendors selling computer hardware and software, designers who help plan forms for the system, the programmers working on the system, and a host of others, such as carpenters who must adjust room space for the installation of new equipment.

Communication.   Analyst communication takes both oral and written forms. The analyst, in the process of understanding the existing system and planning the new one, usually spends a great deal of time interviewing and meeting with the users. The analyst may be called upon to make oral presentations to various audiences of clients, users, and managers involved with the system. The analyst also provides a variety of written reports—documentation—related to the various phases of the system's progress. Depending on the system, these documents range from a few pages to a report that is a few inches thick. An analyst usually spends a hefty percentage of time on the documentation function. The subject of communication deserves a chapter by itself, so we will discuss it in detail in Chapter 4.

Planning and Design.   The systems analyst, with the participation of the user organization, plans and designs the new system. This function involves all the activities from the initial investigation to the final implementation of the system. It is the bulk of the analyst's work. Although it takes just this short paragraph to give a general description, most of this text is devoted to these activities. Chapter 2 provided an overview of the functions, and they are described in detail in Chapters 5 through 13.

Analyst Qualities

Most analysts are hired for their technical expertise and experience, but it is their personal attributes, especially *communication skills*, that become critical if they are to advance. If you cannot communicate, the rest becomes irrelevant. Imagine the reaction of a user to an analyst who throws up clouds of jargon or, perhaps worse, appears to absorb information but really is not listening. Communication arenas include not only one-to-one conversations but also oral presentations and written documentation.

What other qualities might be especially desirable in a systems analyst? An *analytical mind* should top the list: an analyst should be good at understanding and solving problems. A related quality is *creativity*, a talent needed for designing new systems. And, since an analyst is often quite independent, he or she must be *self-directed*. Related to this is the fact that the analyst must be the type of person who is *well-organized*. All of these characteristics are important because the analyst may be thrown into an atmosphere of confusion and, in fact, is usually called in because of that confusion. The analyst is there to solve the problems and, possibly, bring order out of chaos.

The analyst must also be a *leader*, able to guide a team of people through the ebb and flow of the analysis and design processes. As a leader, the analyst must be able to steer a steady course despite setbacks such as: A user inadvertently withholds vital information, the key programmer quits without notice, the special terminals ordered are two months late, a critical file has been converted incorrectly. Although all these problems may not occur on a single system project, they are typical.

Since the analyst is also a *mediator* among users, managers, programmers, and many others, the diplomatic skills required, it sometimes seems, could secure peace among nations. As a leader and a mediator, the analyst maintains a calm posture and appropriate business demeanor. He or she is perceptive but not quick to jump to conclusions.

The most elusive skill is the one that is most difficult to describe: the ability to *work long periods of time without tangible results*. Most analysts come to their profession by way of programming. There is a physical interaction with the program that you look at, touch, talk about, pore over, despair over. The program will be keyed, checked, listed, run, changed—on and on, until you have a very specific finished product. Analysts also produce a finished product—an entire system—but it is not so clear-cut along the way. There will be days when it will seem you have endured endless meetings in which not much was accomplished. An effective analyst must have the patience to get through these dry spells.

Another avenue that requires patience is teamwork. You need to be able to work well in a team environment.

◇ TEAMWORK: WHY GROUP DECISIONS ARE WORTH THE PAIN

An interesting phenomenon in the classroom and in the workplace is that people prefer to work individually. Given the choice of working in teams or by themselves, almost everyone elects to toil alone. Does this mean they are antisocial? Not necessarily. The reasons are more complicated than that. The logistics of getting together, politely hearing out other speakers, dissatisfaction with the quality of another's work, lack of control, difficulty reaching consensus, ill-defined authority roles, personality clashes, and the incredible amounts of time it all seems to take—all contribute to people's desire to work alone. But whenever two or three are gathered together the result is not necessarily unrelieved bad news. There can be strength in group problem-solving.

Systems analysts frequently work in a team environment because different people are needed for their skills and expertise. In addition, if a large project is ever to be completed, the work load must be distributed among several people. Faced with this reality, it helps to recognize the methods and advantages of teamwork.

Experts offer these tips for successful group dynamics:

- Present your position logically but listen carefully to others before pressing your point.
- Do not change your mind simply to avoid conflict.

- If you cannot reach agreement on a first choice, find a next-best alternative that will be satisfactory.
- Accept differences of opinion as natural; disagreements can help the group process because a wider range of information offers a greater chance for a better solution.

Groups function as their members make them function. Keep in mind that, in spite of this approach, this is not quite a democracy. The analyst needs to keep the team on track and, in general, has the final say on technical decisions.

A key advantage of teamwork is the result: It's better. Repeated studies have shown that when a group's final decision is compared to the independent points of view members held when they entered the group, the group's effort is always an improvement. In fact, the group effort is often better than the *best* individual contribution.

SYSTEMS ANALYSIS AS A CAREER

A career as a systems analyst can progress along different paths. Some plan their careers carefully, while others take opportunities as they arise. In either case, the assets and liabilities of the field are similar.

BOX 3-2	**Group Decisions on the Moon**

Play astronaut for just a moment. Your spaceship has just crash-landed on the moon. You are confident, however, that you and your crew can make it to the established moon base just a few craters away. What should you take with you on your journey?

These items are available from your spaceship: matches, a mirror, a short-wave radio, a length of rope, two gallons of water, powdered milk, an inflatable life raft, and a compass. Your individual task is to rank these items in order, putting the most important first. Now get together with your crew and repeat the task as a group.

NASA has provided answers for this classic little problem. Matches, for example, would rank dead last, since there is no oxygen on the moon. But the "right" answers are not quite the point of this exercise. Its purpose is to prove—as it unfailingly does—that the group will always outperform any individual in the group.

Career Paths and Directions

For many years a typical career path went from programmer to systems analyst. It was not considered good form to be "just a programmer" forever. The transition was intertwined with prestige, often encouraged by the organizations themselves. One large bank, for example, kept its programmers herded in a linoleum-floored bullpen, right next to the carpeted individual offices of the systems analysts. Windows too. The message was pretty clear. Some programmers who reluctantly left their craft discovered that analysis was not for them, but to go back would have meant a demotion.

This has changed. Most installations now recognize programming and systems analysis as parallel paths, with suitable rewards for good performances. Although the programmer-to-analyst route is still common, it is no longer the only way to go. It should also be noted that many computer professionals branch out in entirely different directions, specializing in areas such as computer graphics, data communications, database management, computer security, hardware evaluation, office automation, and user information centers.

Moving Through Your Career

The career analyst faces many problems, such as stress, but also enjoys considerable rewards. Let us examine these and then consider appropriate preparatory and ongoing training.

Technostress and Other Problems.   Continuing and relentless work in a computer environment involves a type of mental assault that experts have labeled *technostress*. Some call it a computer-generated form of emotional burnout. A study by *Fortune* magazine noted that the number one stress occupation is waitress/waiter (pressure from both the customer and the kitchen), followed closely by the systems analyst. Analysts are prime candidates for stress because of the constant pressure from people and schedules, both of which need constant attention. In addition, the work load is often heavy.

Sometimes the analysis process seems like trying to hug a bowl of Jell-o. Descriptive words like *vague*, *ambiguous*, and *contradictory* apply. And there are times when the job is very lonely. This lamentable list, however, is usually more than offset by the rewards of the profession.

The Rewards.   Most systems analysts have a high level of job satisfaction. This can be attributed to a number of factors: independence, mobility, variety, creativity, people orientation, work environment, the fact that the job is a continuing learning process, and—not least of all—respect. The most fundamental reward, however, is a job well done—the users' new system. Systems analysts are also paid well, but this is not the critical gratification factor. Overall, the joys of systems analysis far outweigh the woes.

BOX 3-3   Analyst Statistics

Dewar's Career Profile of Computer Professionals is a booklet detailing the results of a nationwide survey. Key findings:

- 94% would not leave the field.
- 82% say they have done as well or better than expected.
- 75% work more than 40 hours a week.
- 82% cited language skills as "very important" to their success.
- 35% of their leisure activities involve computers.
- 94% believe that the computer field offers more opportunities than other fields for women and minorities.

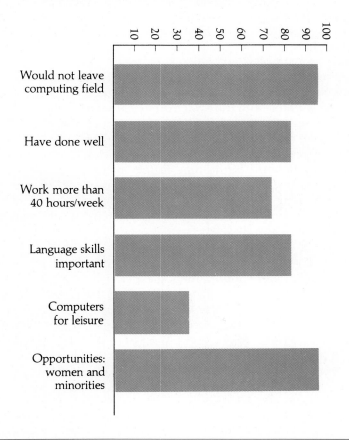

The Right Training.   A popular job description in the computer industry is the hybrid title "programmer/analyst." It could be argued that any programmer should have this title because he or she usually performs some analyst functions. At some point along the way, however, there may be an imperceptible change from *programmer*/analyst to programmer/*analyst*, and then to *analyst*. So, somewhat like instant potatoes, we now have the "instant analyst."

The new analyst is bewildered. Although familiar with the trappings, the analyst has not had sufficient training to be the key player in the analysis and design processes. Would you expect a nurse to be able to step in and perform surgery?

The assumption that goes with this change is that programming skills are comparable to analyst skills. As many chagrined managers have discovered, this is not necessarily so. Some technically competent programmers are not cut out for the people-oriented and organizational chores expected of an analyst. Just as there are natural athletes, there are natural analysts, but this is the exception. One way to develop good analysts is to implement a training plan.

Many consider ideal training to be an informal apprenticeship. The analyst trainee works closely with an experienced analyst. The trainee attends meetings and is exposed to the systems environment. He or she begins to analyze and design small systems. Over a period of time the new analyst is able to assume responsibilities in more complex systems.

Although apprenticeship often produces satisfying results, it is used less frequently today because of the time factor. Instead, new analysts often have to rely on some formal school training and, perhaps, a company-paid class based on some company-used methodology. This type of training teaches analysts tools and techniques, but it is no substitute for participating in the process.

Keeping Up.   Although most analysts launch their careers equipped with a two- or four-year degree, programming experience, and some sort of on-the-job training, background alone will not sustain them. An overt, planned self-maintenance program is necessary to keep current and to keep ahead. It is too easy to get bogged down with the current project—which is followed immediately by another—and lose sight of what is happening in the computer industry.

The well-versed analyst will want to keep abreast of current topics, many of which are directly related to systems work. The vast array of computer magazines can be narrowed down to one general publication such as *Computerworld* and a few specialty magazines on topics such as microcomputers. Of particular importance are the journals of systems-related organizations such as the Data Processing Management Association (DPMA) and the Association for Systems Management (ASM).

Membership in these or similar organizations is another good way to keep current. Typically, they meet once a month for an evening program that includes dinner, a short business meeting, and a speaker. You can learn from both the speaker and the contacts you make. These organizations also sponsor technical seminars for a relatively modest fee. Most seminars last a day or two and feature a recognized speaker on a popular topic. The fee includes written materials you can take with you.

◇ CONSULTANTS

Do you have a fervent desire to be out on your own, to be your own boss? If you look like a walking advertisement for Pierre Cardin, and your friends compare your sparkling conversation to Johnny Carson's, read on. You may have what it takes to be a consultant.

Advantages.   *Consultant* is an umbrella term that covers everything from writing a little program for a friend to running a full-service consulting business. ("Consultant" is also a euphemism used by people between jobs.) The benefits of consulting are many. First and foremost is the opportunity to chart your own course, to participate in the free enterprise system as an entrepreneur. You can get into the business without a large capital investment. The size of the market is limitless: in an endless uncoiling, old problems die and new ones are born. You may have a chance to travel extensively if that is your wish. And, if you are successful, there is an opportunity for big money, far beyond what you could make as an employee of a corporation.

Disadvantages.   One of the reasons consultants are well compensated is that they receive no worker benefits such as insurance and pensions. This problem is minor, however, when compared with the possible negative psychological problems. Employees of the company for which you are consulting may focus on you as the target of their resentment. You could even be made a scapegoat, taking a pratfall for the local manager's unworkable plan. You are never established, you are never a hero, you are never any better than your next job. Despite all this, many people press forward.

What It Takes.   What does it take to be a consultant? Technical expertise, of course, is required. This is the one commodity of which beginning consultants are always proud. The first question you need to ask yourself, however, is whether you can survive financially until you have found and executed some business, sent the bill, and had the payment sent to you. That cycle could be as long as 120 days.

   You must be comfortable with lack of structure; in fact, you must relish it. You must be aggressive enough to hustle business; do not expect clients to flock to you. You must keep current in the computer field. You must be able to become productive quickly in an unknown environment. Lastly, it is essential that you project a professional image. People who can pay for consulting are not attracted to the T-shirt/sandals look, no matter what your brilliance.

◇   THE ANALYST LIFE

Perhaps the most important thing that could be said about life as an analyst is that it is never boring. The business applications, the tasks, and the people provide variety and challenge. Tying all these together is communication, to which we now turn in Chapter 4.

FROM THE REAL WORLD

Thinking Big at Ford

As a budding analyst you may be able to imagine the job in a limited sphere, but can you even conceive of a system designed to analyze 12,000 elements on a car and over 11,000 elements on an engine? To make the task even more awesome, these computers are from the land of computer giants, Cyber 176™ and Cyber 760™.

Part of the engineering task is to analyze the car structure. Powerful computers and software are used to identify weak spots, simulate operations, and analyze results. Software modeling and simulation programs theoretically subject cars, or selected car parts, to all sorts of road conditions. Based on the computer-produced results, engineers can make adjustments quickly and perform another simulation of car activity—and wear and tear—through the computer. As one engineer put it, "We've always been aware of where major stress and weak spots could occur, but could not identify those trouble areas that come with time and extensive use." Now they can. The computer can.

Ford's use of this type of technology serves several hundred engineer users via telephone lines. The system did not develop overnight. Some of the software was purchased and some of the software to meet specific design and engineering needs was developed in-house. These programs support a wide spectrum of structural studies that relate to crash tests, structural effects of thin body parts, placement of engine mountings, vibration studies, and bumper design.

One such application is a software package that uses shades of color to pinpoint areas of high stress and structural weakness. When loads and constraints are applied to a particular part of a body surface, the weak spots become obvious through color coding. The round-the-clock operation is supported by a staff of 50 systems analysts. One does not have to wonder if they are busy.

—From an article in *Computerworld*

SUMMARY

- A systems analyst acts as a change agent when he or she introduces new or revised systems. According to change theory, people often find change difficult but usually adapt eventually. A systems analyst may take the change agent roles of persuader, catalyst, confronter, or imposer.

- Although the basic role for a systems analyst is to find solutions for systems problems, an analyst may take on other unofficial people-oriented roles such as detective, advocate, go-between, and amateur psychologist.

- The primary functions of the systems analyst are coordination of schedules and activities, communication among all project personnel using written and oral tools, and planning and design of the new system.

- Desirable analyst qualities include communication skills, an analytical mind, creativity, self-direction, organization, leadership, mediation, and an ability to work long periods of time without tangible results.

- Although working in teams is sometimes difficult and always time-consuming, a team environment may be used to take advantage of varying types of expertise and to distribute the work load. The key advantage of teamwork and group decisions is that the result is better.

- A systems analyst career may progress along different tracks. Most, but not all, analysts come from the ranks of programmers. The systems analyst has been identified as being in a high-stress career because of the pressures from people and schedules. The reward of a systems analyst career is job satisfaction, due to independence, mobility, variety, creativity, people orientation, work environment, the fact that the job is a continuing learning process, respect, and the fact that the result is a job well done. Ideal training for a new analyst would be an apprenticeship program, in which the new analyst works closely with an experienced analyst.

- A systems analyst must keep current in the field. A plan to keep current could include reading computer magazines, membership in a professional organization, and attending seminars.

- An analyst may choose to be a consultant. The advantages of consulting include independence, the freedom of being an entrepreneur, an unlimited market, travel, and the possibility of big money. The disadvantages are lack of benefits and possible negative psychological problems. To be successful, a consultant needs technical expertise, a financial cushion, comfort with lack of structure, aggressiveness, current knowledge, quick productivity, and a professional appearance.

KEY TERMS

Catalyst	Consultant
Change agent	Coordination
Change theory	Imposer
Confronter	Persuader

REVIEW QUESTIONS

1. Can you see examples in your own life of resistance to change? What deliberate small changes could you make in your daily activities, just to practice adjustment to change?

2. Which systems analyst role could you handle most readily? Which would be difficult for you?

3. Of the three primary analyst functions, which would you expect to occupy most of your time? Why?

4. How many of the analyst qualities listed in the chapter are among your strong points? Do you have an obvious weakness you could work on? Do you have other qualities, not listed, that you perceive as advantages for yourself as a systems analyst?

5. Do you fit the pattern of preferring to work alone rather than in a team? Have you ever worked on a team? What disadvantages do you see for yourself as part of a team? If you are reluctant to join a team effort, do you think you could adjust with relative ease?

6. Has the information in this chapter changed your perception of the systems analyst job? If so, are you more inclined or less inclined to pursue it?

CASE 3-1
RANCHERS BEEF UP THEIR PROGRAM

A group of cattle ranchers in Montana formed an organization, Ranchers Economic Cooperative (REC), to share their business expertise. They concerned themselves with cattle ancestry, feeding patterns, meat quality, and markets for their products. Much of the information they exchanged through their meetings and newsletters was common lore, but they were also quite serious about modernizing their operations.

After some consideration, they used REC funds to purchase a word processor to print the newsletters and copies of seasonal circulars. They engaged a computer services firm and, with the firm's help, purchased a microcomputer, a modem, and a subscription to a network service that would help the ranchers track the commodities markets.

Elizabeth Rodgers, the systems analyst assigned to work with the ranchers, watched them progress from beginners to semi-sophisticated computer users. At this point the ranchers wanted to buy a software package that would track the relationships between amount and type of feed and the resultant meat quality. Elizabeth was asked to participate in the decision process. She located one package that was quite expensive. After checking with other users of the package she decided that it did not provide significant benefits for REC. She was aware of two other promising packages being developed by software houses.

Since the feed/meat issue was new to her, the analyst did not pursue the issue exhaustively. Instead, when meeting with the ranchers, she chose a role that would raise points she felt were pertinent. Elizabeth came as a facilitator, structuring the discussion to help the ranchers reach a decision. They eventually elected to purchase one of the new software packages.

1. Did this group of users find the analyst threatening? Why or why not?
2. What role did the analyst assume? Why was that role appropriate in this situation?

CASE 3-2
PAINT 'N STAIN MOVES AHEAD—
WITHOUT MARSHA

Paint 'N Stain is noted for being a leader in the use of computer technology. Industry watchers credit the company's premiere place in the paint business to its dedication to productivity through technology. Paint 'N Stain maintains a cadre of programmers and analysts who appreciate the opportunity to work with the latest in hardware and software.

COBOL had been the principal language at the company for some time, and as a programmer at Paint 'N Stain, Marsha Blumenthal liked working in COBOL. However, to increase productivity on certain applications, Paint 'N Stain began considering some fourth-generation languages. The committee reviewing the new languages felt that Marsha's new project would be just such an application. Marsha rejected that idea out of hand. No amount of discussion made any difference; she would not budge.

Several months later, Marsha was again asked to be a pioneer, this time on a new split-screen terminal. The company was quite excited about the possibilities of increasing efficiency by updating two related files at the same time. Marsha was not at all excited and, again, found ways to resist.

Later that year Paint 'N Stain decided to pursue the idea of an information center to provide assistance for users. This venture would require many innovative approaches. Even though the beginning stages of this project were rather vague, it was considered a plum assignment. The task did not, of course, go to Marsha. She was disappointed, since she saw her own strength as working well with users.

1. How could a person in the computer field be so resistant to change? Do you think programmers and analysts usually embrace change more easily than other segments of the population?
2. What do you suppose will be the net effect of Marsha's attitude on her career? Is there a place in the industry for programmers who want to stick to what they already know?

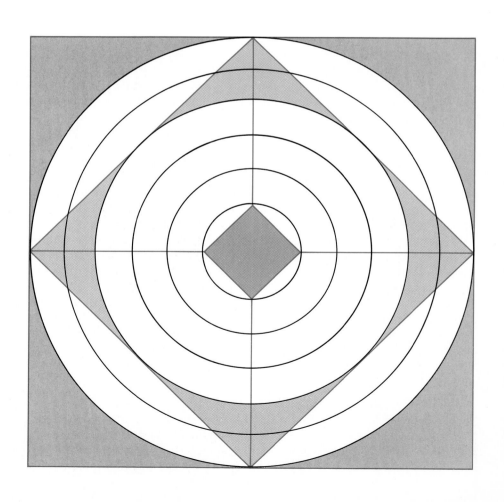

Chapter 4

Communication Skills: Documentation and Presentations

You have just transferred to the quality control group of a large firm, and your first assignment is to use a terminal to retrieve certain reports from the computer. When you ask for instructions on how to do this, you are told that there used to be some raggedy documentation but now it is a word-of-mouth operation. The person who really knows the system, however, is on vacation. You call on your contacts in the data processing department, but no one there seems to know anything about it either. You begin to wonder about quality control right here in the group. Communication on this system is obviously in need of reform.

Hot topics come and go in the computer industry but the subject of communication has had a different history, progressing slowly from orphan to exalted poobah. This change has come from two sources: the managers who got burned trying to handle under-documented systems, and the users who asked pointed questions such as, "Can we do better?"

This chapter focuses on the justification for and overall approach to communication. One-on-one and small-group communication, such as that used for interviewing and design walkthroughs, will be pursued in later chapters. This chapter is divided into two major subsections, one on documentation and the other on presentations. Let us begin here with the need for documentation.

◇ DOCUMENTATION: A MANDATE

Historically, preparing documentation has not rated near the top of favorite tasks. Many analysts come from a programming background, where they develop a distinct preference for the programming process itself rather than for the accompanying paperwork. An interesting phenomenon takes place as the new programmer-turned-analyst eases into the job, however: It slowly becomes apparent to the new analyst that documentation, the written word, is the only tangible evidence of his or her work. The meetings, the interviews, the endless talk-talk-talk leave no lasting mark. Documentation is no longer to be avoided; it is to be embraced. It is not quite that simple, however.

Consider briefly what documentation is. As related to a systems project, documentation includes:

- All written records related to the project phases: manuals of narratives, charts, forms, and so on (These documents are called "the deliverables" because they are presented as each phase is "delivered"—that is, completed.)
- All user documentation: training guides, policies, procedures
- Operations and production instructions: schedules, data, report distribution, security, and so on
- All communication paperwork for the project notebook: minutes, memos, schedules

These types of documents will be considered in more detail later, but first let us develop an appreciation of why they are necessary.

The Need for Documentation: Pay Now or Pay Later

As the computer industry grows more sophisticated, it becomes increasingly clear to management that documentation is a critical element in the development of a systems project. The reasons are many. Before we make that not-to-be-denied list, however, let us review the time-tested reasons why people still try to avoid documentation.

People can find many ways to rationalize not documenting. But those who fail to document systems do so at their peril, even though they usually have some sort of excuse:

- There is not time to document.
- The documentation soon will be out of date.
- Nobody ever looks at it anyway.
- There are more important things to do.

Sometimes managers themselves are guilty of the latter excuse, shifting priorities to more visible tasks: "Say, Carter, why don't you catch up on that documentation later. I'm going to need you for a few weeks on the inventory

BOX 4-1   Obfuscation

Business reports have been criticized for being muddy, pompous, and loaded with jargon. The American Management Association, which presents the following worst-bad-better versions of the same message, is trying to convince businesspeople to write clearly.

Worst

At this point in time the incremental use generated by the various functional areas interfacing with the Data Processing Department has resulted in a cumulative, multi-faceted demand that far exceeds the total working capacity of our present hardware.

Bad

Increasing use by various functional areas within the organization has resulted in a demand that exceeds our present data processing capacity.

Better

The equipment we have does not meet our present needs.

project." Perhaps we should add the ultimate excuse to the list: "I don't want to do it!" Since it must be clear by now that none of these excuses is acceptable, let us move on to the irrefutable arguments in favor of documentation.

The Compelling Case for Documentation

Rather than a last gasp end-of-the-project chore, documentation should be an integral part of the systems project as it progresses through the various phases. As such, time should be scheduled for documentation as for other project tasks. Let us consider the advantages of documentation as part of the systems process.

Documentation Serves as a Safety Net.   Current documentation on a systems project limits the vulnerability of the project to temporary absences and personnel turnover.

Documentation Minimizes Confusion.   As a system develops, many questions arise. Documentation can be used as a tool to help untangle the issues.

Documentation Saves Time.   Rather than being viewed as an onerous, time-consuming separate chore, documentation should be scheduled as an ongoing task. Tackled in a pay-as-you-go fashion, the documentation will prove an invaluable time-saver, both for current personnel who need to cover old ground and for new personnel who need to become familiar with the system.

Documentation Measures Progress.   Documentation lets others know what you are doing. Documentation also serves as a quick measure of what has yet to be done, as well as what already has been done. Finally, documentation is needed as a management decision-making tool; documentation must be approved before the project can proceed to the next step.

Document for Self-preservation.   Picture the progress of your career—not just the first or second system, but the span of years that covers many systems. You cannot accumulate new systems endlessly; you must let some old ones go to pick up new ones. The successful "letting go" process means that your documentation must be in tip-top shape so that those who follow you can understand the system. Unless you plan to leave town with no forwarding address, a poorly documented system could haunt you for years.

◇ THE DOCUMENTS

Documents come in many shapes and forms and serve a wide variety of
audiences. We begin this discussion with the documents that are produced
as a by-product of the systems process: the so-called deliverables.

The Deliverables

The documents prepared by the systems team to record the system's progress and mark milestone events are called the **deliverables**. These documents are noted, in particular, for their need for approval: The cover sheet
of each document will contain a list of people whose signatures signify
acceptance of the document's contents.

Each of the documents is described in a general way here. Their suggested contents will be detailed in the chapters where the related activities
are defined.

Preliminary Investigation Report.   The preliminary investigation report
discusses the nature and scope of the problem. It also suggests the broad
outline of a solution and the resources necessary to make a commitment to
pursue it.

Analysis and Requirements Document.   The analysis and requirements
document contains all the data analysis describing the existing system and
specifies the requirements for a new system.

Preliminary Design Document.   The preliminary design document states
the concept for the new design and the rationale for its suitability. It also
notes the resources needed to complete the plan: hardware, personnel,
money, time.

Detail Design Document.   The thick detail design document is an outgrowth of the preliminary design document. It encompasses all aspects of
the design, describing the input-process-output in detail and covering related
topics such as security and ergonomics.

FIGURE 4-1

The deliverables.
These documents are
"delivered" at the
end of certain sys-
tems life cycle
activities.

Development Document.   The development document consists mostly of program listings and test results. (See Figure 4-1.)

User Documentation Comes of Age

As more and more users enter the world of computers, their initial feeling about systems is strongly influenced by the quality of the documentation. Look what happened to the documentation for personal computer software. Early promoters, targeting the computer professional hobbyist, tossed out slap-dash documentation to accompany their slick software. These same vendors, who now sell to novice users, have a clear understanding that the content *and* the packaging are important. All users, from the naive novice to the computer sophisticate, view documentation as critical. It seems only fair, since they are paying the bills, that they have a reasonable understanding of how their software and systems work.

User documentation once took the form of endless typed narratives slipped into company binders. The written word is still a vital ingredient but its form and substance have changed. In addition, users now are given online documentation that provides assistance right at the terminal.

Written User Documentation.   Since users have become sophisticated documentation users through their experience with microcomputers, they are no longer content with reams of dry prose. They want documentation of substance, documentation that will tell them what they need to know in an easily accessible way. They want charts and diagrams. (See Figures 4-2 and 4-3 for examples.) They want quick reference cards. Above all, they want examples to show them *how* to accomplish their goals. Users need documentation that is pedagogically sound—that is, documentation that can teach.

Consider the following *bad* example. Documentation for a microcomputer word processing package appears in a spiral binder with small type. The type runs from the top to the bottom of the page and almost to the two edges. All the word processing options are described in narrative detail, but without examples. And it gets worse. The average beginner wants to accomplish some small subset of operations, such as enter a document, save it, and print it. Flush with this success, he or she might then wade into deeper waters. This particular document, however, burdens the user with descriptions of dozens of jazzy editing options, most of which the user would not use in a lifetime. And, oh yes, printing is finally mentioned on page 157. The point is that users need basic information before they can or will branch to more obscure topics.

Online User Documentation.   There was a time when programmers took delight in planting earthy messages to appear on the screen whenever a

Control key	Function
	This is the Enter key. You press the Enter key to send the displayed line to the requesting program.

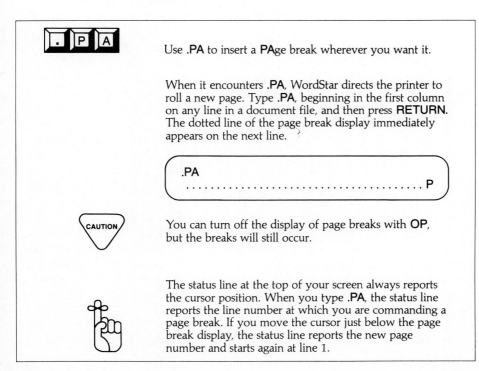

FIGURE 4-2

Written user documentation. This IBM keyboard discussion includes pictures to help the user.

Use **.PA** to insert a **PA**ge break wherever you want it.

When it encounters **.PA**, WordStar directs the printer to roll a new page. Type **.PA**, beginning in the first column on any line in a document file, and then press **RETURN**. The dotted line of the page break display immediately appears on the next line.

```
.PA
. . . . . . . . . . . . . . . . . . . . . . . . . . . . . . . . . . . . . P
```

CAUTION

You can turn off the display of page breaks with **OP**, but the breaks will still occur.

The status line at the top of your screen always reports the cursor position. When you type **.PA**, the status line reports the line number at which you are commanding a page break. If you move the cursor just below the page break display, the status line reports the new page number and starts again at line 1.

FIGURE 4-3

Written user documentation. These instructions in a word processing manual for inserting a page break include several visual aids: the keys to be pressed, a miniscreen version of the results, a "caution" signal, and a "remember" signal.

software user made an error—something like "You blew it again, stupid." It is a measure of industry progress that we would now be horrified at such a prospect. The average user might plead, "Don't tell me how dumb I am, tell me how to fix it."

A typical first-time user is a businessperson who does not really like to read written instructions. Many systems cater to this user by providing a HELP function that will give him or her on-screen information related to the current activity, as you can see in Figure 4-4. Taking this one step further, some systems provide a HELP button right on the keyboard. In fact, one famous cartoon shows a keyboard consisting of nothing but one gigantic button imprinted H E L P !

Online assistance is usually adaptable to a user's level of expertise. The detailed on-screen instructions, to which a beginner might willingly cling, soon become tiresome to the experienced user. Accordingly, a user is given options for various help levels—that is, varying degrees of assistance depending on user need. In other words, the documentation is so flexible that it can change with the user! It should be noted that flexibility is built into the documentation update process, too. Rather than printing revised paper documents, with all the delays attendant to that procedure, a systems person can make changes to online software that can be reflected almost immediately.

Another major source of online assistance is the **tutorial**, a program that teaches a user, step by step, the ways of a software package or system.

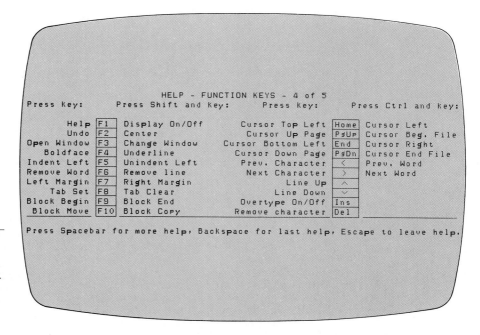

FIGURE 4-4

Online user documentation. This HELP menu for a word processing program provides brief descriptions of function keys.

```
                        HELP - FUNCTION KEYS - 4 of 5
Press Key:           Press Shift and Key:        Press Key:           Press Ctrl and Key:

        Help  F1   Display On/Off       Cursor Top Left  Home  Cursor Left
        Undo  F2   Center                 Cursor Up Page  PgUp  Cursor Beg. File
 Open Window  F3   Change Window     Cursor Bottom Left   End   Cursor Right
    Boldface  F4   Underline           Cursor Down Page  PgDn  Cursor End File
 Indent Left  F5   Unindent Left        Prev. Character   <    Prev. Word
 Remove Word  F6   Remove line          Next Character    >    Next Word
 Left Margin  F7   Right Margin              Line Up      ^
     Tab Set  F8   Tab Clear                Line Down     v
 Block Begin  F9   Block End            Overtype On/Off  Ins
 Block Move   F10  Block Copy         Remove character   Del

Press Spacebar for more help, Backspace for last help, Escape to leave help.
```

Again, we have microcomputer software to thank for this thriving technique. What could be more comforting to a computer-wary executive than to sit down with a "show me" disk in the privacy of his or her own office? Tutorials for popular software packages such as Lotus Symphony have paved the way for the industry. The drawback, of course, is the cost of preparing the tutorial software. Not every system can justify such an expense. Note the tutorial menu for CLOUT in Figure 4-5.

The bottom line on the subject of user documentation is that you must provide enough information so that users can manage without you. Long after you have moved on to other ventures, the user must carry on. Is what you left behind sufficient?

Operations Documentation

If the planned system is to be run in a large-computer environment, then operations and production personnel must be informed about the system. They will need some detailed knowledge:

- Scheduled run times
- Expected input sources and media
- Special output forms
- Delivery schedules, routing, and distribution
- Data file use and retention

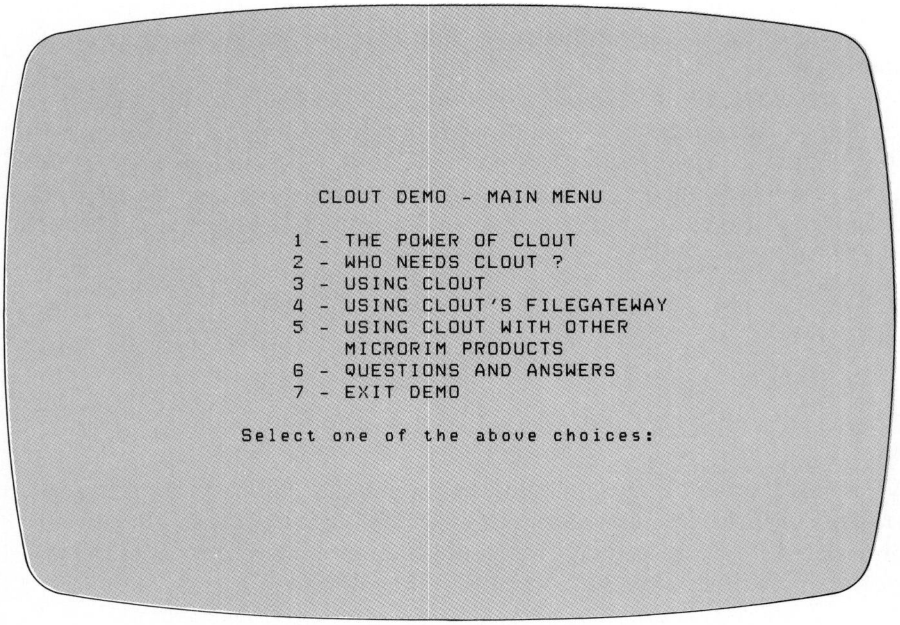

```
          CLOUT DEMO - MAIN MENU

     1 - THE POWER OF CLOUT
     2 - WHO NEEDS CLOUT ?
     3 - USING CLOUT
     4 - USING CLOUT'S FILEGATEWAY
     5 - USING CLOUT WITH OTHER
         MICRORIM PRODUCTS
     6 - QUESTIONS AND ANSWERS
     7 - EXIT DEMO

   Select one of the above choices:
```

FIGURE 4-5

Tutorial. This is the first menu of a user tutorial for a software package named CLOUT, which uses natural language commands to store and retrieve data.

- Libraries
- Planned update schedules for programs and databases
- Backup and recovery procedures
- Security controls

Perhaps the most important item is whom to call if the job fails. This list is typical but not necessarily inclusive.

Much of this data will be prepared in advance in meetings with operations and production staff, who will consider the existing systems and schedules. The form and content of the operations document will depend on established procedures in the data processing organization.

The Project Notebook

If you cannot figure out what to do with it—that piece of paper in your hand—it goes into the project notebook. If it seems to break new ground, then perhaps you should make a tab for it. If you see it as an isolated but indispensable item, then toss it behind the tab marked "Miscellaneous." The project notebook is not as frivolous as those last words may imply, but it is perfectly true that it will accept papers that would be suitable nowhere else.

The primary contents of the **project notebook** are minutes, schedules, budgets, memos, and reports: the people-to-people communications that keep the project humming. Two primary requirements of the project notebook are that it be (1) up to date and (2) accessible to any legitimately concerned person. Any project person or manager should be able to check the status of the project or the latest minutes—and find them in the project notebook.

The importance of keeping written records cannot be overstated. Imagine the consternation of the homeowner who engaged a carpenter for a $6,000 kitchen face-lift and was then presented with a bill for $10,000! You guessed it—nothing was in writing. As sorry as this true story is, businesspeople can ill afford to get caught in such a trap. As one cynic put it, "If it wasn't written down, it wasn't said." Be advised.

◇ PUTTING THE DOCUMENTATION TOGETHER

We have considered what documents are needed, and we discussed the contents of that documentation. Now we need to look closely at an important related topic, document preparation. Many factors contribute to making a document an effective communications device.

Word Processing: Get It if You Can

Would you consider writing a set of documents for a major project with a quill pen? It does not make much sense to use a regular typewriter either, not if word processing is available. If you have not yet used word processing, be assured that it is relatively easy to use and as critical to your career as typing is said to be.

The principles are easy, and much like the text editing procedures you may have used to save and revise program files. After a text file (document) is keyed in, it can be saved on secondary storage, usually disk. When the document needs to be changed, it can be recalled from the disk, revised, resaved, and printed as if new. The point is that only the changes have to be keyed. This is a far cry from retyping an entire document.

Most offices have dedicated word processors—that is, computers that are used for word processing software only. Their particular advantage is that they have special function keys that make certain tasks easier. For example, you might be able to hit the SAVE key rather than typing the letters S-A-V-E. Word processing software is also available for more general-purpose computers.

You may be in an office where documents and reports are regularly submitted to word processing personnel. This is fine. But, since this service may not always be available or convenient, it is urgent that you develop expertise in word processing yourself. Many word processing classes are available through public and private sponsorship. Or you may have the opportunity to teach yourself (using the documentation provided!) on a work or home computer.

What Makes Documentation Good

If you have ever tried to assemble a tool or a toy using the "simple" directions provided, you know that some directions are better—a lot better—than others. Good directions, and good documents, do not appear at random. A certain amount of planning goes into their success. Consider the following pointers.

Content. The actual material is our first concern, for even the most beautifully wrapped misinformation is not acceptable. The document must be *complete* and *accurate*. The material presented must be appropriate to the subject matter and to the audience. And, of course, it must have a table of contents and page numbers, the better to help readers follow the material and discuss it among themselves.

Style. The document should be *well-organized* and *clear* from beginning to end. A well-organized document flows logically and has the added advantage of being easy to update in the future. Clarity is often related to simplicity: Use frequent paragraphs, short sentences, and simple words. Mumbo-jumbo jargon is out; plain English is in. Another important feature is *closure*. The reader should be given information in digestible clumps; that is, each idea must be "closed" before a new one is begun. Finally, especially in user documentation, *examples* and *illustrations* should abound. All these stylistic considerations are important to help the reader understand the material in the document.

Physical Attributes.   The physical appearance of documents has taken a great leap forward since the advent of word processing. Another significant influence has been weary users who no longer tolerate hit-and-run documentation. Physical attributes can increase the appeal of a document. Begin the list with these standards: a *consistent typing font*, and lots of *white space*. If the audience for the finished product warrants it, consider *heavier paper weight*, paper and print *color*, *variable type faces* for headings and examples, *tabs*, and high-quality *binders*. You would probably not use these for in-house documentation. As you can guess, these selected items are probably more likely to appear in commercially sold documentation products, such as manuals that accompany home computer software.

Documentation Standards

It would seem that after all the exhortations to produce quality documentation, an appropriate next step would be to define documentation standards. Many managers have examined that idea. After several meetings, their collective wisdom usually sounds something like this: "If we establish a company-wide documentation policy, write a document to tell how to document, and make every project turn in paperwork that follows these guidelines, we will have compliance and consistency."

These plans do not always work well. No uniform standards apply to all projects. The biggest problem is that the show-and-tell document is so all-encompassing (in a misguided effort to cover every conceivable case) that it fits none of the projects. In particular, documenters of small projects glumly fill in DNA (Does Not Apply) on line after line; they feel silly. When the full ramifications of this policy become clear to almost everyone, the policy is quietly scrapped. So, where does this leave us? Probably with documentation guidelines that are broad in scope but short on detail—that is, standards that are loose enough to be readily applied to a variety of projects.

Far more important than any of these trappings, however, is management's commitment to documentation. More than a memo is needed here. Management must demonstrate over a period of time that documentation will be scheduled, completed, and reviewed.

BOX 4-2	**Problem Documentation**

Guess what your next assignment is! You know that big system, the one with 27 programs that pumps out the engineering design changes? It was written in the late '60s and—uh—it doesn't have much in the way of documentation, just some old flowcharts and listings. And, of course, the programs weren't structured back then—probably about a zillion GOTOs. It would be really helpful if you could write something up so that we don't have so many problems when one of the programs blows up.

Welcome to the land of dinosaurs. *Do not* settle down and try to track all the logic through the programs. Begin by determining the purpose of each program. Then identify all the data files, and draw charts that show the flow of data from one program to another. Then, if you have time to devote still more of your life to this undertaking, tackle the program logic.

The Technical Writer

The technical writer has traveled a long road to find respect. He or she has often been treated more as a clerk, or perhaps even a nuisance, than as a full-fledged project team member. Only recently, with the recognition of documentation as a critical element of the systems process, has the technical writer become an equal partner. The technical writer has the added advantage of serving as liaison between computer people and users. Not every organization, or even the majority, use technical writers regularly. But the trend is obvious and growing.

◇ PRESENTATIONS: YOUR FACE TO THE WORLD

An effective analyst uses presentations as a communication vehicle among systems and user personnel. As a systems project progresses, it is usual to give several presentations—some quite formal, others less so. The development of a major new system, especially one that crosses organizational lines, may require a veritable traveling road show, with the analyst repeating the presentation for several audiences over a period of time. If you are not quite sure you want to do this, read on.

Why It Should Be You

Not everyone relishes the role. A common problem, which will be addressed in some detail later, is nerves. That can be fixed. Perhaps your self-image precludes making a spectacle of yourself. That also can be set aside. The point is this: If you duck this opportunity/responsibility, you will seriously impede your career. Can you possibly be the lead analyst on a project and refuse to give a presentation on the very subject on which you are the reigning expert? You begin to get the idea. Now, having acknowledged that you must do this, let us carefully consider what it takes to be successful. It takes not just a "personality type"—there is almost a formula that you too can learn.

Presentations to Mark the Milestones

Before we discuss how you are going to get through all this, let us briefly consider the content and occasions of presentations. Each will be discussed in detail in subsequent chapters, as the related activities are covered.

Preliminary Investigation Presentation.   The preliminary investigation phase may be concluded with only a written report, but sometimes—especially if the project crosses organizational lines or has political ramifications—a presentation may be appropriate. In this case the analyst will report the initial problem findings and suggest a course of future action, probably leading to a full-blown analysis.

Analysis and Requirements Presentation.   The analysis and requirements presentation is a major one, partly because it may be your first opportunity to show various levels of the user organization your commitment and credibility. And, in order to establish credibility, you must be able to demonstrate that you understand their organization and problems. (You doubtless have done this at the user level; now is the time to build your base to include user management.) Finally, you need to be able to set forth the requirements of their system, so they know you understand their needs before you plan a new system.

Preliminary Design Presentation.   The whole point of having a preliminary design phase is to avoid getting into details until the overall concept is agreed upon by all concerned. The best communication vehicle for such agreement is the presentation, where the design plan will be presented and discussed. Higher management often attends.

Detail Design Presentation.   The audience members who can appreciate the details of the detail design presentation are the same people who have been working on the system all along, and the same people who can understand the detail design document. The presentation describes the details of the concept first presented as the preliminary design plan.

There may be other types of presentations, particularly a final report after implementation and evalution, but these listed are the most common.

Presentations: Secrets of Success

All experienced analysts have favorite tips for giving presentations. The following represent some of the collected practices.

Know Your Audience. Your audience for the presentation must be your first consideration. You cannot understand what they need to know or how to present your material appropriately if you do not know who will be there! From a practical viewpoint, you usually do have a pretty good idea who will be there: the people you have been working with on the project. But even though they may form the core of the group, one or more high-level managers may also be invited. Occasionally, the purpose of your presentation will be to keep peripheral people informed, so the group could be quite different from the usual crowd.

You will want to tailor your presentation to the level of your audience. Upper management may want only an overview of the system. Actual users of the system want a short overview, followed by more detailed information.

Or you may be asked to give the same presentation to several different groups. In this case, you may want to vary the presentation somewhat to suit the audience, especially if your basic purpose changes from informing to explaining to selling. Now, how much of the politician is in you? If you really want to ensure success, do your best to understand the needs and desires of the members of the audience before you prepare your content. Obviously, you would not alter the content significantly just to please, but it cannot hurt to be aware of audience needs.

Plan the Logistics. A smooth-running presentation is a minimum expectation, but it does not happen without preplanning. You must check the room out in advance to determine that the spacing, furniture, and lighting are adequate. You will also want to make sure your equipment works, and take care of such details as locating working outlets. If a microphone is to be used, be sure to test it in advance. If there is to be more than one presenter, you need to work out seating and standing arrangements and plan a smooth transition between speakers.

Use Visuals. Audiences cannot absorb masses of technical material through their ears, so visuals are a must. Since visuals are a key link between you and your audience, make sure they are classy and professional in appearance. A classic approach, and an easy one to use, is the flip chart. Make sure that the letters are large enough and that each page is simple and uncrowded. Consider using the combination upper-/lowercase that people are used to reading rather than all uppercase letters. If you use color, note that yellow cannot be seen from a distance. Similar principles apply to overhead foils and graphics.

More than one vendor has complained about being undercut by a rival sales representative who displayed outstanding graphics. But, as any trip to a computer fair demonstrates, graphics are here to stay. Computer graphics are not a new art form, but they have recently become inexpensive and easy to use, thus making them more available for presentation purposes. "Ease of use" does not mean ten managers gathered around a terminal to view graphics on the screen. Your advance preparation must include translating the graphics to paper and then using these to prepare viewfoils.

Focus the Audience.   Picture yourself walking into a room to be part of an audience. There is a flip chart in the front of the room with some writing on it. There is also a written handout at your place and you idly thumb through it. As the speaker begins, a second person ambles to the other side of the room to control the overhead lights. You now have four directions to look at once and probably are not giving the speaker your undivided attention. As the speaker, you will want to direct the attention of the audience—that is, to focus attention on one item at a time. For example, you would not give audience members handouts until you are ready to take them through it. Nor would flip chart pages be unveiled before use. Also, when using a flip chart, stand near it so the audience does not have to shift its gaze back and forth. Use a pointer to draw listeners' attention to the printed lines or drawing you are addressing.

Communicate Effectively.   Perhaps the most important point here is that most audiences resent being read to. You must be sufficiently well prepared that you can speak and maintain eye contact with the audience. Glancing at your notes is fine, but being glued to them is not. Do not talk too fast—staccato speech is distracting. Jargon, of course, is not acceptable.

A fundamental premise of presentations often is stated as follows: Tell them what you are going to tell them, then tell them, then tell them what you told them. This is a standard teaching technique and is very effective in helping an audience follow the presentation. To put it another way, your presentation could begin with an agenda, be followed by the body of the presentation, and end with a summation.

Maintain Poise.   Poise is grace under pressure. Begin with a businesslike appearance. If you want to be taken seriously you must make a serious presentation. Humor has limitations when you are selling a system. Note these small but potentially distracting activities to avoid: moving around excessively, keeping your hands in your pockets, jiggling coins or keys. Be enthusiastic but not giddy. Do not be disturbed by interruptions. Handle questions with ease. Act as if you know what you are doing. If you are well prepared you can make it look easy, and perhaps that is what being professional is all about.

Stage Fright: The War of Nerves

Would you like to guess what people's number one fear is? Snakes? Heights? Enclosed places? None of these. Survey after survey shows that the Number One Fear for most people is public speaking! Most of us would go to considerable lengths to avoid getting up in front to "say a few words," not to mention giving a full-blown presentation for the brass.

In a perverse sort of way, this is an advantage for you, even if you are one of the fearful. The reason is this: Since most people prefer to avoid making presentations, you can single yourself out by being eager—or at least willing—to give presentations. If you are career-minded, nothing will give you quicker exposure than looking good in front of the decision-makers. Now all you have to do is figure out how to control your nerves—a difficult but not impossible task. Here are some hints.

Control Your Physical Environment.   Most of us are familiar with the signs of nervousness: trembling, cotton mouth, wobbling voice. The physical symptoms can be controlled or hidden. If you have notes, place them on a podium or other surface. Do not hold your notes in your quivering hand where they will rattle noisily. Try to stand behind something if you need to hide knocking knees. Concentrate on projecting a clear voice.

Practice.   Introduce yourself to situations where you will have a chance to practice handling your nerves. A simple technique is to use a tape recorder, which gives your performance a certain urgency. A tape recorder is uncritical but it does not miss much. A video recorder is even better, since you can see the visual impact, too. Practice your speech alone in front of a mirror. Do not write it out word for word. Just put down a few key phrases to remind you of your major points.

View Nerves as Excitement.   Nerves are not all bad. Many entertainers feel that if they are too cool in the wings they will not have that sharp "edge" on stage. Anticipate the best.

Stage fright can be brought under control, but many people never get over that case of nerves just before they are "on." Perhaps the final and best advice about nerves is to concentrate on your audience and on getting your message across. In the process your nerves may diminish.

BOX 4-3	Overcoming Stage Fright

Consultant Jean Withers offers the following tips to reduce stage fright when making presentations.

1. Remove the unknown from your speech.
2. Be well prepared.
3. Remember that you were invited.
4. See the audience as your ally.
5. Focus on the task, not yourself.
6. B-R-E-A-T-H-E.
7. Control your nervousness. Don't deny it.
8. See nervousness as excitement.

◇ SELLING THE SYSTEM

The element common to these vehicles, whether writing or speaking, is communication. You are communicating with all who are involved in the systems process. As a by-product of communication you are selling yourself and the system.

Managers ultimately make the go/no-go decisions, but they must depend on others for their expertise. As a systems analyst, you are assigned tasks based on your expertise, and your ideas will carry some weight. After you have studied and worked on the system for a period of time, you will form quite definite opinions on how the new system should be developed. And on the basis of these opinions you will be selling the system through your communications.

Studying the system and determining new solutions is a step-by-step process. In Chapter 5 we begin a detailed look at the preliminary investigation phase of the systems life cycle.

FROM THE REAL WORLD

Stocks Online

If you are anxious to call up your stock portfolio on the computer, step right up. Major stock brokerage firms are unveiling online networking systems that provide investors with instant information and—of course—an extra nudge to buy more stocks. Investors are charged for this service, even though the brokerage houses do not profit directly from their online services.

To gain access to these systems, an investor needs only a personal computer and a modem. Software is provided by the brokerage firm or recommendations are given as to the most appropriate communications program to buy.

E. F. Hutton, the nation's second largest retail brokerage house, was the first brokerage firm to announce an online system, the Huttonline.™ Huttonline offers its clients access to daily statements of their accounts, electronic mail to and from the broker, research information and analysis, and investment opportunities offered by Hutton. Other brokerages have followed suit with similar packages.

Brokers view customer access to online information with mixed feelings. Some believe clients' online access will cut the time wasted giving quotes over the phone, allowing more time to be spent on research. Customers will be better informed and thus take a more active interest in their accounts, generating more business. Other brokers are not so sure of the windfall, fearing that the lessened client interaction will reduce opportunities to "make that sale."

— From an article in *PC World*

SUMMARY

- Documentation is an integral part of the systems project. The advantages of documentation are as follows: it serves as a safety net in case of personnel turnover, it minimizes confusion, it saves time, it measures progress, and it helps "self preservation" because you can become free of old projects.

- The deliverable documents are those that are delivered as a by-product of the system phases. They are the preliminary investigation report, the analysis and requirements document, the preliminary design document, the detail design document, and the development document.

- Written user documentation has become clearer and easier to read over the years, now including many illustrations and examples. Online user documentation may include HELP functions that give a user on-screen information related to the current activity. A tutorial is an online documentation tool that teaches a user step by step how to use a software package or system.

- Systems to be run in a large-computer environment need operations documentation that describes scheduled run times, inputs and outputs, delivery schedules and routing, files and retention, libraries, update schedules, backup and recovery procedures, and security controls.

- The project notebook contains minutes, schedules, budgets, memos, reports, and other pertinent project-related communications. The project notebook must be up to date and accessible to any authorized person.

- A document may be prepared using word processing. The document is keyed in and saved on secondary storage, usually disk. When the document needs to be changed, it can be recalled from the disk, revised, resaved, and printed as if new.

- Good documentation may be characterized by content (complete, accurate), style (well-organized, clear, containing examples and illustrations), and physical attributes (consistent typing font, white space, paper weight, color, variable type faces for headings, tabs, binders).

- Documentation standards vary by organization. If presented as a detailed document to be used company-wide, such standards may not work well. Guidelines that are broad in scope are more appropriate.

- The status of the technical writer has improved along with the general recognition that documentation is valuable. A technical writer may be an equal partner on the systems team, taking a major role in preparing documentation.

- An analyst can use presentations as a communication vehicle among systems and user personnel. An analyst must be able to give presentations with ease.

- Presentations are usually given at certain times in the systems life cycle, namely, at the completion of these activities: the preliminary investigation, the analysis phase, preliminary design phase, and detail design phase.

- Presentation secrets of success include: Know your audience, plan the logistics, use visuals, focus the audience, communicate effectively, and maintain poise.

- During presentations, nerves can be reduced by controlling your physical environment, practicing, and viewing nerves as excitement.

KEY TERMS

Analysis and requirements
   document
Deliverable documents
Detail design document
Development document
HELP functions
Operations documentation
Preliminary design document

Preliminary investigation
   report
Project notebook
Technical writer
Tutorial
User documentation
Word processing

REVIEW QUESTIONS

1. As a programmer, have you ever postponed the related documentation? Are you convinced that documentation is a necessary part of your career as a systems analyst?

2. How has documentation for personal computer software influenced user documentation?

3. Have you ever used a tutorial for a new software package? If not, find one to try so you will understand their potential value to your users.

4. Can you think of papers not mentioned in the text that would belong in the project notebook?

5. Have you ever used a word processor? If so, describe some of the features of that system. If not, where can you learn/practice word processing?

6. Study some of the user documentation for microcomputer software at your school or in your local store. Do you think it qualifies as "good documentation"?

7. Have you ever given a presentation? If so, were you nervous? Do you think the advice given in this chapter might help you minimize the effects of nerves?

8. Given the choice of media available, which kind of visuals do you think you could use most effectively? Do you think two different kinds of visuals, say viewfoils and flip charts, can be used for different parts of the same presentation?

9. Make a list of the kinds of things you would want to take care of in the room before the audience arrives for the presentation.

CASE 4-1
THE GENIUS

The new microcomputer operating system is on the shelf now, literally. That includes several boxes of program listings and some notes scratched out by Jack Mitchell, the project lead. Jack is considered a genius, the kind of guy who can conceptualize and pump out programs in record time. The group thought they were lucky to get him, even if he did seem a little unconventional.

Just how unconventional he was soon became obvious. His boss, Al Kennedy, knew from the beginning that Jack dressed in his own style, mostly jeans, safari shirts, and open-toed sneakers. He was prepared to live with his work habits too, coding sleepless for three solid days at a stretch and then disappearing for several more. The unforeseen was that Jack had no patience with fancy documentation reports, or any reports for that matter. He was even less interested in making presentations to higher management. Al tried to get around this by assigning a technical writer to work with Jack, but communication broke down completely. Finally, Al decided to give Jack a pep talk, then leave him alone and hope for the best.

After several months Jack declared the programs ready. They did not mesh well with existing system programs. No one really seemed to know how to make the whole package work effectively. Marketing personnel were puzzled about how to describe the new operating system in their sales pitches. Higher management demanded an accounting from Kennedy who tried desperately to get something formal on paper from Jack.

Jack got tired of all the "hassles" and decided to leave the company and start his own business at home. The project was eventually abandoned. Al Kennedy decided to move to a company in another state.

1. How could Al have handled Jack more effectively?
2. What are the prospects for the success of Jack's new business?

CASE 4-2
A PRESENTATION GONE AWRY

Analyst Diana Prewitt spent several months analyzing and designing a new online system to provide shop-floor data collection for management control at Magee Manufacturing. The early work had gone well. Diana had a workable plan to capture source data from process control, heat treatment, plating, machine shop, and assembly. She anticipated saving the company many thousands of dollars.

The preliminary design presentation audience included managers from all the affected organizations. Many of them had never met Diana before. They did not have an opportunity this day either, because Diana was the last one to arrive, nine minutes late. Her appearance was somewhat disheveled and harried, but she apologized, passed out a thick handout, and quickly proceeded to set up the overhead projector and place her foils in order.

She scrambled on her knees behind a table to find an outlet only to discover that it was dead. She did find another outlet that worked, but it was too far away so she had to duck into a nearby office to borrow an extension cord. After turning the viewer on she realized that several people in the audience would not be able to see around the viewer and suggested that they move to a different part of the room, which they grudgingly did.

A woman in the back was able to find the switch to turn out the room lights and the presentation began. (Diana had been too flustered at this point to remember to introduce herself.) On the third foil there was a little pop as the light on the viewer burned out. Diana made a weak joke about the "best-laid plans" and left the room in search of a bulb. The audience had been in the room an hour by this time. Three of the managers stood up and left. The others heard the presentation but had little to say afterward. Diana left with the sinking feeling that it would be very difficult to recover from this fiasco.

1. Hindsight is a wonderful thing. Even though this story is exaggerated, can you identify ten preventable errors that Diana made?

PART 2

INITIAL PHASES OF THE SYSTEMS LIFE CYCLE

The initial phases of the systems life cycle are preliminary investigation and systems analysis. Both of these phases involve background work that is necessary before any action can be taken. The net result of these two phases is that the analyst defines the problem, establishes a relationship with the user organization, and gains an understanding of the system that can be used as a basis for systems design.

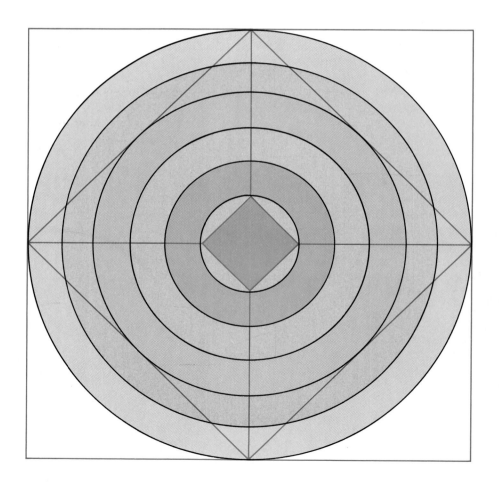

Chapter 5

Preliminary Investigation

Do we have a problem? Can the computer be used to solve the problem? Can the computer be used to create opportunities? In some way, problems and opportunities will be brought to the attention of the systems analyst.

Whose idea is this? The **impetus for change** through systems comes from a variety of sources, as you can see in Figure 5-1. These sources may be within or outside the organization considering a change. A user may feel that an existing automated system is too antiquated to be responsive. A manager may see that problem situations are not being properly reported. Government regulating bodies may decree changes that affect systems, such as the IRS demand for reporting earned interest. Perhaps a manager reads an article in *Fortune* magazine containing a new idea for growth. The introduction of a new product by

the competition may spur your marketing division to establish a comparison-and-impact system. The sources are endless.

It is common for such systems-related requests to pour into the data processing department in a steady stream. In fact, most companies or organizations with established computer expertise have a considerable backlog of requests. These requests usually are filtered through management or some sort of committee, where they are evaluated and assigned to systems analysts for investigation. So, either directly or indirectly, an analyst must have the necessary **management authority** to proceed with the assignment. This is where you come in, at the beginning. The preliminary investigation phase is the first phase of the systems life cycle.

But first we must make a short detour to explain King Books.

◇ THE KING BOOKS CONNECTION

Before we begin the systems life cycle in earnest, we must introduce the King Books case study. As we progress through the next few chapters, we will illustrate various techniques and tools. Most of these examples and illustrations will be linked to a single case, King Books. Although the case is not intended to be comprehensive, the various examples can be tied together with a single thread and do make a coherent package. You can recognize the items related to King Books from the title and the logo. One more note: King Books is not associated with the usual self-contained cases at the end of the chapter. Take time now to read "King Books: The Story." Now you are ready for your first concern, your commitment to the user.

◇ THROUGH THE USER'S EYES

Any study of the systems analysis and design process places continued emphasis on the user. The user views this process from a different perspective than the analyst and with a different kind of knowledge. The analyst needs to see the system through the user's eyes in order to provide adequately for user needs.

The user, like the retail customer, is your reason for being. Without the user there is no system. And, of course, the user pays the bills. As you begin this first phase, it is a good time to establish your commitment to the user.

User Involvement

The phrase **user involvement** is not used lightly—it is the key to the success of a project. Users must be involved from the very inception of the project.

[handwritten margin note: Q: When the author refers to users is that all users or just applications programmers, coding?]

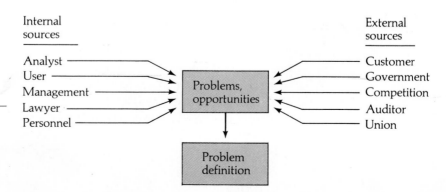

FIGURE 5-1

Sources of requests for system changes. Requests for system changes can come from internal or external sources.

Internal sources: Analyst, User, Management, Lawyer, Personnel → Problems, opportunities ← Customer, Government, Competition, Auditor, Union (External sources)

Problems, opportunities → Problem definition

KING BOOKS

The Story

Although Glen Neilson earned his degree in business, his main interest had always been books—all kinds of books. Glen went to work for a large insurance company but he found time on evenings and weekends to browse in bookstores. His favorites were the small, older bookstores with a musty smell and an atmosphere of reverence for books. He spent time with the bookstore owners, expressing appreciation for their stock and adding to his own growing collection.

Glen also frequented the more popular stores, dazzled by their large and glossy inventories. But he somehow felt that they were not "real" bookstores. Over a period of time, however, he began to change his mind. This change was based on two things. First, the cherished classical bookstore was dying out. Owners admitted to continuing losses, and he watched more than one go out of business. Second, he saw that the public was attracted to the fresh, modern atmosphere of the new stores, and to their lower prices. The idea of more people being exposed to more books appealed to Glen.

Glen gradually formed a plan for his own store. It would be a discount bookstore, and openly advertised as such. He would have a large, varied stock. The key would be fast turnover, a concept that meant emphasis on mass appeal books and a significant advertising budget. He knew the slim profit margin all too well, but counted on his business skills to make the survival difference.

In three years, Glen's plans came into being as King Books. He combined borrowed money with his own savings to revamp an old warehouse in a newly fashionable part of town. He stocked it with a large and attractive selection of books. He did not forget his vow to advertise heavily.

The store was wonderfully successful. Patronage was steady, turnover was rapid, and the King Books staff grew to 87. Glen's head was spinning with new ideas. He considered adding new related sales lines, such as stationery supplies. Further, he envisioned new King Books stores in the suburbs and nearby towns. This rosy picture was marred by Glen's monthly nervousness about paying the store bills and making enough profit to live on. He knew that the best tool to help cut costs was the computer, and he drew up an outline of in-store applications for its use.

The first item on his list was an automated accounts payable system. On the advice of a colleague, Glen called Universal Computer Services, a broad-based consulting company that could handle a variety of needs. Analyst Dave Benoit worked extensively with Glen and his staff over a period of several months on the accounts payable system. Together, they

defined the problems and then moved through the steps of the systems
life cycle. Although we shall not discuss this project in detail, many
examples in the next few chapters relating to the systems life cycle will
involve King Books.

Many users today are quite sophisticated about automation and accept
computer-related assignments readily. Others, with little or no experience,
may stall and even resist the changes. Whether at one of these extremes or
somewhere in the middle, the user must be an early and permanent participant in the process.

Most analysts understand the theory of user involvement. They need
the user early on for authorization and for finding out how the existing
system works. The problems usually crop up later, when the analyst is well
into design and development. Users seem less critical in the computer-
oriented part of the project. But you must not let the user drift into disuse;
the latter parts of the project are just the times when you need the user
alongside to help with testing and training. The user should be carried as
a full partner through the entire systems life cycle. Only then will he or
she feel full confidence in the implemented system.

The User Organization

The need to understand the user organization you are dealing with is paramount. This issue is sometimes lost on analyst trainees, who may invest
large amounts of time with people who do not impact the system or who
are not decision-makers. Or, in a worst-case example, they may feel comfortable talking with the clerical workers and not realize that the project is
doomed because funds are being held captive by a company partner who
finds the project threatening. As tiresome as this sounds, analysts may face
such situations.

So . . . how is a person supposed to recognize a political scenario? It is
not easy. Begin with the organization chart. To weigh their words, you need
to know personnel functions and how the personnel officially interact with
one another. One of your first tasks, then, when beginning a new project
is to obtain a copy of the organization chart. If none exists, you will need
to draw it yourself. Note the example in Figure 5-2. An **organization chart**
is a hierarchical drawing showing management by name and title. As you
talk to various people on the chart you will gradually come to understand
their informal relationships as well. You begin to understand the political
layout. Many people, it should be noted, are appalled at the thought of
political influences on a technical process. They prefer to avoid or not

BOX 5-1	**Project Success Factors**

- Management support
- User involvement
- Realistic scope
- Realistic expectations
- Suitable budget and schedule
- Project tracking and control

acknowledge politics. Unfortunately, office politics is a fact of life, and ignorance is not bliss.

Once you are in the thick of things, politically rife situations will be brought to your attention, often in confidence. It is sometimes your unhappy task to sift innuendo and hearsay from fact, and to discourage recitals that border on gossip. These words are merely cautionary—it is certainly not unusual for analysts to avoid politics altogether on a project.

Having taken the first steps toward user involvement, you need to consider what to do to get yourself and the project off to a good start.

GETTING ORGANIZED

The preliminary investigation phase is similar to the entire systems life cycle, but much less rigorous. You will have to perform some subset of the analysis and design functions, as you determine what the problem is and what to do about it. The net result will be a rough plan on how—and if—to proceed with the full-blown project.

Your preliminary investigation study will not be detailed. It is not usually necessary at this stage of the project to get into particulars of the workaday world, only to get enough information to understand the problem and the organization's objectives.

You should feel a strong urge to make a to-do list at this point. You will find that all phases of the project require a series of activities, but, particularly in the beginning, a list is helpful. It will include having meetings with your own management, having meetings and interviews with users, establishing the problem definition (nature, scope, objectives), assessing the feasibility of proceeding with the project, planning schedules and budgets and personnel, considering general alternatives and their benefits, preparing the preliminary investigation report, and making a presentation of your findings. These and more we now shall pursue in this chapter.

FIGURE 5-2

Organization chart
for Kings Books.

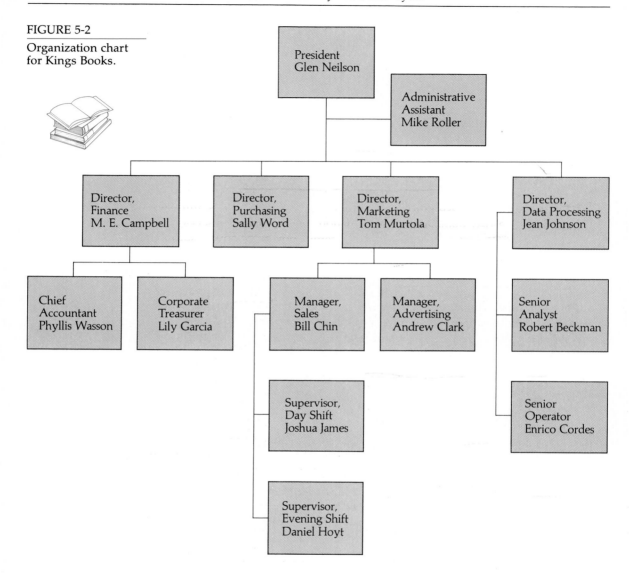

PROBLEM DEFINITION: NATURE, SCOPE, OBJECTIVES

Your initial aim is to define the problem. You and the users must come to an agreement on these points: You both must agree on the nature of the problem or opportunity and then designate a limited scope. In the process, you will also determine the objectives of the project.

The Nature of the Problem

As an analyst, you will think of yourself as a problem-solver. You will see a new or revised system as a response to a problem. You must be sure, however, that the system solves the actual problem—which may be somewhat different from the problem as originally perceived. Your objectivity will help here. As you investigate the existing system you will be free of preconceived biases. (See Figure 5-3.)

The pervasive nature of computers has caused a change in their use, and this change has spilled over into systems planning: Computer systems are seen not only as problem-solvers but also as a means for seizing opportunities. The opportunities come in categories familiar to businesspeople, who habitually use these kinds of words to describe their goals: improved productivity, refined management information, better customer service, and the like. The "Opportunities Through Systems" box gives some examples.

Scope: Where Do You Draw the Line?

In the beginning the analyst and user must agree on the scope of the project —on what the new or revised system is supposed to do and not do. If the scope is too broad, the project will never be finished. If the scope is too narrow, it may not meet user needs. The project scope will be clearly stated in the preliminary investigation report. A project's scope expands both because of new user insights and analysts' helpful tendencies. An industry expert, Dr. Kyu Lee, is famous for coming to the rescue of overdue projects. Asked his secret of success, he replied, "I just trim them down." Is it really that simple? Are troubled projects really just projects grown too fat? It seems to be true that a return to basics—the original purpose of the project—does make a substantial difference. Many of the add-ons can be saved for the maintenance phase of the project. And—an interesting phenomenon— sometimes the add-ons just fade away before maintenance is started. Time can decrease the seeming importance of some features.

Objectives

You will soon understand what the user needs—that is, what the user thinks the system should be able to do. You will express these needs as objectives.

Examine the objectives for the King Books accounts payable system. Note how simple and straightforward these words seem. The people who run the existing accounts payable system already know what such a system must do. It remains for you and them to work out how this can be achieved on a computer system. In the next phase, the systems analysis phase, you will produce a more specific list of system requirements, based on these objectives. For now, let us review our relationship with the user.

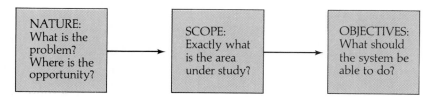

◇ EVALUATING PRELIMINARY INVESTIGATION INFORMATION

Is it *feasible* to proceed with this project? Measure your findings against these criteria: technical feasibility, operational feasibility, and cost effectiveness.

Technical Feasibility.   Is it possible to provide an automated solution that will meet the system requirements? Does the analyst have—or can he or she get—the resources to fulfill the hardware and software needs of the suggested solution?

Operational Feasibility.   Can you suggest a solution that will be workable in the user organization? Do you and the users agree that a suggested solution could meet the objectives of the organization? Can all these things be done within reasonable time constraints?

Cost Effectiveness.   The bottom line, of course, is whether the user organization can afford the system under consideration. Both development costs and operational costs (the day-to-day monies needed to keep the system going) must be considered.

Based on these criteria, four possible courses of action are available:

1. *Do nothing.* The current system cannot be improved by any course of action considered, or the cost of doing so is prohibitive.
2. *Modify the existing system.* Many systems today are already automated, but, if the system was developed more than five years ago, chances are good that new technology will enhance the system significantly.
3. *Develop a new system.* This may involve a manual system that can be automated or an antiquated automated system that needs a complete replacement.
4. *Refer project to the information center.* It may be appropriate to have the user pursue the project using the software tools and assistance available at the company information center. This option is becoming increasingly significant. We will consider the information center in detail in Chapter 17. *Commercial software pkgs*

BOX 5-2	**Opportunities Through Systems**

New or changed systems may be inspired by opportunities envisioned by management or others. Businesspeople are reaching out beyond the traditional heavy-paper-flow applications for new systems. Some typical opportunity categories and examples are as follows:

- Increase productivity

  Speed up supermarket checkers with point-of-sale terminals.

  Introduce robot-controlled factory systems for tedious tasks.

  Set up multifunction cash machines and let the customers provide their own teller services.

- Improve management decisions

  Help managers access files with microcomputer spreadsheet applications.

  Provide cross references among related systems such as shipping, warehousing, marketing, and sales.

- Strengthen system control

  Introduce a bar code badge system to identify marathon runners and accurately post their race times.

  Provide exception reports on products selling below projections.

- Improve customer service

  Institute a word processing system that will speed "personalized" replies to a senator's constituent inquiries.

  Identify fast-paying customers for discounts.

  Balance volume of newspaper ad readership with stocks of advertised goods to avoid rainchecks.

Assuming that you are going to proceed with some plan that requires resources, you will need to consider schedules, budget, and personnel.

◇ THE PLAN: SCHEDULE, BUDGET, PERSONNEL

As you assemble information in anticipation of the preliminary investigation report, you will need to support your plan with some figures for the schedule, the budget, and personnel. At this time, remember, your main concern is the next phase. Knowing that management can scrap a project

KING BOOKS

Problem Definition

True Nature of Problem

The nature of the problem is the existing manual accounts payable system. One part of the problem is the unnecessary expense due to errors related to a manual system of check preparation:

- Paying an incorrect amount
- Sometimes paying the same bill twice
- Writing several checks to the same vendor

The other principal problem relates to the failure to track vendor invoices:

- Failing to take advantage of discounts by paying within specified periods *Make it company policy to demand vendors indicate the purchase/order # on the vendor's invoices*
- Lacking information on vendor account status

Scope

The scope of the project will be limited to the normal accounting and reporting activities associated with an Accounts Payable System.

Objectives

The objectives of the new system are to provide King Books with an automated and flexible Accounts Payable System with the following features:

- Improved accuracy and timeliness of accounts payable records
- System-generated transactions to take advantage of vendor discounts
- Better management control of the accounts payable system
- An easy-to-use orientation
- Reduced operating costs of the accounts payable function *from (Red accuracy) (len labor)*

anywhere along the way, the normal procedure is to present numbers in detail for the next phase only, the analysis phase. You need to decide how long it will take (in weeks or months), how much it will cost (mostly labor), and who will be needed from the data processing and user organizations for the analysis study.

Depending on the nature of the project and the organization you are working for, you may also be expected to determine these figures in rough form for the entire project.

Where do you get all these figures? Although there is a certain amount of jesting about pulling numbers out of thin air, this is rarely what happens. Experienced analysts can sometimes prepare the necessary figures without assistance. Some organizations expect all analysts to work carefully with finance or accounting departments. Others have established formulas for estimating time and dollars. These approaches and other planning aids for analysts are discussed in more detail in Chapter 14 on project management.

Note also that even the "do nothing" option has costs, such as continued inefficient processing, missed product growth opportunities, and the like.

◇ BENEFITS

At this point you have only the most general idea of the solution to the problem or the method of seizing the opportunity. But still, you do have an idea of what benefits the new system will bring. One reason you know is that the benefits are often the flip side of the original problems. Some benefits, such as saving time and money, are common to many systems. Even though you have no specific plan in place, you can already see that any system you embrace will include certain benefits.

Benefits fall into two categories: **tangible** and **intangible**. Tangible benefits are usually easy to measure. In the benefits list for King Books, for example, the number of vendor discounts actually used can be counted and compared. Benefits relating to user productivity or employee morale, however, are difficult to measure and thus intangible.

KING BOOKS

Anticipated Benefits

Benefits are usually closely tied to the system objectives. Note the following benefits of an automated Accounts Payable System for King Books:

- Improved check-writing accuracy
- Vendor discount opportunities
- Improved management information
- Increased user productivity
- Improved employee morale

The benefit list , it should be noted, may change somewhat as the specific design emerges later. For now, recognize that you must have a general benefits list early on, in order to spur the decision-makers forward.

◇ THE PRELIMINARY INVESTIGATION REPORT

Before we get to the actual report, consider this story. Alan Wood, who considered himself something of an intrepid analyst, had just spent three weeks on a preliminary investigation of the Alexander Sports chain. He planned to recommend that they upgrade their existing inventory system to include point-of-sale terminals and an online inventory control system. Although he knew he was talking about over $100,000 for development costs and new hardware, he was confident that the proposed solution was the answer to the user's needs. He had been a junior analyst on a team that had proposed such a solution in a similar situation.

Alan made an appointment with his boss and earnestly described what he envisioned, concluding, "Let's go for it." His boss, recognizing the folly of such a casual approach, insisted that Alan make a full report *on paper* and have it approved by the user. Alan is not the first to try to circumvent paperwork in the name of efficiency, but it would be nice if he were the last. An organization cannot commit vast resources on the basis of some glib remarks, nor can the data processing department enter into a work agreement without some paper being processed.

The **preliminary investigation report** is the first major piece of deliverable documentation. That does not necessarily imply long length. For a brief study, a page may do. More likely, the report will be of medium length and contain several parts. The following sections are not always used, but they are representative of what might be in such a report.

- *Title page*. Include names of key personnel, lines for approval signatures, and an abstract of the project subject matter.
- *Project overview*. Also known as an executive summary, the project overview is a one-page condensed version of the report. It is designed to give busy readers quick information until they have time to study the whole report.
- *Problem definition*. Description of the problem or opportunity, the scope of the system, and the system objectives. Make sure your description shows that you clearly understand what the problem is. Mention exclusions from the scope of the project if they have been disputed.
- *Main text*. This is the heart of the report and typically includes a summary of findings, expected benefits, and recommendations.
- *Costs, schedules, personnel*. Anticipated costs for development and operation of the recommended system. Also include a general timetable and needed project personnel.

- *Appendix*. Charts, graphs, forms referenced in the report. After the report has been prepared and approved, you may feel it wise to make a presentation on the project. Your audience will probably consist of representatives from the user organization and possibly your own manager. Keeping details to a minimum, you present an assessment of the project.

If you and the users and management agree to develop a new system or modify an old one, you will need to assemble a systems team and then proceed into the analysis phase of the project, in which you will study the existing system in some detail.

◇ THE SYSTEMS TEAM

The size and composition of the **systems team**—also known as the **project team**—depend on the size and nature of the project. The team for a small project may be mostly you, the analyst, together with a selected user. In fact, you may handle more than one small project simultaneously. A large project may be handled by an entire team of analysts as well as users. Although this text, for the sake of simplicity, speaks of "the analyst" as if it were always just one analyst, keep in mind that analysts often work together in teams.

Large projects often cross departmental lines, so it is important to have a user representative from each department on the team. Since you are establishing system requirements, this original team may be user-heavy. As the project progresses into the phases requiring more technical expertise, team composition will change, as some users are replaced by data processing representatives. The project leader is usually a systems analyst, a person with both technical and communications skills.

◇ AND NOW TO PROCEED

You have now completed the preliminary investigation phase. The results of your investigation clearly show that the project has some merit. Now you begin your work in earnest, turning to the next phase, systems analysis.

FROM THE REAL WORLD

Computers on the Campaign Trail

Taking the computer along on the latest campaign swing might be as important as taking the candidate. Computers, once considered by political hopefuls as useful for producing address labels, now are an integral part of campaigns at every level.

Campaign support software is blossoming in every direction, but it seems especially right for personal computers, because of their low cost and portability. Two commercial programs that have actually been credited with campaign victories are called *Campaign Manager*™ and *Solon*™, both written for the IBM PC®. *Campaign Manager*'s first victory was the election of a dark horse candidate for mayor of Stamford, Connecticut. The professional politicians did not give him a chance, but they were unaware of the "computer connection." *Campaign Manager* is a menu-driven program that keeps track of polling, media time and space purchases, press relations, candidate scheduling, fundraising, campaign budgeting, and reports to government monitoring agencies.

Solon was planned as the Cadillac of political programs. It is designed to go beyond the campaign itself and help the winner stay in office. The program includes standard campaign features such as voter and contributor targeting, direct mail, candidate scheduling, polling and election district analysis, reporting of campaign finances to the required government agencies, media and volunteer management. The greatest strength of the program is its database retrieval ability. Candidates can sort records in virtually any combination—to print, for instance, all members of a political committee of a given income and who are concerned with a particular issue.

This allows that group to receive a special mailing about that particular issue. (One national candidate figured that his best potential financial supporters were over 40, drove foreign cars, read *The New Republic,* and lived in suburbia—that's pretty specific!)

Security is very tight in most political campaigns. It is for this reason that some campaigns have no communications equipment. High-tech dirty tricks are to be avoided—it would be so very easy to wipe out the opposition's contributor database.

Where will it all end? Obviously, with more and more sophisticated computer applications for candidates of all political persuasions. The way campaigns are run has been changed forever by computers.

—From an article in *PC Magazine*

SUMMARY

- Preliminary investigation is the first step of the systems life cycle. The impetus for change can come from many sources. The analyst needs management authority to begin the process.

- User involvement is the key to success in any project. Users must be involved from the very beginning of the project and continue that involvement until the project is complete.

- It is important that the analyst understand the user organization because he or she needs to know who impacts the system and who the decision-makers are. If a hierarchical organization chart of formal relationships does not exist, the analyst needs to draw one.

- The preliminary investigation phase is like a microcosm of the entire systems life cycle, but much less rigorous.

- Problem definition includes determining the nature of the problem, its scope, and system objectives. The first task is to ascertain the true nature of the problem, as opposed to symptoms. The next is to establish the scope, the limitations of the project. Then the general objectives of the system can be listed.

- To decide if a project is feasible, consider it from technical, operational, and cost aspects. Technical feasibility refers to the availability of hardware and software resources.

- Operational feasibility refers to the possibility of a plan to meet the stated objectives. Cost effectiveness means that the user organization can afford the development and operational costs.

- The plan the analyst develops for the project will need to have supporting data on schedule, budget, and personnel needs.

- Even though the analyst has only the most general idea of the solution to the problem, it is still possible to list the anticipated benefits of the system, which can be tangible or intangible.

- The preliminary investigation report is the first major deliverable document. Depending on the size of the project, this report could be just one page, or possibly much longer for a complicated system. Typically, this report contains the following items: title page, project overview, problem definition, main text, costs, schedules, and personnel. The analyst also may give a presentation at this point.

- The systems team—also known as the project team—usually consists of one or more analysts and several users. If the project crosses department lines, there will be a user representative from each department. The composition of the systems team will change as the project progresses, with fewer users and more technical personnel.

KEY TERMS

Benefits
Cost effectiveness
Feasibility
Impetus for change
Intangible benefits
Management authority
Nature of the problem
Objectives
Operational feasibility
Organization chart
Preliminary investigation

Preliminary investigation
  report
Problem definition
Project team
Scope
Systems team
Tangible benefits
Technical feasibility
User involvement
User organization

REVIEW QUESTIONS

1. The impetus for change can vary significantly with the type of organization. How would you expect the impetus for change in a government agency to vary from the impetus for change in a small, privately owned business?

2. This chapter has stressed the importance of user involvement, but some thought must be given to who the users are. Who do you think would be the key users in these situations:

   a. A proposed computerized system to smooth the flow of operations among marketing, ordering, and shipping in a publishing company.

   b. An investigation of the various microcomputers used by the five scouts for a major baseball franchise, with an eye to sharing player data through a network.

   c. Changing the automated accounts receivable system for a major food brokerage from batch to online. (There are 77 employees in the accounting department.)

3. A hospital billing department is having trouble billing people properly with the existing manual system. There seem to be patient record identification problems and some careless record-keeping. Also, the department needs to be able to bill insurance companies directly, with patient authorization. Finally, it needs to be able to keep accurate records in order to receive government monies for patients on Medicare. Although this does not reflect an actual investigation and the information supplied here is sketchy, can you write up a possible problem definition—nature of the problem, scope, and objectives? Include a list of possible benefits also.

4. Can you think of situations that might fail the technical, operational, or cost feasibility criteria?

5. What advantages can you list for giving a presentation for the preliminary investigation phase? Under what circumstances might you feel it was not necessary or appropriate?

CASE 5-1
BIG IDEAS, SMALL BUDGET

When Alex Melson opened his first sports store, The Athlete Elect, his motto was "Think Big." He admonished his employees to: "Think big, think really big." Following this dictum, the store ordered clothes and equipment in large lots, used saturation advertising techniques, maintained high visibility in the community, and captured a large and loyal following in the first year of business. The store was operating at a profit by the seventh month, and each succeeding month produced a rosier financial picture.

Alex could see control problems, however, as he realized that he was not tracking inventory adequately, not billing his customers in a timely manner, and not taking advantage of discounts from his own suppliers. Alex saw computers as a way to solve these problems, so he invited several consultants and software vendors to make appointments to hear his story. Alex envisioned the solution as some sort of inventory system, one encompassing adequate supplies and taking into account his records for both customers and suppliers. He budgeted accordingly.

Hearing the key word *inventory* from Alex, the software people all came prepared to discuss just that. As the discussions progressed, however, it gradually became clear to Alex that their offerings did not match his vision. The fourth vendor recognized that Alex was really talking about much more than inventory control—he wanted accounts payable and accounts receivable as well. After absorbing this information, Alex realized that what he needed was not possible to achieve in the limited time frame and budget he had established. After discussions with his managers and some pencil scratching, he was able to come up with an acceptable plan for all these items that would cover a two-year period.

1. The main issue could be summarized in one word, a word central to the preliminary investigation process. What is it?
2. Shall we place any blame for this delay/misunderstanding? Was anyone inexperienced, negligent, or stupid?

CASE 5-2
THE PERILS OF POOR PLANNING

Its very name conjures up pleasant images: *Merry Go Round. Merry Go Round* is, in fact, a current events magazine for elementary school readers. When Wayne Stempel took over as president and publisher of *Merry Go*

Round, he had a mandate to cut costs. Two years after he decided to use a computer to help cut circulation costs, Stempel is still grappling with an in-house circulation program running on a microcomputer. Wayne does not smile when he says, "It has not been a pleasant experience."

The system is built around a hard disk and a series of custom programs prepared for use with a common database software package. Leland Dial, hired on a contract basis, created the programs to print subscriber labels, update addresses, delete expirations, request demographic information, and track accounting for the 55,000 subscribers to *Merry Go Round.* The programs have been in place for more than a year, and running, more or less, once a month. Wayne wearily admits, "This has ruined every one of my weekends for the last six months."

Wayne describes Leland as brilliant and says that he has a great sense of "conceptualization." But he looks baffled as he continues, noting that the advice he would give to other small businesses is to find someone who will stay with you, someone who has an innate sense of a business deadline. Wayne says that, in his opinion, amateurs—meaning himself as the user—cannot be taught to run the programs. In desperation he turned to a free-lance programmer, Lila Christie, who was more available than Leland. Lila ended up sort of running along behind Leland, patching the programs as problems occurred. And Wayne added yet another problem to his list: ref-ereeing between the two technical people without having the skills to declare a winner.

Each of the three has a different opinion of what went wrong. Wayne says that Leland does not really understand magazine circulation and that he is a dreamy artist rather than a businessperson. Leland thinks Wayne's expectations were not realistic, that "he wants to run the ship at full speed while I'm still working on the engines." Lila declares that Leland does indeed understand what is required and has written a sophisticated application. She is more critical of his method of implementation—lack of systematic deadlines, no real audit of the output, poor user communication, and little coherent documentation. Wayne, summing up months of anger and despair, says that micros have been oversold to American business.

1. This sad story is not far from some true stories. Although the details are somewhat sketchy, it is still possible to pick up the thread of mistakes. Can you list several?
2. What could have been done to avoid these mistakes?
3. What people are actually hurt by these events?

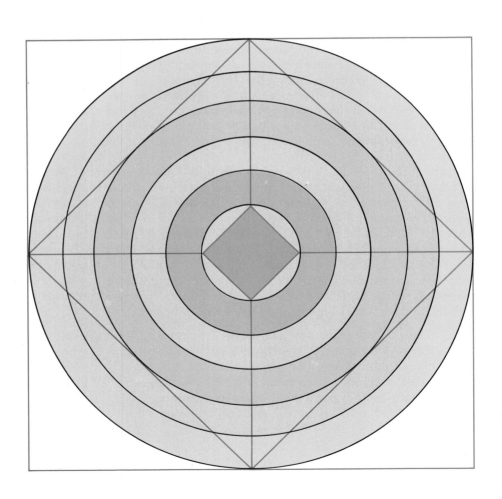

Chapter 6

Systems Analysis

The beginning of the analysis phase can be an unsettling time for a systems analyst. Typically, he or she has just completed a whirlwind of activity, convincing management and the user organization that it really is feasible to proceed with the project.

The analyst has, in a sense, made a promise. Now, to deliver on that promise, the analyst must plunge into the user area to understand it well enough to be able to plan an appropriate solution. Imagine waking up one morning and realizing that today you must begin to learn how a major insurance company processes auto claims. Or perhaps today is the day you start looking into how aerospace engineers process paperwork for space shuttle designs. This is not to say that you will become an insurance adjuster or an engineer, but your exposure to the processes will be extensive.

The purpose of the systems analysis phase, you recall, is to understand the existing system and, based on that understanding, prepare the requirements for the system. How will you do your job? You begin by gathering data, and then you analyze the collected data. Finally, the system requirements are prepared. Let us consider these in turn.

◇ DATA GATHERING: JUST THE FACTS

As the analyst—or as one of a team of analysts—you need to collect data relevant to the system being investigated. "Data gathering" is a common industry phrase for the process of finding out everything you can about the existing system. There is no standard procedure for gathering data because each system is unique. But there are certain techniques that are commonly used: written materials, interviews, questionnaires, observation, and sampling. The data gathering techniques you choose will depend on the nature of the system, time constraints, and minimum impact on the system users.

Regardless of the techniques used, the data gathered must pass two tests: validity and reliability. **Validity** means that the questions asked were appropriate and unbiased. **Reliability** means that the data gathered is dependable enough to use. A questionnaire for union members, for example, will probably reflect the bias of the writer if it is made up by management negotiators, and thus will lack validity. On the other hand, if some of the respondents are union members in name only, their answers may lack reliability. (See Figure 6-1.)

Let us now examine these data gathering techniques: written materials, interviews, questionnaires, observation, and sampling.

Written Materials

Gathering written materials is not a routine activity, because each system is different. On the one hand, there may be very little documentation, and thus no trail to follow. On the other, there may be masses of written material for the analyst to sift through. Much of this material may be incomplete, out of date, and ignored by the very people who are supposed to live by it. Procedures manuals may have been superseded by memos or word of mouth; it is even possible that the original documentation was poorly prepared at the start and therefore useless even when new. Written materials that do exist may be spread all over the organization—on a shelf here, in a drawer there. Locating the data is just the beginning.

The written manuals give guidelines for managers and employees and say how the system should be run, but they may differ a great deal from the way the system actually *is* run. Here is a sample list of written materials you should review if they are available:

- Policy and procedures manuals
- Operations manuals that tell how the system should run
- User manuals for existing automated systems
- Job descriptions
- Forms

Forms are so important that they deserve a separate discussion. Forms are carriers of data and come in all shapes and sizes. They are used for every imaginable purpose, from ordering restaurant supplies to balancing

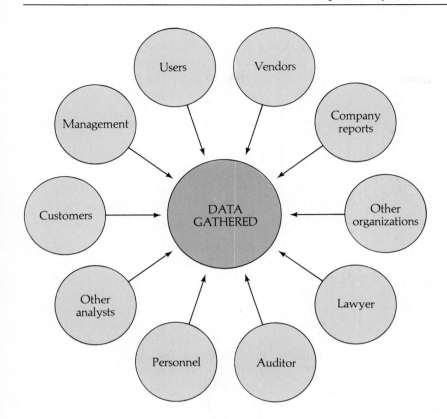

FIGURE 6-1

Sources of data gathering. Data about the system being analyzed can come from many sources.

a bank teller's account to submitting a factory report of parts that meet quality standards. You should take time to obtain a copy of each form used in the system, whether it originates within the system or comes from another organization. You will want to know the following for each form:

- What is the purpose of the form?
- Who fills in the form?
- Where is the form routed?
- Who uses the form?
- What authority does the form carry?

The best way to see how a form is used is to obtain both a clean and a used copy of the form from your contacts in the user organization. In the case of forms that have been filled in, identifying information may be removed from the form for security reasons.

Interviews

"Because we've always done it that way" belongs in a bad joke. Most businesspeople would be reluctant to make such a statement, but that is exactly what the file clerk told analyst Jane Leech when she inquired about the auto assembly reports. The accumulation was astonishing. Three-inch-thick computer-produced reports, delivered twice a week, were stacked on metal shelves that went from floor to ceiling around the room and down three aisles. Worse, no one ever looked at them! Jane stepped gingerly along a line of questioning that eventually yielded this information from the reluctant clerk: The reporting system had been developed three years ago, at the insistence of her boss, at a cost of $80,000. He was, naturally, reluctant to dismantle the useless system.

Most interviews do not expose such covert motives, but an analyst never knows when skillful techniques may be needed. Interviewing is unique among data gathering methods because of the person-to-person interaction. The best way to learn interviewing skills is through practice on the job, but we can make a start here.

Advantages of Interviews.   Talking face to face with someone has several advantages over reading written words. One advantage is that you can evaluate not only what is said but also how it is said. The words people say may be of minor significance compared with the tone of voice and body language used. Also, some people tend to be more candid if they only have to talk, as opposed to committing themselves on paper. An interviewer can offer open-ended questions and let the interviewee answer in his or her own style. Interviews provide a flexible format for the participants—the entire conversation can shift if you discover a fertile area of investigation. A by-product of the interviewing process is that you can establish rapport with the user and lessen resistance to change.

Disadvantages of Interviews.   The disadvantages of interviewing are essentially time and money. Neither is available in abundance, so the people you interview must be chosen carefully. Information gathered from the masses must be acquired through some other technique.

Start at the Top.   To get a complete picture, you will want to interview several people at different levels. The prevailing philosophy is to begin the process with the people at the top of the organization. There are several reasons for this. Starting with the boss will enhance your credibility with users at lower levels and lend you the authority you need to get their attention. Also, the boss is the person most apt to see the big picture, the person who can give you a frame of reference in which to work. The boss also can direct you to the best people to interview at other levels.

Structured versus Nonstructured Interviews.   A **structured interview** is one in which questions are preplanned and no deviations are permitted. Questions are asked as stated, with no rewording or explanatory comments. This is not designed for the free flow of ideas and is, in fact, very much like a spoken questionnaire. A common application of the structured interview technique is in a job interview when, by law or custom, each interviewee

BOX 6-1	**Successful Interviewing**

- Listen carefully to everything the interviewee says rather than thinking ahead to what you will say next.
- Use plain English. Avoid technical jargon.
- Dress and behave in a businesslike manner.
- Respect the interviewee's schedule.
- Avoid office gossip and discussion of the interviewee's personal problems.

Interviewing. Interviewing is an art. One develops expertise with practice.

is asked identical questions. Otherwise, the practical uses of the structured interview are limited to street and phone interviews of the general public.

A **nonstructured interview**, however, can vary considerably from the original plan. This is not to say that you ever just sit around and chat with a user, hoping something interesting might turn up. Even unstructured interviews are planned in advance. But it is not necessary to stick to the script if the conversation heads in some unexpected direction. An analyst can feel free to probe a little, asking the interviewee to expand and explain.

Planning the Interview.   In general, you need to decide whom to interview and what information you want from them. Only then can you plan what

BOX 6-2	The Reluctant Note-Taker

You can get into rather heated discussions about whether or not to take notes during interviews. You wish to be unobtrusive and non-threatening, but there is a deep-seated need for accuracy as well, and not all of us have total recall. To complicate matters further, there may be occasions when you think a tape recorder would be appropriate, but you have visions of your user clamming up completely. Here are some guidelines that may help:

- If you think you may need to take notes during the interview, carry a clipboard with a fold-over cover, so that the notes, once taken, will not be exposed.
- If the interviewee appears nervous or reluctant, try to post-pone taking notes until a subsequent interview when some level of trust has been established.
- Use the answer to a specific statement, such as the number of employees in the organization, to begin note-taking. Ask permission.
- Take notes judiciously, briefly recording key points, rather than trying to prepare a verbatim transcript.
- If it seems appropriate, mention your confidentiality policy briefly. Do not make a big deal out of it or you may get results that are the reverse of what you intended.
- Tape recorders should be saved for people you know rather well. Even so, you should ask in advance.

You may want to vary these guidelines if, in your view, other approaches are justified. These are merely guidelines, not hard-and-fast rules. You are the one on the scene and you are paid, above all, for your expert judgment.

the questions will be. Since most interviews are unstructured, you will probably include other questions as the interview progresses. First interviews are often quite general. When you are better acquainted with the interviewee and the subject matter the questions can be quite specific. Interview questions, unlike those on questionnaires, can be open-ended ("What suggestions would you make for improving the order processing system?" and so on).

Interviews must be by appointment, scheduled at the convenience of the interviewee. Normally, the analyst goes to the work area of the interviewee. Ideal conditions would place the interview in an office or a conference room where you can be free from interruptions. In practice, however, the person you have chosen to interview is likely to be too busy to leave the phone or work area. If you are working with a particular user who has a great store of information, several short interviews are better than one marathon interview.

Follow-up.   Every interview is documented in some way, even if it is just saving the notes you took during the interview or made immediately afterward. In theory, the analyst prepares a memo for the interviewee, highlighting the main points of the interview and giving polite thanks. If an analyst works with certain users over a period of time, everything becomes more casual, from the may-I-come-over-now phone call to the distribution of photocopied interview notes.

Questionnaires

There are some very bad questionnaires floating around—you have probably received some in the mail. And you may have wondered how some people have the gall to ask you to take your time to fill them out. Gene Metcalf, who was thinking about offering a business service to doctors, sent a questionnaire to all the physicians listed in the yellow pages. The questions were general and required several-sentence answers ("Describe your office environment," "List your hospital affiliations and your reasons for those choices," and so on). The questionnaire was seven pages long. Gene received three replies. Two had unprintable comments scrawled across the front and the third threatened a lawsuit—something about illegal access to information. There are some guidelines for questionnaire use, but Gene did not know anything about them.

When used appropriately, questionnaires can be a very useful tool. You might elicit data via questionnaire from a group of clerical workers in the quality control section, or from the entire telephone sales staff of a telemarketing department, or from all the employees of the shipping department, or all the patients at a drop-in clinic. On many occasions a questionnaire can yield a lot of data from many people in a short amount of time for a relatively low budget. All this becomes clear as we examine the advantages of questionnaires in more detail.

Advantages of Questionnaires.   The key advantages of questionnaires are intertwined: their use for *high-volume* responses and the fact that they are relatively *inexpensive* to administer. It would be prohibitively expensive, for example, to personally interview each of the 250 retail clerks in a major department store, but a well-planned questionnaire would elicit useful data. There are also questionnaires, notably political or product surveys, when *standardized wording* is needed to reduce bias. A written vehicle also eliminates any influence from a questioner's voice. In addition, information received from questionnaires is likely to be candid, since respondents may be *anonymous*. Finally, the use of a questionnaire can be *fast* and *efficient*.

BOX 6-3	The Right Word

Questionnaire construction involves choosing the right word to convey meaning. Some words have more impact than others. A classic example is a question first used by pollster Elmo Roper in 1940. The "free speech" question was phrased in two different ways and produced radically different results. When asked "Do you think the United States should forbid public speaking against democracy?" about 50% of the interviewees said yes. But when asked "Do you think the United States should allow public speeches against democracy?" about 75% said no. No one seems to be able to explain this phenomenon. Words—in this case *forbid* and *allow*—seem to have connotations that may intrude on data gathering in unexpected ways.

Disadvantages of Questionnaires.   Perhaps the most obvious disadvantage of using a questionnaire is that it has *limitations on the types of questions* that can be asked. A questionnaire is not suitable, for example, for probing questions, nor can you pose follow-up questions inspired by the ruminations of the respondent. Based on the written word only, both the questions and the answers are subject to *misinterpretation*. And both of these problems are based on the assumption that you do get the questionnaires back. If you have no control over the answering group, such as store customers, then you may have a *low return rate*, a sample too small to be statistically significant.

Guaranteeing a Response.   How can you guarantee a response? Well, of course, *guarantee* may be too strong a word, especially in view of Gene's troubles, which we just described. But there are ways to improve the expected return rate. If the respondents are in-house, say the employees in the quality control department, then you can use the **double envelope method** for returns. That is, the outer envelope lists the employee name for accountability (so you know all questionnaires are returned) but the inner envelope is unmarked and will be grouped with the other unmarked envelopes to preserve anonymity. For groups who are not a captive audience, a little ingenuity may be necessary. A mail order house, for example, feeling a need for more demographic data on its customers, offered a 25% discount on any item in its catalog in return for a completed questionnaire. Likewise, a textbook publisher surveying college professors offered a choice of appealing fitness books as a bonus for filling out a classroom-needs questionnaire. These enticements may seem faintly like bribery, but people are inundated with requests for "just a few moments of your time." As usual, you must make the decision about using such inducements: Is it worth it?

Questionnaire Construction.   This is the most critical aspect of questionnaire use. Your questionnaire must be brief, appealing, clear, and tested. Why *brief*? Woe to the questioner who submits open-ended questions requiring long answers ("Describe your job functions."). You can bet that the return rate will be abysmally low—everyone will be "too busy." Furthermore, it is difficult to draw conclusions from high-volume narrative answers. Such questions are best left for interviews.

You may puzzle momentarily over *appealing* as a criterion. You do want the questionnaires answered without malice and you do want them returned. Appeal involves three factors. The first is overall appearance. Brevity has already been considered. Other appearance factors are sufficiently large type, lack of crowding, and white space. The second is an appropriate format—see the descriptions in the box and note the examples. The third is to avoid "loaded" questions—that is, questions that (1) make the respondent feel uncomfortable or (2) suggest the "correct" answer. Questions such as "As a bus driver, do you belong to the Amalgamated Transit Union?" (he doesn't) or one beginning "Do you agree that . . ." (she doesn't) will convince a squirming respondent to skip the whole thing.

Making your questionnaire *clear* means gearing your questions to the intended audience and removing ambiguity. If your knowledge of the respondents is shallow or if you let your own bias dominate, the questionnaire may be so far off the mark that the recipients do not even recognize themselves. For example, a health center sends a questionnaire to potential clients. A single urban career woman may not identify with a heavy percentage of questions on children's illnesses and vaccinations. Or imagine a politician sending his or her constituents a mailing with multiple questions about foreign problems when they are most concerned about crime and inflation at home. Why would you answer a questionnaire that did not seem to be related to what you actually do or are? Those who feel they do not "fit" the questionnaire will not return it, thereby distorting the results. Removing ambiguity is not a trivial task. You must scrutinize the prepared questionnaire for unintended meanings.

Preparation of the questionnaire must be followed by thorough *testing*. After you have planned the subject matter and written the questions, you should try the questionnaire out on a colleague and then on a small subset of the respondents. The answers will be revealing and aid you in making revisions.

Computer-assisted Questionnaires.   It makes sense to use the computer as a data gathering tool. It is obvious that a computer can be used to prepare the text of a questionnaire. It is less obvious that the questionnaire can be read and answered on the computer screen. In essence, such a questionnaire is an **online survey**. Furthermore, the answers can be tallied by a computer program using the answers as input. This method would be very helpful for gathering data from large groups of people, but it is based on two assumptions. The first assumption is that microcomputers or terminals

BOX 6-4	**Questionnaire Formats**

Keeping in mind that a questionnaire should be concise and easy to tally, there are several possible formats to consider. Note that all of them present some limited set of choices, as opposed to open-ended questions, which invite the respondent to ramble.

RATING SCALE

Describe an activity or facet of the organization that the respondent can rate (poor-adequate-good-excellent) or make a statement with which the respondent can choose some level of agreement or disagreement (always-sometimes-never or agree-neutral-disagree). A simple Yes/No is a subset of this category.

	Strongly agree	Somewhat agree	Neutral	Somewhat disagree	Strongly disagree
Promotions should be made from within the company.	____	____	____	____	____
Most overtime is unnecessary.	____	____	____	____	____

MULTIPLE CHOICE

Present a list from which to make a choice.

Which method is your primary method of written office communication:

    a. Hand-written notes   c. Typed memos
    b. Formal letters   d. Computer messages

Various work hour methods are being considered for this organization. Which do you prefer:

    a. Fixed hours for everyone
    b. Flex time with midday core hours for everyone
    c. Flex time with two-hour-range starting time

Approximately what percentage of your time is spent on paperwork?

    a. Under 25%   c. 40%–50%
    b. 25%–39%   d. Over 50%

CHECK-OFF

Respondent is invited to choose as many as appropriate. You will usually need an "Other" category.

Check all professional magazines you read regularly:

_____ InfoWorld	_____ Computerworld
_____ Byte	_____ Data Management
_____ Systems Management	_____ Business Computer
_____ PC Magazine	Systems
_____ Personal Computing	_____ Datamation
_____ Popular Computing	_____ Computer Graphics
_____ Online Today	_____ Microcomputing
_____ Other _____	_____ Other _____

are available and convenient for such a task. A second assumption is that respondents can use the automated questionnaire readily. There are many organizations today where these conditions exist, however, so the computer-assisted questionnaire might be ideal. One more point should be made about a computer-assisted questionnaire: it is more expensive to administer than the standard written questionnaire.

Sources of Error.   The possibility of distributing a questionnaire whose questions are largely irrelevant to the recipient has already been cited as a factor in incomplete returns. Questionnaires that are returned, however, can still be subject to error. Rarely does the person from whom data is being gathered see an advantage to be gained by supplying information. In fact, such data could conceivably be used against that person. Combine this with the natural tendency to try to look good on paper, and the accuracy of the results can be in doubt.

One source of error is the _perceptual slant_ of the respondent. All of us have known and/or unknown biases that may creep into our answers. We also are subject to occasional _faulty memory_, as we innocently rewrite history. A more severe problem is _fear of consequences_, making people hesitate to report actual occurrences, to protect themselves or others.

Despite these cautionary notes, the use of questionnaires is a rewarding technique, yielding quick insights for relatively low effort and expense.

Observation

As a technique, **observation** is particularly suitable for a paper-flow application. You, the analyst/observer, watch the flow of data: who interrelates with whom, how paper moves desk to desk or station to station, and how paper comes into or leaves the organization. Types of data gathered this way cannot easily be collected by other techniques. The key advantage of observation is firsthand information: You can take notes that describe the activities as they occur.

This is not a clandestine operation, so you will not be lurking behind the potted palm. Rather, the purpose of your visits is made known to the members of the organization by their own manager. "Observation" usually means just that—watching and seeing. You would rarely interrupt workers or distract them from their tasks. When people are being directly observed in their jobs, they have a tendency to enrich their own functions and performance. It may be necessary for you to make several visits so people become accustomed to your presence and return to their normal habits.

Suppose, for example, that you are analyzing an existing loose system of office microcomputers. Management suspects the microcomputers are underused. The first stated objective is to increase the productivity of the microcomputer users. Before you and the managers can decide how that might best be done, you must gather data on how the microcomputers are being used now. Based on a few interviews and questionnaires for all the users, you learn that usage seems to make up about 40% of the day, with that time split as follows: 50% spreadsheets, 30% word processing, and 20% other.

But you get hints from the interviews, and from managers, that those figures may be inaccurate. So you decide to see for yourself. After observing the microcomputer users on and off for a week, you find that the actual figures look more like this: 25% spreadsheets, 25% word processing, and 50% other. Unlike the first set of numbers, these figures may indicate that some people need training in the use of spreadsheets and that the "other" category needs closer examination.

One form of observation is **participant observation**; in this method the analyst temporarily joins the activities of the group. This practice may be particularly useful in studying the activities of a complicated organization. The theory behind participant observation is very much like "learning by doing." At the risk of feeling like an intruder, you can learn a lot in a short time. A side benefit may be the enduring respect of the user organization, whose members may admire your bravado as you risk looking a bit foolish.

Sampling

You may need to collect data about quantities, costs, time periods, and other factors relevant to the system. How many subscription orders can be handled by a clerk in an hour? What is the typical cost for spare parts for an engine assembly? How long does it take a pharmacist to process the paperwork for a prescription? These and other questions may be best answered through a procedure called **sampling**: You need not gather all the data, only a certain representative subset. For example, instead of observing all 75

BOX 6-5	Only Some of the Folks

A famous sampling error put a big smile on the face of Harry Truman. Before the 1948 presidential elections, pollsters predicted a runaway victory for rival Thomas Dewey. The pollsters, however, had made a fatal error in their sampling choice: They queried voters by phone. In those days lots of people had no phone. But they did vote, and they voted for Truman. The Chicago *Tribune* compounded the error by printing the DEWEY DEFEATS TRUMAN headline before all the votes were counted.

clerks filling subscription orders for an hour, pick a sample of 3 or 4. Or, in the case of high-volume paper output, such as customer bills, you could collect a random sample of a few dozen.

Work sampling is particularly appropriate when there is a large volume of data. It would be too expensive to check all the data, but the same information can be obtained by using sampling. How much data is enough? Knowing how much and what data to select is a job for specialists who use statistical techniques. The actual techniques are beyond the scope of this book. Some analysts are trained in statistical sampling; more often, the analyst will work with a statistician to obtain the needed information.

| BOX 6-6 | The Analyst as Artist |

Perhaps it is another member of your family who has the artistic bent. In fact, perhaps one of the reasons you elected the computer field is that you are not the artsy type at all. Well, all thumbs or not, you must become a competent artist. Your brilliant logic will be dimmed if it is presented in a messy or slapdash format.

You need to learn to use rulers, ink pens, templates, lettering guides, professional lettering tape, and other materials available in your company supply depot or the local office supplies store. Learn to draw without smudges or, failing that, learn how to cover the smudges you do make.

These manual tools, although still necessary, can be supplemented or even replaced by modern tools such as computer graphics software specially designed to help you prepare presentations.

Consider this exercise an investment in your future. When it comes to documentation, neatness counts.

◇ DATA ANALYSIS

Data analysis may take place along with data gathering. More likely, however, you will acquire so much data that you will have to take time from data gathering to sort through and examine what you have so far. The narratives you prepare as part of data analysis should be supported by tools such as charts and diagrams.

There are a variety of tools for analyzing data, not all of them appropriate for every system. You should become familiar with the techniques, and then use the tools that suit you at the time. It should be noted, too, that many of the tools used for analysis are also suitable for design purposes.

The reasons for data analysis are related to the basic functions of the analysis phase:

- Assemble data gathered in a meaningful way.
- Show how the current system works.
- Provide easily accessible reference material.
- Set forth the current system as a basis for future comparisons with the new system.

As you analyze the data you find, keep in mind that you are searching for a better way. Should some parts of the system be eliminated because

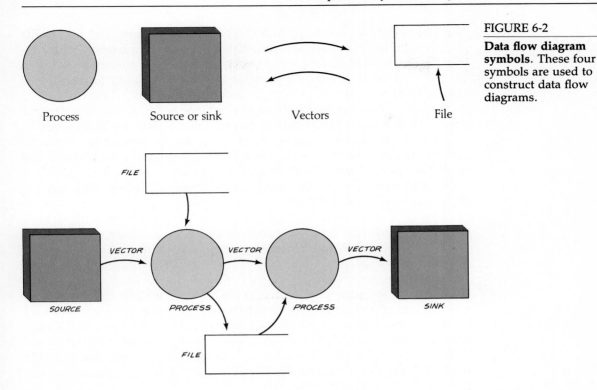

FIGURE 6-2

Data flow diagram symbols. These four symbols are used to construct data flow diagrams.

they are counterproductive? Can the system be streamlined? Can a certain process be simplified? Can reports or forms be combined? Can we make the system more responsive to user needs? These are some of the questions you will consider as you analyze the data.

Data Flow Diagrams

A **data flow diagram**, as its name indicates, shows the flow of data through a system. Although data flow diagrams (DFDs) can be used in the design process, they are particularly useful as a communication vehicle between you and the user during the analysis phase. Suppose, for example, you are to study the paper flow in a large order-processing organization. You would probably spend several hours interviewing the manager and other workers, your eyes glazing over as you make copious notes on what goes where. But that is only the data gathering function—now you must somehow analyze your findings. One method is to summarize the results in narrative form and have the user read it for accuracy. In fact, this method was used exclusively for years.

But now there is a better way. Instead of generating prose, you convert your notes to a diagram, a picture of what was said about the data flow.

Now you approach the user with the completed picture. The reaction is gratifying because the user can immediately spot errors and misunderstandings. The beauty of this device is that, although it takes some training to be able to draw data flow diagrams, it takes little training to read them. Users like that.

Data flow diagrams were introduced by Gane and Sarson as a structured analysis tool in 1979. DFDs have been promoted heavily by Tom DeMarco, who wrote his own book and is constantly on the lecture circuit. The notation used here is DeMarco's, because it is informal and easy to draw and read. Figures 6-2a and 6-2b illustrate the notation.

Four elements are used in data flow diagrams: processes, files, sources and sinks, and labeled vectors.

Processes.   These are the actions taken on the data—compare, check, stamp, authorize, file, and so on. Processes are represented by circles, often called bubbles.

Files.   A file is a repository of data. This is not necessarily the meaning of file that comes to mind when the subject is computer systems. You may think of a tape or disk file, or even an ordered set of papers in a file cabinet. Any of these are valid, but so are papers on a clipboard, mail in an in-basket, or blank envelopes in a supply bin. You begin to see that the DFD method is flexible enough to handle all sorts of office practices. Files are represented by an open-ended box.

Sources and Sinks.   A source is an origin for data outside the organization. To know what is "outside the organization," you must first define the organization. If the organization is a certain department, then any other department, or any person or organization in the world, could be considered a source. Examples would be an x-ray authorization memo sent into the radiology department from a doctor in another clinic, or a payment sent to a department store by a charge customer. In these cases, both the doctor and the customer are outside the organizations; since they are sending data into the organizations, they are sources of data. A sink is a destination for data going outside the organization, such as the bank to which money is being sent from the accounts receivable organization. A source or a sink is represented by a square.

Labeled Vectors.   Vectors are simply arrows, lines with directional notation. A vector must come from or go to a process bubble. The important rule about vectors is that they must be labeled with the data they carry. They must be labeled since the whole point of a DFD is to show the *flow* of data, and this way readers will know what data is being carried. This seems like a straightforward rule, but it is the greatest source of error on DFDs drawn by beginners. Now, having said that, there is one exception to the labeling rule: Arrows going to or from a file may not need to be labeled because the file is already annotated. However, if only a certain item is being extracted from the file, the vector will need to be labeled. One more thing—it is acceptable to have several vectors, each carrying different data, enter or leave the same process.

Note the example in Figure 6-3. The "organization" in this case is just an instructor grading papers. The source is the student who provides papers to be graded. There are two processes, one to grade the papers and one to record the grades. The action verbs *grade* and *record* indicate what is happening to the data. Note that the name of the data changes after it goes through a process, because the process changes the data and this change should be reflected in a new name. So "papers" become "graded papers" and then "recorded papers." The file used in this example is the gradebook, a repository of the grading data. Finally, note that, in this case, the source is also the sink, since the student gets the recorded papers back.

A little advice on putting all this together is in order. First you must dispel some notions associated with your programming experience. Begin with a list of what a data flow diagram does *not* do. A data flow diagram does not include any decisions or any information on the circumstances under which data may be sent here or there—just that it *is* sent, at least some of the time. No timing or sequence is included. And one more note of interest: Programmers are accustomed to the straight, neat lines of other kinds of charts, whereas the vectors used on DFDs are flexible, curved lines.

Assume you are working from a sheaf of notes acquired through interviews. Begin—in pencil!—drawing the processes, and then tie them together via labeled vectors. Add sources, sinks, and files as appropriate. A problem you may encounter is a temptation to cross lines, which is not allowed. Instead you must rearrange the components, which is also what you must do if you find that two processes you have drawn far apart need to use the same file.

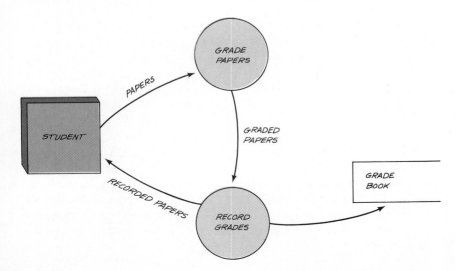

FIGURE 6-3

Simple data flow diagram. This drawing shows an instructor's grading system.

BOX 6-7	Data Flow Diagram Labeling Conventions

- Label each process with an action verb that shows what it does to the data.
- Label the vectors with names that show the transformation of data. That is, the data coming out of a process will not have the same name as it did when it came in, because it has been changed by the process.

When you must prepare a complicated diagram, begin with a simple drawing. In Figure 6-4, a sketch of the existing Accounts Payable System for King Books, only the most basic processes are drawn. This drawing shows a very high-level version of the system. It is a good start, but a user could tell you right away that there are a few things missing. For example, the voucher (the authority to write a check) would not be written until the invoice was compared with the original purchase order. Otherwise, any vendor could send a bill and it would be mindlessly paid whether anything was ordered or not. So you would change the DFD to show that someone matches the purchase order with the invoice. This is reflected in process 3 in Figure 6-5, the final version of this data flow diagram.

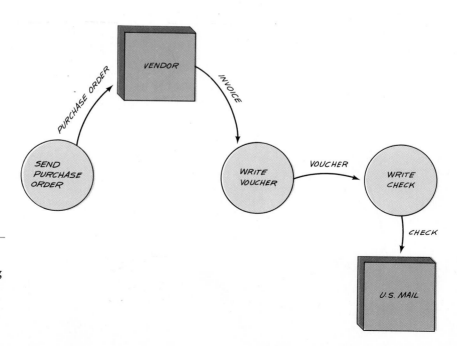

FIGURE 6-4

Initial data flow diagram sketch. First, "bare bones" drawing of data flow diagram for King Books current Accounts Payable System.

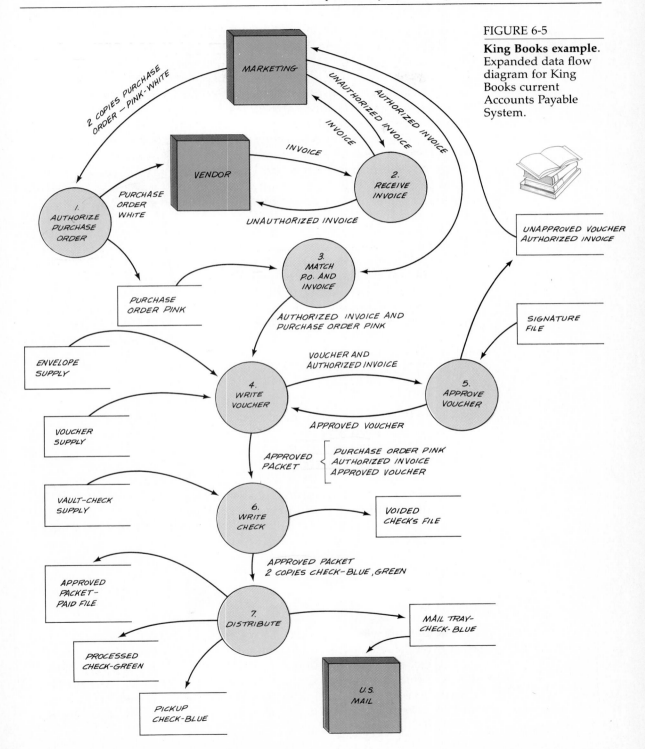

FIGURE 6-5

King Books example. Expanded data flow diagram for King Books current Accounts Payable System.

If you find that a complicated system does not fit on one sheet of paper, you must turn to leveling. **Leveling** is a system of breaking down data flow diagrams into smaller components that can be represented on separate sheets of paper. The processes on the initial diagram, the context diagram, are numbered, although in no particular order. Each process may be further subdivided at a lower level, and each process on those lower levels may likewise be subdivided. You may, in theory, continue indefinitely with this hierarchical practice. Process numbering is the key to keeping track of diagram leveling. Note the example in Figure 6-6. Process 3 in the context diagram could be represented on Diagram 3, whose processes are then numbered 3.1, 3.2, 3.3, and so on. If process 3.2 were to be subdivided further, it would be done on Diagram 3.2, with processes 3.2.1, 3.2.2, 3.2.3, and so forth.

One more issue must be raised in connection with leveling—the concept of **balance**. A balanced data flow diagram is one in which any process can be replaced by the entire subdiagram, if there is one, that bears its number. This is a theoretical approach to make sure the lower level diagrams are set up properly. That is, you would not actually want to replace a process by its subdiagram because it would not fit on the page. The point is that, in theory, you *could* replace it. To balance a DFD, be sure that the data going into and out of the lower diagram are the same data going into and out of the original process.

Structured English

Structured English is a tool often used in conjunction with data flow diagrams because it explains all the things that the DFD leaves unanswered. **Structured English** describes in detail the processes that were drawn in the DFD—what action takes place and under what conditions. Structured English has been described as "plain vanilla" English, but it also resembles pseudocode. At the heart of structured English are imperative verbs—no wishy-washy statements here. You can also skip adjectives, adverbs, compound sentence structures, and most punctuation. Overall, the rules are pretty loose; you can make up your own as you follow your data flow diagrams. Note the structured English example to match the King Books data flow diagram.

Data Dictionary

A **data dictionary** is a file of data about data. In its simplest form, it is just a list of all data names associated with the system. Names of files, records, data items, processes and, in some organizations, anything else relevant can be included. At minimum, a data dictionary serves as a repository for information about the data in the system. In many organizations, it takes on a more formal purpose, serving as a central clearinghouse for data names.

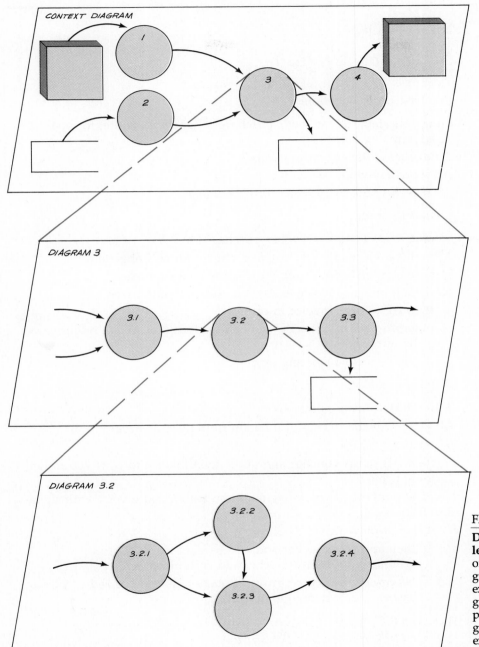

FIGURE 6-6

Data flow diagram leveling. Process 3 on the context diagram has been expanded to diagram 3. Likewise, process 3.2 on diagram 3 has been expanded to diagram 3.2.

KING BOOKS

Structured English

Purchase order initiation:

Marketing department sends purchase order (pink, white) for books to Accounts Payable System.

AP assigns purchase order number.

AP sends PO-white to vendor.

AP files PO-pink by purchase order number.

Invoice Receipt:

Vendor sends invoice for books purchased to AP.

AP sends invoice to marketing department for authorization.

Marketing adjusts dollar amount if some books returned.

Marketing returns invoice to AP. If invoice not authorized

   then AP returns invoice to vendor

   otherwise AP matches PO-pink and authorized invoice by purchase order number and

      AP begins vouchering process.

Voucher initiation:

AP assigns voucher number.

AP writes voucher using information from purchse order.

If voucher > \$5,000

   then AP sends voucher and authorized invoice to chief accountant for approval

Chief accountant compares signature on authorized invoice with signature from file

   If signatures match

      then approve voucher and

         return approved voucher and authorized invoice

      otherwise place unapproved voucher and authorized invoice in basket to be returned to marketing

         and exit routine

   otherwise approve voucher as is.

AP sends approved packet of PO-pink, authorized invoice, and approved voucher in an envelope.

Check preparation:

Typist receives approved packet.

Typist retrieves numbered blank check.

Typist types 2-part check (blue, green) using data from approved voucher.

Typist enters invoice number on check stub.

AP files approved packet in paid file.

AP files check-green.

If check is to be picked up

    then check-blue is placed in pickup tray

    otherwise check-blue is placed in mail tray.

The data dictionary is begun in the analysis phase by the analyst as he or she begins gathering data. Since the dictionary will be updated as data gathering continues, it is kept on an easily accessible automated file. Some organizations use word processing or text editing to maintain the file. Others use a data dictionary software package, usually associated with a database.

The data dictionary produced in the analysis phase will be used as the basis for the data dictionary for the new system. The two dictionaries, though possibly similar, will not be any more alike than the old and new systems are. The data dictionary for the new system may be monitored by a formal organization in the company, whose functions include approving of any new data names. If you are working on a system that has already been automated, you may be frustrated by several rounds with the bureaucracy as you try to win data name approval.

In our first cut at a data dictionary, for the analysis phase, there are few rules. The most important one is that entries must be in alphabetical order, although fields will be in the order in which they appear in the record. The other rules are notational, as noted in Box 6-8. Note the sample data dictionary for King Books.

BOX 6-8	Data Dictionary Notation

=	Equivalent
+	And
[ ]	Select from choices
{ }	Repetitive
* *	Comment

KING BOOKS

Data Dictionary

CHECK-FORM = CHECK-NUMBER +
                    CHECK-DATE +
                    CHECK-AMOUNT +
                    CHECK-PAYMENT-NAME +
                    CHECK-VOUCHER-NUMBER

INVOICE-FORM = INVOICE-NUMBER +
                  INVOICE-DATE +
                  INVOICE-VENDOR-NUMBER +
                  INVOICE-VENDOR-NAME +
                  INVOICE-VENDOR-ADDRESS +
                  INVOICE-VENDOR-CITY +
                  INVOICE-VENDOR-STATE +
                  INVOICE-VENDOR-ZIP +
                  INVOICE-PURCHASE-ORDER-NUMBER +
                  INVOICE-DUE-DATE +
                  INVOICE-AMOUNT-DUE +
                  INVOICE-DISCOUNT-PERCENT +
                  INVOICE-DISCOUNT-DATE

PURCHASE-ORDER-FORM = P-O-NUMBER +
                            P-O-DATE +
                            P-O-VENDOR-NUMBER +
                            P-O-VENDOR-NAME +
                            P-O-VENDOR-ADDRESS +
                            P-O-VENDOR-CITY +
                            P-O-VENDOR-STATE +
                            P-O-VENDOR-ZIP +
                            P-O-DELIVERY INFORMATION +
                            {P-O-ITEM-DESCRIPTION} +
                            {P-O-ITEM-QUANTITY} +
                            {P-O-ITEM-PRICE} +
                            P-O-AMOUNT

VENDOR-FORM = VENDOR-NUMBER +
                   VENDOR-NAME +
                   VENDOR-ADDRESS +
                   VENDOR-CITY +
                   VENDOR-STATE +
                   VENDOR-ZIP +
                   VENDOR-PHONE *INCLUDES AREA CODE*

VOUCHER-FORM = VOUCHER-NUMBER +
                    VOUCHER-VENDOR-NAME +
                    VOUCHER-VENDOR-ADDRESS +
                    VOUCHER-VENDOR-CITY +
                    VOUCHER-VENDOR-STATE +
                    VOUCHER-VENDOR-ZIP +
                    VOUCHER-DATE +

{VOUCHER-INVOICE-NUMBER} +
{VOUCHER-INVOICE-AMOUNT} +
VOUCHER-AMOUNT +
VOUCHER-APPROVAL

Data Element Charts

Sometimes it is useful to show data on charts. Although you may find the data dictionary sufficient for most systems, it is helpful to be familiar with chart forms and suitability.

Data Element Listing.   A separate chart is made for each form and report, listing all the data elements in each. The meaning of each data element is recorded with it, in order to avoid misunderstandings.

Data Cross-reference Chart.   Sometimes it quickly becomes obvious that the same data elements are being transcribed again and again onto various forms; that is, those data items are recurring. In such cases it is useful to

KING BOOKS

Data Element Listing

Document: Invoice Form

Data Elements:

1. Invoice number. Established by vendor.
2. Invoice date. Date invoice sent. Month, day, year.
3. Invoice vendor name.
4. Invoice vendor address. Street and number.
5. Invoice vendor city.
6. Invoice vendor state.
7. Invoice vendor zip. Full nine digits.
8. Invoice purchase order number. PO that caused the invoice.
9. Invoice due date. Date after which account becomes delinquent.
10. Invoice amount due. Total to be remitted to vendor.
11. Invoice discount percent. Apply to reduce bill if paid before invoice discount date.
12. Invoice discount date. Date by which bill must be paid to take advantage of discount.

chart the data, cross-referencing data elements against the forms on which they appear. This chart identifies duplicate data so it can be consolidated in the new system. Note the recurring data in Figure 6-7.

Decision Tables

A **decision table** is a standardized table of the logical decisions to be made regarding the conditions that may occur in a given system. Decision tables are useful for a series of interrelated decisions. Their use helps to ensure that no alternatives will be overlooked.

As you can see from the decision table format (Figure 6-8a), the table is divided into conditions on the top and actions on the bottom. Any given rule will combine certain conditions, which then yield certain actions. In

Data Cross-Reference Chart					
Data element \ Form name	Check Form	Invoice Form	Purchase Order Form	Vendor Form	Voucher Form
Check number	✓				
Check date	✓				
Check amount	✓				✓
Vendor name	✓	✓	✓	✓	✓
Voucher number	✓				✓
Invoice number		✓			✓
Invoice date		✓			
Vendor number		✓		✓	
Vendor address		✓	✓	✓	✓
Vendor city		✓	✓	✓	✓
Vendor state		✓	✓	✓	✓
Vendor zip		✓	✓	✓	✓
Purchase order number		✓	✓		

FIGURE 6-7

King Books data cross-reference chart. As you can see, many data elements appear on more than one form.

other words, "If these conditions exist, then take these actions." On the left of the chart, the condition stub states the possible conditions, while the action stub states the possible actions. On the right side of the chart, Y (for Yes) is entered in the rule column for each condition invoked, and X is entered in the same column on the line for any action resulting from the conditions. Study the example in Figure 6-8b.

Decision tables may be useful in assembling information about a given system you are studying. Decision tables are also useful in the design process. They can be used to establish a set of conditions and actions from which the programmer can write code.

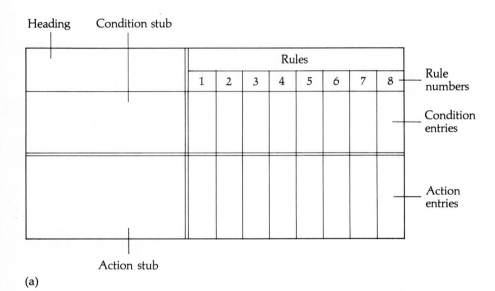

(a)

Data processing course placement	Rules							
	1	2	3	4	5	6	7	8
Taken survey course	Y	Y	Y	Y	N	N	N	N
Passed entrance exam	Y	Y	N	N	Y	Y	N	N
2 years industry experience	Y	N	Y	N	Y	N	Y	N
Any advanced course	X				X			
Structured design course		X	X			X		
Entrance exam				X			X	X
Survey course								X

(b)

FIGURE 6-8

(a) **Decision table format**. The number of rules (R) will vary with the number of conditions (C): R = 2 to the power C.
(b) **Decision table example**. This decision table shows the rules for data processing course placement. Note that three conditions yield eight rules (2 to the 3rd power = 8).

Decision Trees

A **decision tree** is a graphic way of noting combinations of conditions that lead to certain actions. Decision trees are easy to read but less compact than decision tables. Notice the decision tree in Figure 6-9 for the voucher approval process. At the left is a box that represents an **act fork**. Proceeding to the right are two branches, indicating that a voucher may or may not be written. If a voucher is written, then we encounter an **event fork**, a decision based on whether the voucher is greater than $5,000. If the voucher does exceed $5,000, then an authorized signature is required before a check can be prepared. Each branch of the tree produces some eventual **outcome**, such as "write check."

Physical Layout Chart

If your observations make it clear that the movement of people and paper within the organization is complicated or inefficient, a **physical layout chart** of the work area may be appropriate. Such a chart shows the workspace, desks, workstations, filing cabinets, and the paper flow therein. An example is shown in Figure 6-10.

A new system will change this flow. It is important to document the work relationships and anticipate problems resulting from the changes.

◇   REQUIREMENTS DEFINITION

The most common computer-related lawsuits involve a dissatisfied user filing suit against computer system personnel or vendors. Should we be talking about malpractice insurance for analysts? The subject no longer brings guffaws. Doctors thought such insurance was preposterous only a few years ago. The point is that analysts can give themselves a measure of protection by keeping the users actively involved in the system.

There is the old refrain again: user involvement. You have certainly been involved with the users through the data gathering stage, and they were needed to confirm your data analysis. Now you need to work with the user to draw up the requirements list. Begin with the system objectives you established in the preliminary investigation phase. Now you can refine them, in terms of your expanded knowledge of the system.

It is common to quantify requirements when possible—that is, to attach actual performance numbers to an activity. In particular, when timing is critical, exact timing definitions must be included in the list of requirements. An online system that is used while customers are waiting, for example, might require a response time of no more than two seconds from a query to the computer. Other requirements may have no numbers attached to them.

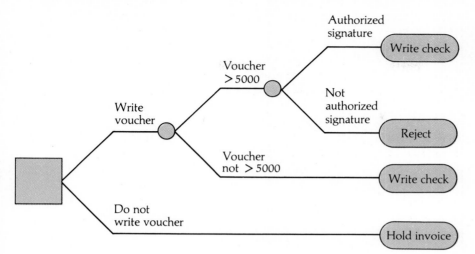

FIGURE 6-9

Decision tree. This example shows a decision tree for a small part of the King Books Accounts Payable process.

The objectives described, in a general way, what the system should be able to do. The requirements describe the functions of the system, without going into detail about how they should be accomplished. Notice the objectives-to-requirements transition for King Books.

ANALYSIS AND REQUIREMENTS DOCUMENT

Completion of the analysis phase is a major milestone, one marked by this significant deliverable: the **analysis and requirements document**. It is with this vehicle, accompanied by a presentation, that you show your own management and the user organization that you understand the problem, the user organization, and the requirements for a new system.

The analysis and requirements document will be fairly hefty. In fact, this becomes a burden as you make copies for each person involved and, especially, as you make changes to all those copies. It is usually fattened further by an appendix, where you stash data to support your conclusions stated in the main body of the document.

Suggested content for the analysis and requirements document follows. Remember that it is not practical to present rigid documentation standards. You are expected to deviate from this or any other outline whenever, in your judgment, it is warranted.

- *Title page*. Include names of key personnel and an abstract of project matter.
- *Project overview*. This is a one-page condensed version of the document.

FIGURE 6-10

Physical layout chart.
This example shows
the physical layout of
the facilities of the
King Books Accounts
Payable System. The
arrows mark the key
movements of
people.

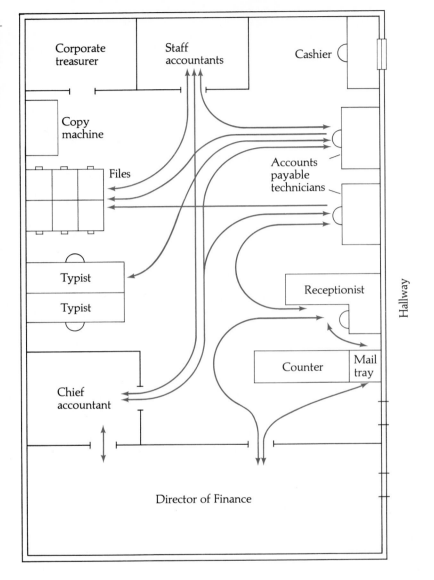

- *Problem definition*. A repeat of what you wrote for the preliminary investigation report. Readers need the background and you cannot assume that they remember—or have handy—what you wrote before. Just lift it right out of the other document.
- *Requirements*. Although these are the result of your study of the current system, and have yet to be described, readers like to hear them first. You can tell them how you got there later.
- *Benefits*. General list of benefits of a new system, as mentioned in the preliminary investigation phase.

KING BOOKS
Requirements

You may want to review the objectives for the King Books Accounts Payable System that were presented in Chapter 5. Briefly, they referred to accurate records, discounts, management control, user orientation, and cost savings. The requirements build on those objectives, but provide more detail, based on further information acquired in the analysis phase. Note also that several of the requirements are quantified.

The requirements for the King Books Accounts Payable System are as follows:

- Update vendor accounts within 24 hours of when an invoice is received.
- Generate computerized checks.
- Combine monthly invoices for the same vendor and write a single check for that vendor, reducing the total number of checks generated by 15%.
- Get early-payment discounts on 35% of invoices.
- Provide both on-demand and scheduled management reports.
- Provide security controls throughout the system.
- Provide a user-oriented system whose online usage can be learned by a new user in less than two hours.

- *Description of current system*. Include any data analysis tools used:
  Organization chart (picked up from first phase)
  Data flow diagrams
  Structured English
  Data dictionary
  Data element list
  Data cross-reference list
  Decision tables
  Decision trees
  Physical layout chart
- *Costs, schedules, personnel*. Anticipated costs of the design, development, and implementation of the system. Include activities schedule and personnel needed.
- *Appendix*.
  Current forms and documents
  Questionnaire results
  Sampling results
  Anything else germane

The analysis and requirements document must be complete and delivered to all interested parties before any presentation, so that attendees will have a chance to look it over beforehand. The timing is never easy, so try to plan ahead.

A presentation at this point will probably be required. In any case, you will certainly want to give one to demonstrate your progress and get approval to proceed. General presentation guidelines were presented in Chapter 4. Specific content for this presentation is included in Box 6-9.

◇ HOW TO KNOW WHEN YOU ARE DONE

When is the analysis phase complete? It has been remarked, tongue in cheek, that the analysis phase is over when time is up. The schedule is, in some ways, a handicap. You will naturally tend to draw to a close as you see the deadline approaching, whether you are truly finished or not. However, if you finish early, you will doubtless spend the remaining time hand-wringing and second-guessing.

There is one more category of schedule confusion: "paralysis by analysis." This applies to the people who are so determined to get it right the first time—to get it all—that they dither endlessly, constantly searching for new ways to extend the schedule. As mightily as you try, you will probably not succeed in 100% data gathering. We have already indicated that the systems life cycle tends to be iterative, so accept the fact that you may have to do a little backtracking. Meanwhile, move on to design.

BOX 6-9	Analysis and Requirements Presentation

Although your presentation could vary from this, here is a general approach to consider:

- Introduction to presentation
- Background—whose idea is this?
- Problem or opportunity description
- Hint of a new system—what can be done about the problem
- Anticipated benefits
- Full description of problem in context of the organization and how it works
- Schedules, costs (These should be toward the end so your audience is not distracted by dollar figures.)
- Summary—end on an upbeat note
- Questions

FROM THE REAL WORLD

Crime Online

The fascination with being a systems analyst lies, in part, in the endless variations available. In addition to the diverse subject matter, your role can vary. As a programmer/analyst, you can be the whole show: analyzing, designing, developing. Or you may be the only analyst, interfacing with programmers who produce programs from your specifications. Or you may be one of several analysts on a complex, time-consuming project. This last mode is the one used by the district attorney's office in the borough of Brooklyn.

District Attorney Elizabeth Holtzman was elected on a platform that promised a coherent strategy for fighting crime. The crime-fighting system was clogged with paperwork in a useless jumble. The chances in Brooklyn for "beating the rap" were good, simply because of prosecutor error.

District Attorney Holtzman set about to change all that. She knew she needed help from computers. If you want to improve the management of a very large office of this nature, she noted, then you've got to have computers—to keep track of the information so you can see where the problems are and make changes and improvements.

Holtzman resolved to bring the private sector in to review management practices. She turned to the New York Partnership, a philanthropic organization that mobilizes private corporations to share the latest expertise with the public sector. Systems analysts—this is where the large-complex-project comes in—from 12 major corporations spent several months in the district attorney's office. They thoroughly surveyed all work methods. They scrutinized the daily operations of the agency's 700 employees, including the 300 lawyers who do the actual prosecuting. After a thorough analysis, they concluded that, based on today's standards, the current system was primitive.

On the strength of the analyst's recommendations, Holtzman asked for city funding to support a computerized criminal-justice system. Her request was granted. A computer service firm specializing in law offices was hired to do the job. A new system, including hardware and software, was in place in 18 months. Although the system components do everything from word processing to monitoring grand jury records, its most important feature is a trial-tracking program that provides prosecutors with up-to-date information for monitoring cases. Dismissals for failure to meet speedy trial requirements have been reduced by an amazing 80%. The computer system now makes the judicial system function the way it was intended to function.

—From an article in *Business Computer Systems*

SUMMARY

- Data gathering means finding out everything about an existing system. The data gathered must be valid (questions asked were appropriate and unbiased) and reliable (data gathered is dependable). Data gathering techniques include written materials, interviews, questionnaires, observation, and sampling.

- Written materials include, as available, policy and procedures manuals, operations manuals, user manuals, job descriptions, and forms. When possible, the analyst should get a clean and a used copy of each form.

- Advantages of interviews include evaluating how words are said, candidness, open-ended questions, flexibility, and user rapport. Disadvantages of interviews are time and money, since interviewing is a lengthy process. If you interview people at different levels, begin at the top. A structured interview is one in which questions are preplanned, and no deviations are permitted. A nonstructured interview can deviate from the original interview plan.

- The advantages of questionnaires are high volume, low expense, standardized wording, anonymity, speed, and efficiency. The disadvantages of questionnaires include limited types of questions, possible misinterpretation, and low return rate. The double envelope method, which promotes anonymity, may be used to improve the return rate.

- A good questionnaire will be brief, appealing, clear, and tested. Some organizations use a computer-assisted questionnaire called an online survey. Sources of error on questionnaire responses include perceptual slant, faulty memory, and fear of consequences.

- The technique of observation is one in which the analyst watches how a system works: personnel interactions and the flow of paper through the organization. Participant observation means that the analyst temporarily joins the activities of the group.

- Sampling means that only a representative part of a set of data needs to be gathered, as opposed to all of it. This is useful for high-volume data.

- Data analysis includes narratives with supporting charts and diagrams. The reasons for data analysis are to assemble data in a meaningful way, to show how the current system works, to provide easily accessible reference material, and to set forth the current system as a basis for future comparisons with the new system.

- A data flow diagram shows the flow of data through the system. The four elements used in a data flow diagram are processes (actions), files, sources and sinks (data origins and destinations, respectively, outside the system), and labeled vectors to carry the data.

- Leveling is a system of breaking down data flow diagrams into smaller components that can be represented on smaller sheets of paper. A balanced data flow diagram is one in which any process could be replaced by the entire subdiagram.

- Structured English describes in detail the processes in the data flow diagram—what action takes place and under what conditions.

- A data dictionary is a file of data about data, a list of all data names associated with the system.

- A data element listing can be made for each form and report, noting each data element therein. A data cross-reference chart cross-references data elements against the charts on which they appear.

- A decision table is a table of logical decisions regarding the conditions that may occur in a given system. A decision tree is a graphic way of noting combinations of conditions that lead to certain actions.

- A physical layout chart is a diagram of the organization work area and the paper flow within it.

- The analysis and requirements document usually contains the title page, project overview, problem definition, requirements, benefits, description of the current system, costs, schedules, personnel, and appendix.

KEY TERMS

Act fork
Analysis and requirements
  document
Balance
Data analysis
Data cross-reference chart
Data dictionary
Data element listing
Data flow diagram
Data gathering
Decision table
Decision tree
Double envelope method
Event fork
Files
Interviews
Labeled vectors

Leveling
Nonstructured interview
Observation
Online survey
Participant observation
Physical layout chart
Processes
Questionnaires
Reliability
Sampling
Sinks
Sources
Structured interview
Structured English
Validity
Written materials

REVIEW QUESTIONS

1. Give an example to differentiate validity from reliability.

2. Which of the five data gathering techniques would you be most likely to use in each of these situations? Explain why.

   A group of 283 phone solicitors key customer data using a key-to-tape system. You need to know how long each average customer session takes, not counting no-answers and hangups.

   Something is wrong in the accounts receivable department but you are not sure what or why. There are two managers and seven clerical employees, and they seem cooperative.

   A manager in a college registration office thinks employees spend too much time away from their desks in unproductive tasks and would like some data on just how much time is actually spent at their desks.

   In order to increase time usage of the company's 46 sales representatives, the head manager first wants a breakdown of the activities they spend their time on.

3. Would it ever be appropriate to let an unstructured interview go in a completely different direction than originally planned? Explain.

4. What questionnaire format do you favor, and why?

5. Under what circumstances do you think it would be practical to use a computer-assisted questionnaire?

6. As a user, what do you suppose you would like or not like about reading data flow diagrams prepared by an analyst?

7. Make a data flow diagram of a system with which you are familiar, such as the registration process or lab procedures at your school. Write the structured English to go with it.

CASE 6-1
EMPTY HOTEL ROOMS

A major hotel chain used a phone network to handle its reservation system. This system was unresponsive. In many cases, hotel rooms were left empty that could have been rented, causing lost revenue. In other cases, hotels were overbooked, resulting in angry former patrons.

   The potential benefits of an online computer system were obvious to management. Data processing manager Rob Stearns was hired to direct the project. He in turn hired 7 programmer/analysts and quickly expanded that staff to 34.

Rob convinced the managers that little time needed to be spent on analyzing the existing system, because the entire concept would be changed as the new system replaced the old. The new system was developed rapidly and was soon implemented to run side by side with the old system. But the old system was kept running because the new system was not functioning well. In fact, in a few weeks the new system was eliminated entirely, and, as a direct consequence, the staff reduced from 34 to 5.

After an evaluation, the cause of the failure was determined to be the inability of the new system to handle out-of-the-ordinary transactions. Most exceptions had not been planned for. The loss to the hotel chain was in the millions of dollars.

1. Design or analysis? Word went out to the company that the new system had been poorly designed. What really happened?
2. Where should the blame for this failure be placed?
3. What would you have done differently?

CASE 6-2
SEND US YOUR REQUIREMENTS

An aerospace firm developed an automated system to process engineering drawings. Since management had structured the company into divisions according to airplane model, it was decided to bring up the new system one model at a time. This plan worked successfully for two models, but work on the third model could not seem to get started.

Analyst Donna James was pulled off another job to tackle the "third model" problem. The user offices were in a somewhat distant suburb, so Donna scheduled several interviews and planned to spend the day there. The prior analyst warned her that the users were uncooperative, and she did sense a little anxiety when she called for the appointments.

To her surprise, the interview sessions went well. After some initial hesitation, the users were able to describe what they were doing and Donna was able to communicate what the new system could do for them. This first day was only one small step, but it seemed clear that the transition to the new system would be successful.

The user manager, whom she had interviewed first, took her aside as she was leaving and candidly expressed his relief at the way the day had gone. He said he had been mystified by phone calls from the original analyst, a man he never met, who repeatedly said "Send us your requirements."

This rather bizarre true story has some interesting twists. The third model was brought on board in record time. Its users became enthusiastic supporters. Donna James was given a raise.

1. Why do you suppose the original analyst tried to communicate only by phone?

2. What was the user's reaction to the send-us-your-requirements request? What is probably the reason the user did not challenge the analyst who made the request?

CASE 6-3
ALL THE MARRIED WOMEN FROM ST. ANNE'S

Saint Anne's College, a private liberal arts college for women, prides itself on academic excellence and the diversity of its graduates. As the Class of '66 prepared to converge on the campus for their 20-year reunion, class scribe Pamela Lucas decided to send a where-are-we-now questionnaire to her classmates around the country. Pam was well known; she lived in an eastern suburb where her activities centered around volunteer work, her husband's business, and her four children.

Here are some typical questions from Pam's questionnaire:

Who handles the money at your house?
_____ Self
_____ Husband

In what direction do you think the church should go?
_____ More conservative
_____ More liberal

What volunteer activities do you pursue?
_____ Hospital
_____ Church work
_____ Children's charities
_____ Girl/Boy Scouts
_____ Sports—coaching, etc.

Pam received a 43% response. She was amazed at how homogeneous the class was; all were married with children and most were active in the church and doing volunteer work. When the classmates gathered for their reunion, however, the actual statistics did not support the questionnaire results.

1. What could we say about Pam's skill in constructing a questionnaire?
2. What went wrong here? Why were the results skewed?

PART 3

SYSTEMS DESIGN

Systems design is the most challenging of all of the phases in the systems life cycle. Rather than studying a system that already exists, the analyst must plan and design a new system, one that will meet the needs of the users. The first step is preliminary design, in which the analyst formulates a general design plan. After the design plan is approved, the analyst and the rest of the systems team prepare a detail design for system output, input, processing, and files.

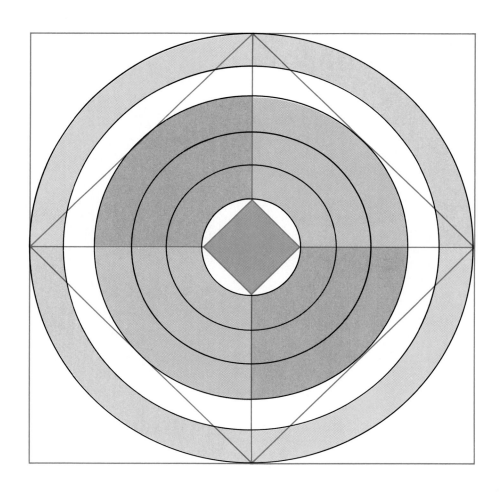

Chapter 7

Preliminary Systems Design

You have come to the point where you must design a system. This makes you, for the time being, a designer. But if someone asked what you do for a living, would you say "designer"? If you did, you would probably be mistaken for a person who works with fabrics and clothing styles. "Systems designer"? That's getting closer, but the title would still be lost on the general public. No, you are still a "systems analyst," albeit a multi-faceted one. Whether you are about to design a system for books, burglar alarms, or bicycles, you are entering the next phase of the systems life cycle.

◇ THE TRANSITION FROM ANALYSIS TO DESIGN

This is an appropriate time to review just where you are in the systems life cycle. You have passed through the preliminary investigation and analysis phases and are ready to begin the design phase. The focus of your work now changes from analysis to design. Your role shifts: Instead of asking questions, you will be answering questions. Now that you understand the problem, it is time to solve it.

The design phase, you recall, has two subphases: preliminary design and detail design. Preliminary design sets up the big picture, so that you and the users can agree on a general way to proceed. We will study preliminary design in this chapter. Detail design takes more time, as well as more study; it will be presented in four separate chapters on output, input, processing, and file processing and databases.

Changes to the Systems Team

One of your first actions on entering the design phase is to make changes in the **systems team** (also called the project team). Formed at the beginning of the project, the team originally was top-heavy with people from the user organization(s). This was appropriate because they were the people who could provide information about the existing system, which was being studied.

Now that the analysis phase is complete, or largely so, user roles diminish somewhat—not in importance but in focus and in time commitment. The emphasis of the team effort is now on designing a new system, and for that purpose, more technical people are needed. Depending on the size and nature of the system, the team may take on another analyst or two and possibly a lead programmer. These people may devote full or part time to the project.

Goals of Systems Design

There are general design goals that can be applied to any system. Perhaps the most important is *suitability*—making sure the design meets the system requirements. Certainly the system designed must do what the user wants, but another major characteristic of the new system must be *reliability*. Users must know that they can count on the system to work as specified and to produce valid and accurate output. A third design goal is *ease of use*. No system will be fully used, not to mention tolerated, if it is complicated or difficult to use. A related goal is *simplicity;* an unnecessarily complex system is one just waiting for trouble and costing extra money in the process. Finally, as always, a system must be designed with *economy* in mind—if the users cannot afford the system then its wonderfulness becomes irrelevant.

This list of desirable system characteristics may seem rather obvious. If they truly are obvious, then surely they would have been heeded by the analysts who have preceded you. However, this has not always been the case. Recheck the list once in a while.

System Constraints: What's Getting in the Way

Imagine going ahead with system design plans and discovering—too late— that there is an insurmountable barrier in your way. Let us say, for example, that the system will need disk space and that disk space for the organization is saturated and there are no equipment funds available. Are you to be blamed for this? Probably. No one is going to hold you accountable for a tornado, but basic considerations such as disk space are really your responsibility. That is, you must look ahead and be aware of problems that may impact your system. Your statement of constraints is basically one that says, "The system will work as long as we do not run into problems such as . . . ." This is not an open invitation to cover yourself from all possible contingencies, just a warning to look for the ones that may come your way.

Constraints usually fall into these categories: operational, psychological, political, and economic.

Operational.   Operational constraints are most easily understood. They involve anything specifically needed to keep the system running. General operational categories are hardware and software. Problems you might anticipate in these areas involve available storage capacity, a suitable database management system (DBMS), availability of computer run time, adequate data communications equipment, and availability of software development tools.

Psychological.   Although less concrete, psychological constraints are also an important consideration. The usual ones are the unwillingness of new users to accept the system and the overall resistance to change in the organization.

Political.   Now it gets a little sticky. In practical terms, political constraints may be the most troublesome because pointing to a powerful personage as a constraint on the proposed system is difficult to do. In some organizations, the person with the most clout can impact everyone else by opting for high-priority treatment on everything from computer test time to office space. You probably will not have trouble recognizing the problem when it occurs, but it may be difficult to predict. Plan a flexible schedule when possible.

Economic.   And lastly, never far from our thoughts, are economic constraints. The bottom line is really very simple: If the user cannot afford the system, that constraint takes priority over all others. The "best" system for the user may carry a high price tag. You must work within the financial limitations of the user organization.

It is common to present a list of constraints as part of the preliminary design document. As you can see from the preceding discussion, however, it is not politically expedient to put everything you know on paper.

◻ THE BIG PICTURE: WHAT DO YOU DO NOW?

The object of preliminary design is to determine, in a general way, how the system will be implemented. There are actually treatises and books that recommend that your next task is to sit down and *think*. Well, of course you will think, but not in some dreamy reverie at your desk. You already have guidelines in the form of the system requirements, and these give you some coherent notion of what will work. You know, for instance, that a system to track football plays for the coaches will not be very useful if it produces only monthly batch reports. And so on. The point is that you are not lost—the groundwork has already been laid.

Now you need to consider what steps you will take and in what order. The following suggested topics provide a framework for preliminary design (see Figure 7-1):

- Review the system requirements.
- Consider alternative systems and choose the one that most closely matches the requirements.
- Define hardware requirements.
- Define major system outputs, inputs, and files.
- Describe processing requirements.
- Prepare high-level system charts to show the relationship among processes, outputs, inputs, and files.
- Describe the benefits from the proposed system.
- Prepare cost studies.

It is at this point, then, that you make the major choices defining what the system will be. This is followed—or perhaps paralleled—by your preparing, in the form of high-level documentation, concrete evidence of your choice.

We shall now give our attention to the suggested topics just listed, and then look at another approach altogether—prototyping.

◻ REVIEW OF SYSTEM REQUIREMENTS

The system requirements are a statement of what the new system should look like. You may not think you could forget the requirements that you worked so hard on, but the memory can dim fast. Many analysts are distracted by other subjects and other tasks that need their attention, so a review now of the requirements is probably a good idea.

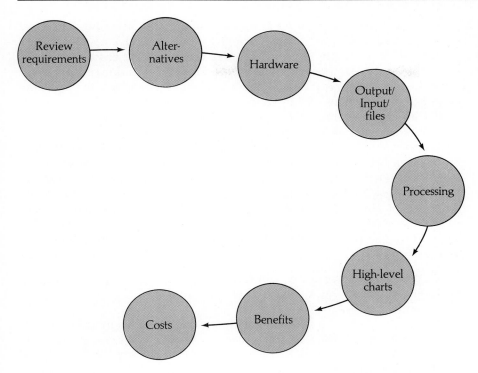

FIGURE 7-1

Preliminary design. These are the general steps in the preliminary design process.

◇ MAKING CHOICES

Any time you have to make a decision, you will find that the alternatives seem endless. Even so, as the search begins, you can address, in a general way, some of the major choices to be made. This chapter focuses first on two major issues: Should the new system be online? Might purchased software packages be the best answer? These two questions are not the only ones you will have to answer, but they are important enough to consider separately. Let us begin with these issues and then go on to a general discussion of how to set up and weigh alternative systems.

Online Systems

Just as early analysts were most likely to be converting large manual paperwork systems to automated batch systems, so you are likely to be converting these old batch systems to new online real-time systems. Large manufacturing and service companies, in particular, are abandoning the paperbound batch systems they grew up with.

A batch system, by definition, accumulates data and processes all of it at one time, as a group or "batch." A real-time system, on the other hand, processes data and gives output in a time period that is fast enough to affect the work environment. In most systems, "fast enough" usually means "immediately." Real-time systems are desirable because customers order and complain in real time, workers work in real time, and inventories and other assets accrue in real time. Today's marketplace demands the real-time answers that a real-time online system can provide. (An aside: It is possible to have an online system that is not real-time, but this is not common.)

Online systems allow an organization to be more responsive to customers and workers, support timely control of assets and liabilities, and provide quick, up-to-date information to managers. The heart of an online system is often transaction processing. In its most common form, **transaction processing** is the use of an online database for data retrieval and update by a group of concurrent users. Successful implementation of such a system means that everyone who uses the system can count on the system data being up-to-date at all times. Transaction processing is appropriate for many types of systems:

- A pharmacist keys a prescription transaction, receives a no-conflict (with other current medications) report, and updates the customer's record.
- An army lieutenant in a simulated field exercise keys in the unit's position and receives on the screen the relative positions of the nearest units.
- A supermarket point-of-sale terminal sends a product-identifier transaction, receives the product price in return, and updates the product inventory record.
- A social worker with a seriously ill mental patient keys in a transaction to get a screen reply on the availability of a temporary hospital bed.
- A hotel manager with unexpectedly high guest bookings keys in a transaction request to determine which employees are available for overtime work.

Note that in each of the systems mentioned in this list other system users could be performing the same activity at the same time.

Online systems have both advantages and disadvantages. The key advantages of choosing an online system design are that data is available sooner (as soon as it is entered to the system) and that it can be dispersed immediately to user departments. There are two key disadvantages. The first is that, for security reasons, there must be a total backup system; transaction "journals," files of records made from the generated transactions, must be maintained and audited. The second disadvantage—perhaps it should be called a difficulty instead—with online systems is the hardware/software interface, especially as it relates to data communications equipment. As we shall note in Chapter 10 on detail design processing, the connections between a computer and its peripheral equipment (not to mention other computers) are not standardized and often present a host of problems.

Online systems provide advantages that batch systems cannot, but they are not appropriate for every application. Online systems are more expensive, so their extra costs must be justified. Other aspects of online systems, including the important issue of controls, will be discussed in Chapter 10 on detail processing.

BOX 7-1	**The Rush to Go Online**

United States Navy Captain Grace Hopper, famous in computer circles for her role in developing COBOL, enjoys telling a story about "keeping up with the Jones"—online. As soon as one department developed an online system, she says, it was a foregone conclusion that other departments would follow suit. Faced with a series of departments wanting to climb aboard the online bandwagon, Grace was skeptical of their needs. The departments were all accepted, but Grace played a little trick on them.

Each record in the new system included a counter. The system incremented the counter by 1 each time the record was used. In six months Grace performed an audit of the system, printing each record in which the counter was still zero—that is, each record that had never been used by anyone since being added to the file. As she explains it, "One whole department fell out." In other words, one department had so little need for the system that they never used it, as evidenced by the inactivity on their records.

| BOX 7-2 | **Some Software Shopping Hints** |

As always, let the buyer beware. For some applications there are literally hundreds of software choices; for example, there are over 300 payroll programs. If you have decided to purchase software, here are some guidelines to help you choose the right package.

- Keep in mind that any given software package is compatible only on certain hardware and with certain operating systems. This point alone will narrow your search significantly. Do not get enthusiastic about software that will not run on your equipment, unless you are—as sometimes happens—purchasing software first.
- Evaluate each likely package by matching your requirements to its functions.
- If you find software you like, find out how many times that particular package has been sold and installed. If the numbers are low or the answers nonexistent, you may find yourself working out the bugs in a new system.
- Examine the documentation carefully. Have your users look it

Software. This selection of software would be of interest to anyone who has business applications on a personal computer.

over. Can they understand it easily? Or does it remind them of an imagined thesis on Sanskrit? Buyer beware.

- Is the product easy to use? Try it out!
- What kind of support is available from the vendor? Will the vendor provide you with source code so your technical people can make fixes?
- If you are considering a set of packaged software, will the programs interface with each other? Purchasing an integrated software package alleviates this problem.
- Does the vendor offer training? How much? How long?
- Does the software manufacturer offer a price break on subsequent editions of this software item?

Packaged Software: Make versus Buy

It is much less expensive to buy software than to write it from scratch. Although this is a simply stated fact, the to-buy-or-not-to-buy question is not so simple. The options are many. You could purchase a single software package for a single microcomputer user or as part of a larger system. You could purchase a set of software prepared for your user's particular industry; for example, many packages are available for banks, insurance companies, credit unions, and the like. Or you could contract with a service bureau to develop and manage the entire system. These are merely samples of the options available to you.

If you are considering buying software, you have two basic decisions to face: (1) Will buying a package be better than developing software in-house, and (2) if so, which specific package is the most suitable?

There is no magic in a purchased software package. It may fit your needs; it may not. Most users think that their system, the one to be automated, is unique. No existing package in the world could satisfy their requirements. This may be true, but often there is a package close enough to be used. Another possibility is a compromise—to purchase the basic package but pay for modifications. If these options are being considered, you will want to make a cost comparison chart for in-house development, purchase, or purchase with modifications. The chart will clearly show that homemade systems are the most expensive. This fact alone may move recalcitrant users to consider packaged software if it could be made to meet their needs.

The advantages of packaged software are as follows:

- Reduced cost.
- Reduced time.
- Proven system.
- Less staff.

The main disadvantage is that packaged software will not fit user needs as exactly as custom software will. Another disadvantage is that there may be no on-site expertise to fix it. Other disadvantages could be categorized as risks that, with some effort on your part, can be minimized: package search time (the time it takes to locate suitable software), poor documentation, unknown program bugs, program inefficiencies, and hardware configuration variances. Note the growth in packaged software in Figure 7-2.

If the proposed system is to be implemented through purchased software packages, perhaps it is a suitable application for the information center. That is, you can be the initial liaison between the user and the "user service" department, which would then take up most of the work. This option will be discussed more thoroughly in Chapter 17.

Other Choices to Be Made

You may recall that in Chapter 2, which contained an overview of the systems life cycle, several preliminary design possibilities were presented. Now it is time to formulate a scenario to fit the requirements of this particular system. You have already considered some of the big questions: whether the system should be online and whether software should be purchased.

Now look at other aspects of the system. What do the users need and how could the system provide it? How will data be captured? Will new equipment be needed? How will programs relate to one another? Do some

FIGURE 7-2

Packaged software growth. This chart shows the extraordinary growth of the packaged software market, which will increase sixfold between 1984 and 1989.

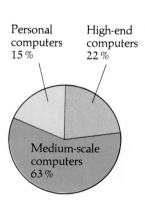

1984: $10.2 Billion

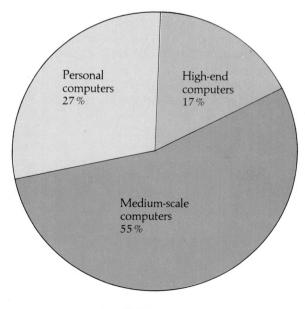

1989: $63.8 Billion

activities need an immediate response? What processes need to take place weekly? Monthly? And so on.

In addition to your own experience and that of your co-workers, you may want to call on the expertise of vendors, the people selling hardware and/or software. It does not hurt to look around and see what others have to offer. Vendors are happy to spend time with you, hoping that the time they invest will lead to a sale, as it might. If you decide that buying is not appropriate, you still benefit from the exposure to other ways of thinking.

Eventually—perhaps sooner than expected—you will develop an approach to the design. In order to communicate with others (not to mention remembering for yourself), you will prepare a written draft, with drawings, to describe what you have in mind. Methods of recording your design are discussed in the upcoming section called "Describing the System." Meanwhile, consider your choices.

Lining Up Alternatives

You may have several choices, each with a slightly different approach, possibly different equipment, and probably a different price tag. Each of these choices is called a **system candidate**, another name for an alternative design. Sometimes managers demand that alternatives be offered. Other times the various candidates seem obvious and it makes sense to offer them. The following list suggests an approach to comparing candidates.

- Consider system requirements in terms of importance.
- Develop alternatives related to the requirements.
- Evaluate the alternatives against the requirements and benefits.
- Choose the alternative best able to achieve the objectives within the allocated budget.
- Evaluate the chosen system in terms of general design goals.

If you want to make a more formal comparison of system candidates, consider the **weighted criteria method**. Suppose you have three system candidates—say, one for a centralized batch system, another for online access but centralized processing, and a third for a distributed system that uses a separate computer in each needed location. Any one of these could do the job, but there are large differences in costs and in services received.

To use the weighted criteria method, you must list desirable system characteristics and then give each a rating, usually 1 to 10, indicating a given item's importance. These ratings are called weights; that is, how much weight does the user attach to each characteristic? This is a bit difficult because anything you might think to write down you usually want very much, so all the weights are high. Think about the advantages of the candidates and write those down, too. If they are not important to you, that alone may provide the clear evidence to favor another candidate. In the King Books weighted criteria example, note that the weights have been written in parentheses next to the items in the left column. (See Table 7–1.)

TABLE 7-1

King Books:
Weighted Criteria

| | Candidates | | |
Desired System Characteristics	1 Centralized Processing	2 Online Access	3 Distributed Processing
Accessibility (8)	2 (16)	8 (64)	10 (80)
Fast processing (7)	2 (14)	6 (42)	8 (56)
Customer service (6)	4 (24)	6 (36)	8 (48)
Accuracy (9)	7 (63)	8 (72)	8 (72)
Independence (1)	2 (2)	5 (5)	10 (10)
Economy (9)	10 (90)	6 (54)	0 (0)
Total weighted score	209	273	266

Now you must assign an expected performance value on the items, by candidate. That is, on a scale of 1 to 10, how good do you think each item is, or how well do you believe it will perform? A 1 or 2 is poor, whereas at the other end of the scale, a 10 is perfect. Each of these will be multiplied by the respective weights, producing another number, related to the item row and the candidate column. (These numbers are in parentheses in the example.) When these numbers are added for the candidates, you have the total number for each candidate. Candidate 1 is clearly the loser, since its centralized processing does not meet the system needs. (The fact that it would be much less expensive becomes irrelevant.) Candidate 3 meets the system needs well and thus has high scores in many areas. If money were no object it would be the first choice. But cost is a significant consideration—a 10—so the winner emerges as candidate 2. Sometimes these methods can produce the "wrong" winner, making you wonder about the validity of the system. The system is, of course, highly subjective. But it is still valid—at least it showed you what you really wanted. Now you may want to adjust your thinking, or the weights, to justify it!

◇ DESCRIBING THE SYSTEM

At about this point, it is time to describe, in a general way, the system you have in mind. You understand the organization, with its needs and limitations. You can formulate a plan and describe it in a way the user can understand. To do so, you must have some idea of the hardware requirements; the input, output, and files needed; and the processing. Specifics about these subjects, plus databases, communications, and security, can wait until the detail phase. Review the design plan for King Books to get an idea of how you might proceed.

KING BOOKS

Thinking Out Loud About the Design

King Books is a fair-sized operation—it would be surprising that it has no computer power if it were not so new. Glen Neilson understands very well that there is a lot of room for automation. In fact, he has big ideas about both company growth and automation. What's more, he wants to do it in-house, with his own staff. He has already hired a skeleton data processing staff in anticipation of the big change. They are currently underemployed, dealing only with a software house to process payroll and working with outside analyst Dave Benoit on the new accounts payable system. Glen is an eager user, but as talks with Dave progress it seems that bringing in a total hardware setup and a complete data processing staff is a giant step. There are just too many problems: lack of expertise, lack of space, and lack of capital funds.

Although the ultimate goal is still a self-contained King Books operation with many systems, it was tentatively agreed that Dave's consulting company, Universal Computer Services, would handle the accounts payable system for two years. Universal's fees include providing some on-site hardware at King Books on a lease-option basis. It is still up to Dave to produce an acceptable accounts payable system design, however, before Glen signs a contract.

A review of the design requirements indicates that users need immediate access to the status of vendor accounts. Such access would make it easy to combine invoices for the same vendor and also to check vendor discount terms. This kind of system would also reduce errors such as duplicate payments. Finally, queries from vendors about the status of their accounts could be handled quickly. All this points to an online system for updating and querying key accounts payable information—purchase orders, invoices, and vendors. This will be the heart of the system.

Not everything needs to be online, however. The checks themselves can be printed weekly. (Discounts can still be used because bills are due monthly and invoices are received with a two-week slack.) Since this is a financial system, Dave correctly anticipated a number of reports. The majority of the reports are weekly or monthly, however, thus lending themselves nicely to batch processing.

As noted in the weighted criteria discussion (Table 7-1), a totally centralized system is untenable, but a computer on-site in the finance department is economically unsound. The middle-ground solution is an online system to Universal's mainframe. Universal will keep database files of accounts payable data on disk. Terminals can be placed in the finance department for instant access and update. Universal will run batch reports on a scheduled basis, producing them on a printer that will reside on-site at King Books.

Since there are over 300 accounts payable software packages available, Dave thought they were worth investigating. He found one from SoftPro that he thinks can be modified to meet King Books' needs.

This, in a nutshell, is the design plan.

Hardware Requirements

Describing the hardware at this point is not too difficult. What do you need in the way of computing power, and associated peripheral equipment? So far, with a general design in mind, you know if you need terminals or disk storage or microcomputers or a fast printer or all of the above. You can describe these needs in narrative form and support them with a diagram in the preliminary design document, as part of the design overview discussion. In the next section, we shall describe the effort needed to acquire hardware, if that is part of the plan.

Outputs, Inputs, Files Description

At this stage you need have only the most general idea about outputs, inputs, and files. Once you have a design plan, you can list the outputs needed, whether screen or printed reports. Detailed content is not needed yet. Similarly, you will have an idea of what inputs are needed to produce the output, and whether the data thus presented to the system should be kept in sequential or direct access files, or on a database system. Note the list of outputs and inputs for the King Books example.

Process Description

Now that you have described the output, input, and files in a general way, you are ready to integrate them into the design process picture. Once you have drawn the processes at a high level (that is, in a general way), you should be able to state acceptance criteria and plan a schedule for the rest of the project. We will take these in order.

Process Design.   At this point you need to present just enough information so that the user can understand the overall design plan. One high-level design tool is the **Leighton** (*lay*-ton) **diagram**, which is simple to draw and easy to understand. It is a high-level system chart, hierarchical in nature, listing the processes by function or processing time (daily, weekly, and so on), and breaking these into ever smaller chunks. At the lowest level, outputs, inputs, and files are attached to the processes. The detail design

KING BOOKS

Outputs and Inputs

Outputs

Screen output:

> Status of purchase orders
> Status of vendor accounts
> Status of invoices

Printed reports:

> Checks
> Vendor name list
> Total vendor status report
> Journal of purchase orders
> Bank reconciliation report
> 1099 report

Inputs

Online input:

> Vendor data
> Purchase orders
> Invoices
> Vouchers

processing chapter (Chapter 10) presents other hierarchical methods that will be appropriate to use for a fuller description of the system. Meanwhile, note the Leighton diagram for King Books in Figure 7–3.

Acceptance Criteria. Acceptance criteria are interesting in that they can be very effective as a political tool. Their power lies in the fact that people are willing to commit to reasonable-sounding actions if the commitment is well in advance of the actual event. Politician or not, do not miss this opportunity. You will be very pleased later.

Acceptance criteria can be a set of very simple statements. In the accounts payable system for King Books, for example, straightforward acceptance criteria could include "The system will be able to produce computer-gener-

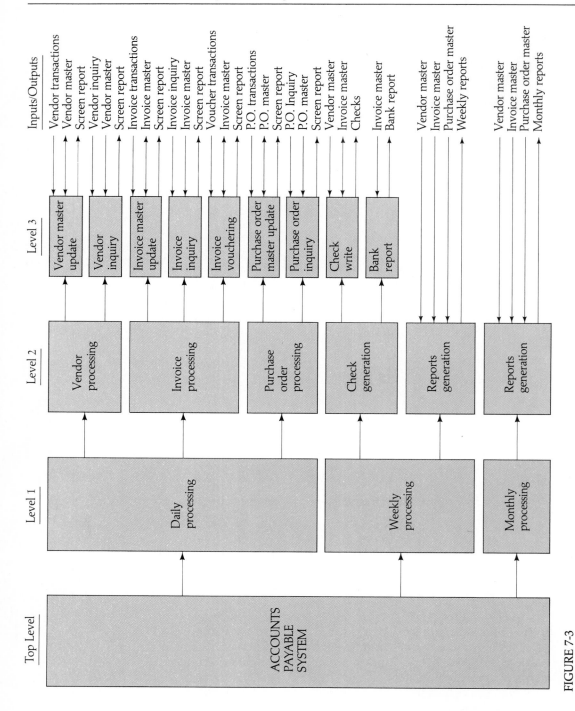

FIGURE 7-3

Leighton diagram. The Leighton diagram shows high-level processing, as well as inputs and outputs.

ated checks" and "The system will be able to enter, update, and query vendor data online."

The payoff is at implementation time, when the system is ready to be put to use. Assume that the system is functioning according to plan. There is your user, however, foot-dragging week after week. Somehow there is always a reason why the system cannot be implemented. You gradually come to realize that you are faced with a bigger problem: reluctance to make the change. But you have a trump card—the acceptance criteria.

The list of acceptance criteria is drawn up and signed by both the data processing and the user organizations in the preliminary design stage. You are promising that you will not ask that the user accept the system until these things are done. The user is promising that, once these things are done, the system will be accepted. It works very nicely for both parties.

Scheduling.   As you continue to put together a preliminary design package, you will need a schedule for the coming work load. A common device for presenting a schedule is the **Gantt chart**. Note the King Books chart in Figure 7–4. Each major activity is represented on a Gantt chart by a horizontal bar that travels across the time line represented by the *x* axis. A Gantt chart can thus be a very simple bar chart, although some organizations use more sophisticated versions of it to indicate progress made and delinquent schedules. At a later time you may wish to have a detailed Gantt chart for each major activity, listing tasks within the subcategory. Other measuring devices will be discussed in greater detail in Chapter 14 on project management.

Benefits

Benefits have already been noted in a general way in the preliminary investigation phase. However, now that you understand the system better and have a clear idea of the overall design, this would be a good time to reexamine the benefits. If you choose to, refine and rewrite them.

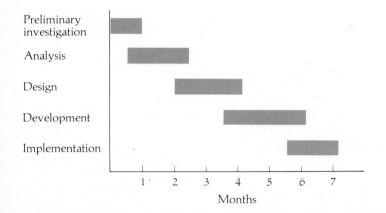

FIGURE 7-4

Gantt chart. This overview of a Gantt chart shows the systems life cycle for King Books.

◇ HARDWARE ACQUISITION

If you are truly starting with nothing and plan to build your own software in-house, then you are talking about a lot of equipment. You are also looking at long lead times—that is, a long time from order to delivery. You need to begin the process as a by-product of the preliminary design. The following are some considerations related to hardware acquisition.

Purchase or Lease. Computers and their associated equipment are expensive. Some users are reluctant to buy costly equipment that they may have to keep using when new and better hardware becomes available. Also, used equipment has little value on the market, so the purchase of a complete system should be made with full utility in mind. That is, you better know that you can get your money's worth out of it because you will not get much when it comes time to resell. Finally, some companies need to conserve capital outlay for other purposes or may have certain tax strategies in mind. For all these reasons, many organizations choose to lease equipment. Equipment can be leased directly from manufacturers or through leasing companies that buy a variety of equipment and juggle it among leasees.

Proposals. You need to prepare a list of what is expected from a new computer system. These needs can often be stated quite specifically. For example, there must be enough ports to support a certain number of terminals, or a compiler for a certain language must be available. In many organizations, computer acquisitions are made through a system that includes a request for proposal. A formal document, the **Request For Proposal (RFP),** states the computer equipment needs of the organization and invites vendors to respond.

Performance. Comparing computers involves more than just reading glossy brochures and talking to vendors. At some point, some serious efforts must be made to compare one machine to another in terms of performance. A **benchmark** is a formal performance test. All kinds of benchmark tests are available, or you can invent your own. You usually want to compare factors such as throughput and response time in an environment that closely represents the systems that will be run on the machine. There are even benchmark programs available to measure printer capabilities.

Site Preparation. Installation requirements for a new computer system are often quite complicated. The manufacturer can provide you with detailed specifications. These include everything from computer dimensions to humidity requirements. At this point you could be coordinating with facilities managers about an appropriate location for the equipment. Details of site preparation are mentioned in Chapter 13 on implementation.

Delivery.   When do you need the equipment? Will all related equipment items be available when you need them? It does little good, for example, to have the mainframe and terminals delivered if no cables are available. What written guarantees can you get from the vendors?

Besides purchase or lease other options are possible. For example, more likely are situations in which no extra equipment is needed, or additions are needed to equipment already in place, or some minimum leased equipment is needed because the processing is being contracted out. You may need to order terminals, modems, microcomputers, extra disk storage, or extra printers. Once the needs of the system are clear, however, you will be making the rounds of the hardware vendors, who will be very attentive, at least until you sign on the dotted line. That last statement sounds cynical, but it is all too true. The lesson here is to get everything you can—in writing—before you are committed. After the signing, the vendor is much less available.

◇ COST/BENEFIT ANALYSIS

In order for the users to get their money's worth, the value of the system benefits—that is, savings—must exceed the system costs. Figuring out whether or not this is the case requires some agility on the part of the analyst. The biggest problem is quantifying intangible benefits. Suppose, for example, that you plan to convert a rather old batch system to an online system. Input data that used to be keypunched will now be keyed directly to the disk through a time-sharing system. Since the data preparers already think they are about to leave the Dark Ages, one of the benefits you anticipate is an improvement in morale. "Improved morale" is certainly a benefit, but can it be quantified? Can you assign a dollar value to it? The answer is yes, but it is not easy. You will need to relate it to more work processed or fewer employee absences or something else measurable. Let us examine the whole issue of money and see how to look at it in relation to justifying the system.

Costs and *benefits* are terms that have general meanings. To really compare them, you need to break them into finite units and hang a price tag on each unit. Then you add up the totals and make a comparison.

System Costs

Begin with the two kinds of costs you will be seeing: development costs and operations costs. Development costs are one-time costs, whereas operations costs are recurring costs. **Development costs** are the monies spent for all resources used to develop the system. Each phase of the life cycle has costs that fall into several of these categories:

| BOX 7-3 | The Second-System Effect |

How would you like the systems you design to be remembered? Or, perhaps more to the point, what kind of analyst reputation would you like to have? This not-so-idle question is worthy of some thought before the fact; it is easier to do than undo.

Would you like to be known as the analyst who designs systems that are reliable, systems that do what you promised? Or perhaps for systems that are clean and lean, getting the most for the money? These sound good, but many analysts become known instead as producers of systems that try to do something for everyone but actually do nothing for anyone. Why is this? Well, partly, because analysts are so agreeable.

In his book *The Mythical Man-Month*, Frederick Brooks describes the "second-system effect," which he attributes to overanxious analysts trying to please the world. An analyst's first system design will be pretty plain vanilla; although frills may be considered, the analyst is too nervous to allow these luxuries to creep in. This hesitation is gone, however, by the time the second system is designed. Flush with success from the first effort, the analyst now feels confident about adding frill after frill, embellishment after embellishment. That is, there is a distinct tendency to overdesign the second system.

The second-system effect has a long history. Perhaps you could learn from this and avoid the pitfalls.

- *Labor:* Analysts, programmers, operators, clerks, data entry, users, consultants, management, technical writers
- *Equipment:* Purchase and/or use of computer time, terminals, printers, disk space, tapes, data communications equipment, office equipment
- *Supplies:* Paper, copying supplies, desk supplies
- *Overhead:* Heat, air conditioning, light, cleaning, maintenance

Operations costs, usually calculated monthly, are the costs of running the completed system. Although money spent falls into the same categories just listed, the emphasis changes. First, the cost of running a system for a month is considerably less than the development cost. Second, the labor costs—always the largest development expenditure—are significantly reduced.

System Savings

Savings are related to benefits. Benefits also fall into categories, but they are not as clear-cut as costs. Benefits come in more varieties, depending on the type of system, and tend to be more difficult to quantify. Benefits can be derived from either cost reductions or increased revenues. Typical cost reductions can come from increased productivity or decreased user personnel—probably by attrition, as opposed to layoffs. Typical increased revenues come in the form of increased business due to improved customer service or better investments due to better access to management information.

Now let us approach the issue of quantifying benefits. The vague phrase "increased productivity" is a good example of a benefit that is difficult to quantify. Consider the directory assistance change that the Bell System has installed in many parts of the country. When the operator keys in the name of the person whose phone number you need, a list of possible names appears on the screen, each with a letter along the left screen margin. The operator then keys only the letter of the correct line, causing the computer to generate a spoken message containing the phone number. Meanwhile, while you are still listening, the operator goes on to serve another customer. The customer/operator interaction time has been reduced. To get some meaningful cost reduction figures, you would have to estimate how many more customers a given operator can serve and, consequently, how many fewer operators are needed to serve the same population. This figure can be applied to salary data to come up with a monthly savings figure. Notice that "increased productivity" turned into decreased personnel in this case.

Taking another short example, a supermarket that converts to an automated bar-code checkout system can expect more business from customers who like the speedy service. You can estimate both the customer increases and the dollars spent per extra customer. In cases like these, there is usually historical data, provided by vendors and other users, to use as a guideline. In fact, historical data is useful when trying to prepare benefit figures on any type of system.

Return on Investment

Now that you have some idea about costs and benefits, we return to the original question: Is it worth it? Managers see system development as an investment in the future. The various system phases all cost money—investment money. At this point management has already spent funds on the preliminary investigation and systems analysis phases. Now they must decide whether to proceed with the remaining phases and actually implement the system.

One measurement they can use is the **payback period**, the amount of time it takes to "pay back" the investment funds with net savings from the new system. Once the new system is running, net monthly savings can be computed by subtracting the monthly operations costs from the monthly savings benefits:

Net savings = Savings − Cost

The development cost is then divided by the net savings to obtain the payback period, the time period it takes to recoup the invested monies:

Payback = (Development cost) / (Net savings)

As an example, suppose that the total development cost of a new system is estimated to be $100,000. If the new system costs $3,000 to operate but has benefits of $7,000, then the net savings per month is $4,000. Dividing $4,000 into $100,000, gives 25, the number of months in the payback period. All figures could be computed on a yearly basis instead if payback could be expressed more meaningfully in years. (There is not much point, for example, in talking about "84 months" when "7 years" expresses the same fact more clearly.) The end of the payback period is sometimes called the **breakeven point**. Note the chart in Figure 7–5, showing the breakeven point at 25 months.

This discussion has been fairly light. Many books devote whole chapters to this subject, giving all kinds of charts and ratios. There are, of course,

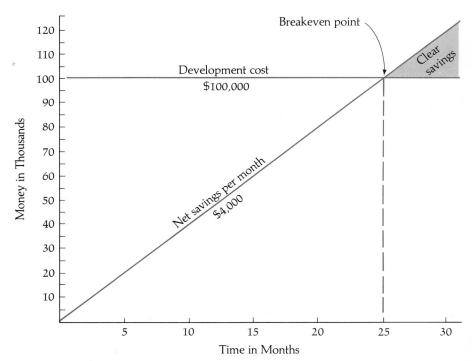

FIGURE 7-5

Breakeven chart. Payback period for an investment of $100,000, with net monthly savings of $4,000, is 25 months.

entire books on business finances, but their audience is not the systems analyst. Before you concern yourself too much about your slim skills as a cost accountant, be assured that you will have help. At the very least, any company has dollar cost figures for various activities, such as X dollars per hour for a programmer. With figures like these, you can estimate the number of hours and multiply it by the hourly rate to obtain the cost figure for programmer services. Some companies go further and provide full accountant services. You will work with these specialists, giving the estimates for your system. In other words, you do not need to know everything, but you do need to know something, so you can talk to the experts.

◇ PROTOTYPING

The idea of building a prototype, a sort of guinea pig model, has a lot of appeal—if you are building an airplane or a car. The prototype is the first of its kind and not much is expected of it. A prototype is a working model, one that can be tinkered with and fine-tuned. The final version of the prototype has been put through the paces and is usually highly satisfactory to all.

Considered from a systems viewpoint, a **prototype** is a limited working system, or subset of a system, that is developed quickly. Can you adopt this approach to systems development? This text speaks forcefully to the issue of planning before acting—that is, doing the steps in the systems life cycle in proper order. Leaping in with a loose test model does not seem to belong in the natural order of things. And yet, some analysts in the computer industry are doing just that. Before you volunteer to try this approach, note the circumstances when such action might be appropriate.

Prototyping Tools

The prototype approach exploits the advances in computer technology, using powerful high-level software tools made practical by inexpensive hardware. These software packages allow analysts to build systems quickly in response to user needs. These systems can then be refined and modified as they are used, in a continuous process, until the fit between user and system is acceptable.

Many organizations use prototyping on a limited basis, for example, for a certain data entry sequence, a particular screen output, or a particularly complex or questionable part of the design. That is, prototyping does not necessarily have the scope of the final system.

You might use prototyping if you work in an organization that has quick-build software and management support for this kind of departure from traditional systems procedures. A prototype system is developed very quickly, often in a few days. The software tools used are usually a time-sharing computer system, a database management system with a high-level query language, an easy-to-use fourth-generation language, and a generalized report writer package. The idea is to get the system into the user's hands quickly, so that the refinement process can begin. This puts the user in a comfortable and enjoyable role, that of wise consumer. Figure 7-6 illustrates the refinement process.

Prototyping and the User

Prototyping is not always appropriate. Situations ripe for prototyping are marked by several of the following circumstances:

- Users can state objectives readily but have trouble deciding on actual requirements.
- An inherent communication gap—time or distance—exists among participants.
- The system is high risk and high cost.
- The analysts available lack expertise in this type of application.
- A system of this type is untried and untested by the users.

Prototyping is used most commonly to help users determine requirements. Users often have some idea of what they want, but are vague on details. Statements like "Maybe it could . . ." or "It would be nice if . . ." are often heard. Prototyping is based on the assumption that the complete set of requirements will not be discovered until the user has had a chance to experiment with a working system.

Prototyping Results

What is the net result of a prototype system? What will it produce for users? If the whole system is being prototyped, it initially will look something like this: minimum input data, no editing checks, incomplete files, limited security checks, sketchy reports, and minimum documentation. But actual software uses real data to produce real output. Remember that prototyping is an iterative process. The system will be changed again and again.

In some organizations, the prototyped system is refined until the end product is acceptable. In other organizations, prototyping is not the end system. In fact, the prototype may be thrown out when it has accomplished its purpose (of, say, helping establish requirements), and an up-to-standards system written to replace it.

Prototyping is not offered as a panacea for everyone, only as an option that is right for some of the people some of the time.

◇ PRELIMINARY DESIGN DOCUMENT

The **preliminary design document** is another major milestone document. This document will be built on the analysis and requirements document you have completed already. It contains some of the same materials in the

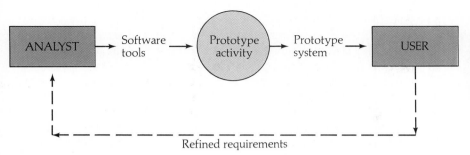

FIGURE 7-6

Prototyping. The prototype process is iterative, with users and analysts working together to refine the system.

beginning to set the stage for the design. Again, this list may vary with your needs.

- *Title page.* Key personnel and an abstract (concise summary) of the project subject matter.
- *Project overview.* As before.
- *Problem definition.* As before.
- *Requirements.* As before.
- *System constraints.* If any.
- *Alternatives considered.* Include weighted criteria if appropriate and a narrative backing it up.
- *Design overview.* Narrative. Why you did what you did. How the design meets user requirements.
- *User diagram.* Diagram of the proposed system from a user, non-technical point of view.
- *Benefits.*
- *Description of new system:*
  General hardware requirements
  Outputs, inputs, files
  Process description
  Software packages, if any
  Acceptance criteria
- *Costs, schedules, personnel.*
- *Appendix.*

This document will be in the hands of interested parties well before the design review.

◇ PRELIMINARY DESIGN REVIEW

The **preliminary design review** may be the most important one you give. You are now convincing both your own management and the user organization that you have a good plan that will solve some problems and be within an affordable price range. However, you probably will not be springing any surprises. This review may be almost a formality if the system is one that many people have agreed upon already in private. You would not, in fact, consider going into such a review unless you had a good idea of what the outcome was going to be.

From a political standpoint, the actual review process is often a vehicle for users to demonstrate to their own management, through you, that they have been wise stewards. The bulk of the work, remember, has been done at lower levels, where analyst and user have formed a working alliance. Although higher management has been kept informed and is supportive of the efforts thus far, they must now be convinced to proceed with the plan.

Management wants—needs—to hear the design plan and understand how it will impact the organization. They need to see a direct link between their goals and the services the system will provide. If there is one overriding bit of advice we can offer at this juncture, it is to keep the presentation user-oriented.

An unstated goal of the formal presentation is to give management another opportunity to size up the analyst and the data processing organization he or she represents. You will not have many chances to look this good, so seize the opportunity.

BOX 7-4	**Preliminary Design Presentation**

As before, topics in the preliminary design presentation may vary. Note some of the similarities to topics in the analysis and requirements presentation. A little duplication does not hurt, however—it refreshes the minds of some and gives needed information to those hearing it for the first time.

- Background
- Problem or opportunity description
- Alternatives considered
- Overview of the new system, strictly from the user viewpoint
- Anticipated benefits, as related to new system
- Full description of new system, but not in excruciating technical detail
- Schedules, costs
- Summary
- Questions

◇ THE END OF THE BEGINNING

So now you have begun. You have a general idea of the design plan and agreement from your own management and users at all levels. Now the work begins in earnest, as you nail down the details of the design. The first detailed task is output definition, followed by definition of the input required to get that output. The final design task is processing, to complete the input-process-output picture. We now turn to these topics.

FROM THE REAL WORLD

Factory Management at Corning Glass

Have you tried to buy an American-made radio lately? Chances that you will find one are slim, because radios are no longer manufactured in the United States. On the other hand, if you check your hardware store's selection of electric drills, you will not find any imports.

Why the difference? Back in the 1960s, Black and Decker, the leading maker of electric drills, successfully implemented an outstanding computerized manufacturing system and became so efficient that imports could not compete. The name of the system is Materials Requirements Planning™, commonly called MRP, and it has had an impact on American manufacturing way beyond electric drills.

Another happy user of MRP is Corning Glass, which manufactures all manner of glassware, including the famous Corningware and Corelle dishware. In the mid-1970s executives at Corning began hearing grumbles from two of its largest retailers, Sears and K-Mart. Corningware and other products were not being shipped on time. In fact, only 50% of the company's orders were shipped on time. When managers heard excuses but saw little improvement in performance, they began to worry that big retailers might discontinue their wares. They brought in the MRP system to be used on their IBM 3083® mainframe, networked to DEC VAX® minicomputers at each of Corning's 61 plants.

MRP works something like this: If a company receives an order for 100 widgets but has only 30 in stock, it should plan to make 70 more. If there is only enough raw material to make 20 widgets, it should order raw material to make 50 more. These sorts of calculations would be simple enough in a small shop, but as a business grows, a computer becomes necessary. Manufacturing systems differ from other types of businesses for two reasons: (1) They are enormous, and (2) they cut across organizational lines—inventory, engineering, purchasing, receiving, shipping, and others. Everyone must cooperate to give the computer system an accurate view of what is going on.

Used properly—and it takes commitment to make it work right—the MRP system delivers big results. Among other improvements that Corning noted were a reduced inventory and overtime pay. The big payoff, however, was a 99.8% delivery success rate. No more grumbles from Sears and K-Mart.

—From an article in *Softalk*

SUMMARY

- The design phase has two subphases, preliminary design, which establishes a general design plan, and detail design, which presents specific information on the design of output, input, processing, file processing, and databases.

- As the design phase begins, the systems team usually shifts somewhat so there are fewer users and more technical people.

- The goals of systems design are suitability, reliability, ease of use, simplicity, and economy.

- System constraints refer to possible problems that may impact your system. Constraints usually fall into the categories of operational (anything needed to keep the system running), psychological, political, and economic.

- The framework for preliminary design includes reviewing the system requirements; considering alternative systems and choosing one; defining hardware requirements; defining major system inputs and outputs; describing processing; showing relationships among processes, outputs, inputs, and files; describing the benefits; and preparing cost studies.

- Many automated batch systems now are being converted to online systems. The heart of an online system is transaction processing, using an online database to retrieve and update data. The key advantage of choosing an online system is that data is available as soon as it is entered into the system.

- Two decisions must be made when considering buying packaged software: (1) Will buying a package be better than developing software in-house, and (2) if so, which specific package is the most suitable? The advantages of packaged software are reduced cost, reduced time, proven system, and less staff. The disadvantages are lack of exact requirements fit, lack of on-site expertise, package search time, poor documentation, unknown program bugs, program inefficiencies, and hardware configuration variances.

- Each possible alternative system is called a system candidate. Alternatives are developed and considered in light of the system requirements. A formal method for comparing system candidates is the weighted criteria method, which assigns weights and performance values to desirable system characteristics and computes a total for each candidate.

- Hardware requirements, if any, as well as outputs, inputs, and files are described very generally at this stage.

- The process design needs to be presented in a general way so that users can understand the overall design plan. One high-level design tool used for this purpose is the Leighton diagram, a hierarchical diagram that shows the relationships among processes and inputs, outputs, and files.

- Acceptance criteria are a set of statements agreed to by the analyst and user, on which the acceptance of the system will be based.

- A Gantt chart is a bar chart used to schedule system activities.

- If you are considering hardware acquisition, the two basic options are purchase or lease. A document called the Request for Proposal (RFP), specifying the computer equipment needs for the system, is prepared for vendors. A benchmark is a formal performance test for hardware being considered. Site preparation may be required before a major equipment installation.

- System costs include one-time development costs and recurring operations costs. System savings are related to benefits, which must be quantified.

- The payback period is the amount of time it takes to pay back the investment funds with net savings from the new system. Net savings equals savings minus cost. Payback is development cost divided by net savings. The breakeven point is the end of the payback period.

- A prototype is a limited working system that is developed quickly, using powerful high-level software tools.

- The preliminary design document usually consists of the title page, project overview, problem definition, requirements, constraints, alternatives considered, design overview, user diagram, benefits, description of the new system, costs, schedules, personnel, and appendix.

- The preliminary design review provides the opportunity for you to demonstrate to management that the design plan is a good one.

KEY TERMS

Acceptance criteria	Payback period
Benchmark	Political constraints
Breakeven point	Preliminary design document
Detail design phase	Preliminary design phase
Development costs	Preliminary design review
Economic constraints	Process design
Gantt chart	Prototype
Goals of systems design	Psychological constraints
Hardware acquisition	Request For Proposal (RFP)
Hardware requirements	Site preparation
Leighton diagram	System candidate
Net savings	System constraints
Online systems	System savings
Operational constraints	Systems team
Operations costs	Transaction processing
Packaged software	Weighted criteria method

REVIEW QUESTIONS

1. Describe, in a general way, the difference between preliminary and detail design. Why is it a good idea to have the design phase split into these two parts?

2. Analysts in training are sometimes nervous about that day when they will be on their own for the first time, actually responsible for planning a new system. Can you think of what aspect of this process would make you nervous? Can you think of resources you could call upon for help if you get stuck?

3. The issue of constraints is often a difficult one for trainees, who envision only available resources. From an operational standpoint, list five conceivable barriers to system success that might become problems *after* the system has been planned.

4. What is the purpose of reviewing the system requirements, which were prepared in the analysis phase, before proceeding with the preliminary design phase?

5. List five examples of systems that could benefit from online transaction processing. Now list five systems that should be batch only because there is no real ecomonic gain from having them online.

6. As an analyst, do you think you would be more likely to consider packaged software seriously if you worked as an independent consultant, as opposed to working within a large company that had its own programming staff?

7. When alternatives are considered for a set of geographically separate stores or offices, the choices usually come down to some variation on (1) centralized processing with remote access through local terminals, or (2) local processing with local computers, or (3) some local processing but with access to a central computer. In general terms, give the advantages and disadvantages of these choices. Do not forget the economic factors.

8. A supermarket manager wants to install a system that will scan bar-coded products, automatically look up prices in a computer's storage, and compute the bill. You have probably seen such a system in your local store and know that the system has additional functions, such as providing a tape for the customer to keep listing all purchases. Try to think of a list of possible acceptance criteria for such a system.

9. The subject of hardware acquisition is uncharted territory for most new analysts. If this is true for you, but you need to know more about it, where can you turn for assistance?

10. The subject of prototyping seems vague when first encountered. Assuming that you do know something about the high-level software tools mentioned, and that they are available to you in your working environment, for what might you use prototyping? Can you think of a particular application?

CASE 7-1
MAKING IT LOOK STORE-BOUGHT

The accounts receivable system at Cameron University was a jumble of manual procedures executed by revolving-door clerical help. The whole operation was held together by an experienced supervisor who joked about using band-aids and chicken wire. Her boss, Controller Mel Schoen, often thumbed through glossy brochures extolling the benefits of automated accounts receivable systems. These systems, he knew, could be modified to run on the university computer, but he lacked funds to purchase such a system.

In an inspired mood, he turned to the Management Information Systems Department for assistance. He knew that graduate students needed "real life" projects and thought he might get some free services. Schoen was able to begin work with a team of three graduate students.

Over a period of two semesters Schoen and the student team plowed through the systems phases, always aided and abetted by the brochures. Schoen wanted all the bells and whistles. The students tried to focus on the needs of the existing system, but they were always brought back to the fancy reports and other extras of the brochure systems.

We have here, alas, a system failure. At the end of the school year the system was incomplete. It not only did not look like the store-bought systems, it was not able to perform even the most basic functions. School professors had planned to pass the system on to a new student team for completion, but somehow that never happened. Cameron University, as of this writing, still has a manual accounts receivable system.

1. This is not really a make vs. buy decision, but a make vs. imagination fantasy. If the students had been more assertive, how could they have alleviated the situation?

2. Assume that the most widely imitated accounts receivable package had a purchase price of $2,500. If the students had been paid entry-level salaries, how would you compare, in general, the costs of the in-house system to the purchase price of the package system?

CASE 7-2
COMPUTERS ARE TAKEN TO THE CLEANERS

Glenda Underwood saw a business opportunity. Computers were helping run businesses large and small—why not her friendly neighborhood cleaning establishment? Although her analyst experience was somewhat limited, she felt she could easily handle a small project on her own. Based on the familiarity of her face and her business card (GLENDA UNDERWOOD, COMPUTER SYSTEMS CONSULTANT—in embossed letters), the owner of the establishment, Eric Thompson, granted her an interview.

Glenda spoke confidently of cost savings, better control, and improved customer service via computer systems. She also spoke convincingly of the increasingly lower cost of computer hardware and how it was now within the reach of every small business and even the home. Thompson was favorably impressed and agreed to proceed.

Glenda, in fairly rapid order, selected a microcomputer and several financial management software packages that would run on that machine. Then she sat down with Thompson and explained how it was all going to work. She worked with Thompson to produce a list of benefits from the system. Glenda also wrote a six-page report on the system and attached a cover letter and her bill to it.

1. What chance does this system have of working?
2. Are things a little out of order here? Rearrange.
3. What necessary system activities are missing altogether?

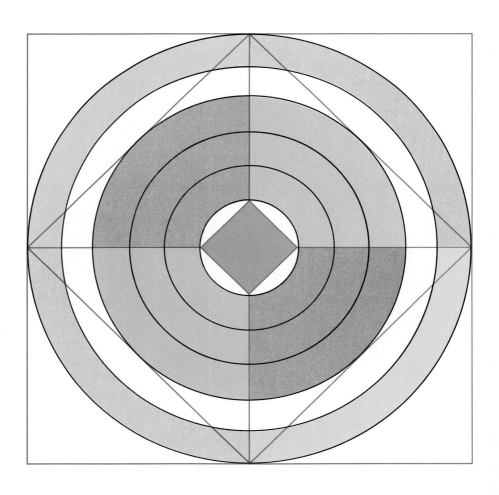

Chapter 8

Detail Design: Output

When analyst Jan Kumasaka began working with the Chelsea Green Hotel, she saw all kinds of possibilities for improving the hotel's profit picture. Since management's main goal was installing a second restaurant to draw in local customers as well as guests, Jan and management discussed types of restaurants in some detail. Jan used a microcomputer integrated software package to make spreadsheet cost/profit comparisons of restaurant types and then convert her findings to terminal screen bar graphs for the presentation to management.

Although she expected the graphs to make a dramatic impact, she knew she could not huddle the nine managers around a single computer screen to view them. The task of getting the graphics into a form usable in the presentation, however, turned out to be surprisingly difficult—she had a problem with software/printer compatibility. She did manage to find a combination that worked, just in time. When Jan vowed that she would take time to become more familiar with graphics options, she faced a fascinating, and increasingly complicated, task—one we shall examine in this chapter. First, however, we begin with a consideration of user needs.

◇ OUTPUT CONSIDERATIONS

A new system is designed to meet user needs. Nowhere are the needs as clear as in the design of output. Users can often be quite specific about what they want to see in front of them. Your approach to the discussion can take on the form of a litany:

> What output information do they need to see? (Purchase order status? Baseball statistics? Letters to constituents? Planting cost per acre?)
>
> How would they like it presented? (Screen? Printed reports? Plotter output?)
>
> What format should it be in? (Columns? Pictures?)
>
> When do they need it? (Immediately? Weekly? Only in exceptional circumstances?)
>
> How much output is there? (Does high volume point to microfiche? Does low volume mean inexpensive media?)

These questions should be part of your early discussions with the user on the subject of outputs. And the answers should form the basis for output planning, including printer and screen output, and report design. Viewing output in terms of user needs, we shall begin with output media and then examine criteria for reports.

◇ OUTPUT MEDIA

Printed reports were the major output medium through three generations of computer technology. Printed output is still important, but its use is diminishing as we rely more and more on immediate responses from terminals. Although most output is produced on paper (hard copy) or screen (soft copy), other exciting media may be suitable for certain applications.

BOX 8-1	**Color Graphics to Go**

Instant slides or instant pictures—they are almost as close as your computer. If you need a quick transition from screen graphics to prints or slides, you can find commercial devices available for this purpose.

The trick involved in getting instant prints or slides is a camera imbedded in a shield that exactly fits over the screen. You place the camera up to the screen, press the button, and within minutes you have a finished print. Slides are a little more complicated. You load the completed slide film into a processor, crank it through, and wait 60 seconds—then you have slides ready for mounting. These are separate cameras, though. You'll have to choose prints or slides.

Since use of this text is predicated on a basic knowledge of computers, this discussion will be fairly brief. Uses and advantages will be highlighted; mechanics and operation will be downplayed.

Printer Output

Although printers are clearly needed for producing reports in batch systems, it is not always obvious to planners that—somehow, somewhere—a printer will be needed for *any* system. Even users of the least expensive home systems soon put printers at the top of their must-have lists. We just do not manage well without the printed page.

There are many varieties of printers, from room-size giants to inch-wide pocket calculator printers. Microcomputer buyers are often surprised to discover that printer options are even more confusing than computer choices, and making a decision about a printer may be as complicated for these buyers as for a major system user. The speed and cost of the printer are usually the deciding factors. In this chapter, it is somewhat practical to discuss specific speeds, but costs vary too much and too often to merit inclusion. Realistically, however, the system will often use a printer already in place in the organization, and you need to recognize and understand its characteristics.

Physically striking the paper with the printing element is the hallmark of **impact printers**. The mechanical nature of impact printers makes them relatively slow. **Nonimpact printers** form figures on the printed page without actually touching it. Nonimpact printers for large computers are very fast, which makes them suitable for high-volume output. Typical nonimpact printers in a mainframe computer environment are laser printers, which print up to 20,000 lines per minute, and ink-jet printers, with speeds of 45,000 lines per minute.

Printers can also be classified as page, line, or character printers. Page printers, which print an entire page at a time, are the fastest. Line printers print a line at a time, at several hundred lines per minute. There is a large variety of impact line printers. Character printers, often used with microcomputers, print only one character at a time and can be as slow as ten characters per second. In fact, there is a slight irony here, in view of what we just said regarding the relative speed of impact and nonimpact printers. A typical microcomputer impact line printer is much faster than a character-at-a-time microcomputer nonimpact printer.

```
r you for your interest in the
onal    ad,    the    QX-10    feature
ware    systems    which    allows
duling,    calculating,    filing,
hs with  just  a  few  minutes  of
  Epson    Business    Package    als
e-pack  of CP/M software, incl
```

Output from dot-matrix printer.

```
K you for your interest in th
onal ad, the QX-10 features t
ware system which allows you
duling, calculating, filing,
hs with just a few minutes of
, the Epson Business Package
e-pack of CP/M software, incl
```

Output from letter-quality printer.

Printers can be further classified as either **solid character printers** or **dot-matrix printers**. Solid character printers produce a fully formed character, much like a character produced by a typewriter. So-called **letter-quality printers,** which make such characters, are used with word processors, primarily for letters and documents that need a fine appearance, usually because they will be used by customers or in other situations where a good impression is important. Dot-matrix printers form characters as a series of dots. Dot-matrix printers are the most common hard-copy output device. Many high-quality dot-matrix printers come with options, such as the ability to form other printed characters, for example, italics.

There are a number of reasons why some applications must have printed output. One is the need to send printed information such as bills to customers. Another is a possible legal requirement to keep printed records. A third is the need to send information to users, such as on-site construction workers, who have no access to a terminal. Still another is the need for output that is physical, such as paychecks for employees and stock reports for owners. In other words, if the output must be permanent, the medium should be paper.

Temporary output can appear on a screen. Some applications need little printed output or possibly printed output only on demand. In these applications the users who need the output have access to terminal or microcomputer screens. Furthermore, their applications—banking, accounting, reservations, insurance, and many more—are suited to pulling data up on a screen.

Screen Output

Output on a screen is appropriate and usual for any kind of online system. Taking this a step further, screen output is necessary if data is needed for immediate use. Data is produced on a screen for an endless variety of uses—hotel reservation inquiries, medical record updating, inventory level checks, musical compositions, law precedent databases—to name just a few.

Screen Output Considerations.   Some considerations for screen output are screen size, screen shape, resolution, and color. The choice of *screen size* is not often a major issue, but sometimes it is related directly to the application. A point-of-sale terminal, for example, has a tiny one-line screen. A terminal or microcomputer that needs to be portable usually has quite a small screen. The *screen shape* can also be related to the application, but is usually the standard rectangular configuration. There are some interesting exceptions, however. One is the type of screen that is letter-shaped, long and narrow, for processing documents. Another different shape is the large circular screen used by air traffic controllers.

The number of addressable points, called **pixels**, available on the screen determines *screen resolution*. The higher the point density, the greater the clarity of characters and figures formed on the screen. Resolution is important for graphics displays. Text character clarity also depends on the size of

Windowing. This screen illustrates the windowing concept, showing different outputs (three in this case) in different sections of the screen.

the matrix used to form the characters. Some word processors, for example, form each character with a matrix 9 dots wide and 14 dots tall; this makes a very clear character. The readability of the screen may be critical for systems in which users must look at the screen all day.

Screen *color* may or may not be important to the system. The importance of color usually increases with the importance of making an impression on the viewer, and that impression is often related to sales of some kind—selling customers or selling management. Color can also be used very effectively in showing engineering or product design. Color monitors cost more but are preferable for graphics displays. If the screen will be used for text only, a monochrome screen is sufficient. Monochrome is usually preferred for word processing because of its superior resolution.

Windowing.   An important development in the use of screens is windowing, the ability to display several different outputs in different locations on the same screen at the same time. Users can run more than one program at once, each displayed in a separate section—"window"—of the screen. Users can move, resize, and stack windows in order to focus on one output while referring to others.

The implications of such software power are significant. For example, a user writing a large document whose sections are in separate files could make a change to one section of the document and be able to modify a related section in another window on the screen. Likewise, an insurance claims agent could check the damage claim report in one file, cross-referencing the customer's insurance record, with both files showing on the same screen. Several software developers offer such packages and the prices are reasonable.

Screen Output Forms.   Screen output can take the form of screen reports, screen query replies, and screen graphics output. Report design philosophy, discussed later in this chapter, applies to both printed and *screen reports*. Because of their transient and immediate nature, *screen query replies* raise special issues related to input. An input query is necessary to produce the brief responses in a real-time environment such as airline reservations. Because this subject is closely related to the input process, however, it will be discussed in Chapter 9 on input. *Screen graphics output* is a fascinating topic that we shall give special attention now.

Graphics Output: Getting the Picture

Output graphics have become very important in the computer industry. Graphs help managers spot trends and relationships more quickly than they could from a sheaf of number-laden reports. Computer graphics can be shown best on a graphics terminal, which accepts lines, shapes, and

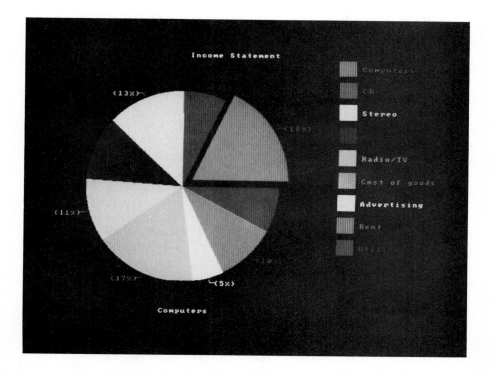

Screen graphics. This pie chart shows the relationships among the figures more quickly than a column of numbers would.

colors. But graphics software packages are available in many varieties for all kinds and sizes of machines. In particular, integrated software packages allow users to cull data from databases for spreadsheets and/or graphics.

Graphics devices now exist that can take the graphics displayed on the screen and transfer them to paper, transparencies, or slide film. These are called **hard-copy graphics** devices. The devices can be grouped into three broad categories: dot-matrix printers, plotters, and devices that read data directly from the screen to create slides.

Dot-Matrix Graphics.   In addition to letters and numbers, most dot-matrix printers can produce bar charts, pie graphs, and pictures. The process of recreating screen graphics on paper is called a **screen dump**; what you see on the screen is approximately what you get on the paper. The key here is compatibility; you must make sure that the software to accomplish these tasks works with the hardware. If you want to use an impact printer, you may prefer a multicolor ribbon, usually containing four colors: blue, red, yellow, and black. Nonimpact printers, particularly ink-jet printers, are fast, reliable, and very quiet. Ink-jet printers can have one color per ink spray nozzle. The nozzles will not degrade over a period of time the way a ribbon does.

Plotters.   You may feel you need higher resolution for your charts than a dot-matrix printer can provide. Plotters have outstanding line resolution and color quality. Plotters can generate all the standard charts and graphs and other complex patterns directly to paper or transparencies. There was a time when an operator had to stand around waiting to change color pens on a plotter, but now plotters, with four to eight pens, are capable of changing colors automatically. You may want to look carefully at plotter speeds. A slow plotter can take five times longer than a fast one to complete a drawing. Historically, plotters have been quite expensive. However, small machines suitable for a business environment now cost under $1,000.

Slides.   When a large number of graphics are to be shown, slides are particularly convenient. Slides can be generated directly from a personal computer at a reasonable cost. Slide-generating devices read video information directly from the computer and process the film.

Special Output

Although most business applications use paper or screens for computer output, other modes are important in the appropriate circumstances.

Microfilm.   Many organizations, notably the federal government, maintain huge files that need to be accessed occasionally. It is not economically feasible to keep this information on disk available online. However, the miles of paper reports for applications such as census data cost money to store and are not easily accessible. To save precious storage space, therefore,

Plotter. This detailed drawing is rendered with great clarity and precision.

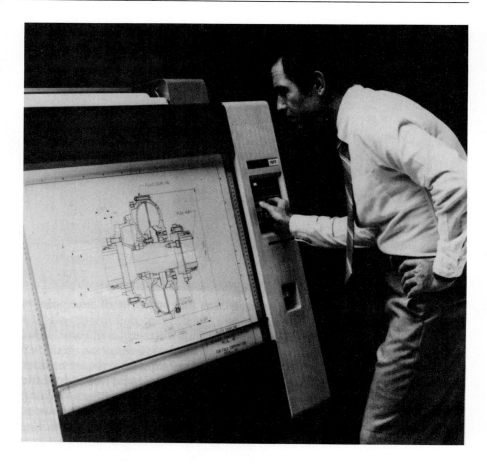

computer output microfilm (COM) was developed. Printed output is photographed as small images on microfilm, which can be in a roll of film or on **microfiche**, 4-by-6-inch sheets of film with 200 printed page images per sheet.

Audio Output.   "Spoken" output from a computer system is becoming common. Stored digital patterns can produce humanlike sounds through speakers associated with the computer. Do not confuse these, however, with actual recorded human voices, which can also be invoked by the computer. Sometimes it is hard to tell the difference. Telephone calls ("junk calls") from a computer and directory assistance replies are both prerecorded voice output. Interactive sessions—between user and computer—would use actual digitized output because the conversation is not predictable. A key example of an interactive session is an audio response terminal for a visually impaired programmer, who can hit an audio release button to get the computer's reply to keyboard action.

Microfiche. A single microfiche is the equivalent of 200 printed pages.

Robotics. Perhaps you never envisioned your career having anything to do with robots. Most analysts deal with office business systems, but in the computer industry anything can—and may—happen. Robots are machines, usually with "arms," that can be programmed to perform humanlike tasks such as those found in a typical assembly line. Their work could be considered a form of output. **Robotics** is the study of the design, construction, and use of robots. If your analysis work is related to factory job automation, you may indeed find yourself planning the work of robots.

◇ REPORTS

The printed word may carry weight, but never so clearly as when printed by a computer. There is a computer tale about a certain budget that was perfectly balanced and beautifully prepared on a fine typewriter. It was rejected. The budget preparer—no fool—produced the exact same budget on a word processor and had it computer-printed. It was accepted. Users long ago figured out that report content *and* appearance count.

Reports People Really Use

Although report content and appearance are important, not all reports are well planned. In fact, many users, especially managers, do not use all the information they receive from computer systems. The common complaint is that most computer reports are too confusing to use. Managers are accustomed to receiving information through informal, personal, oral reports from subordinates. Computer reports, however, are often excessively long, abstract, tabular arrays of figures with few clues to help the manager extract the information needed. In the next section we shall consider some design techniques that make reports really useful.

Report Design

The basics of report design—placing page numbers and dates, suppressing leading zeros, inserting dashes and commas, aligning columns, giving summary totals, and so on—are taught in any standard programming language course. Since readers of this book have some familiarity with a computer language, we shall not linger over these items here.

The first thing that you should understand about report design, whether a printed or screen report, is that it is an iterative process. That is, you will continue to interact with the user, until you—and they—get it right. It would be naive to set about designing reports by yourself. If you expect a pat on the back from the user, you may be unpleasantly surprised. Independence is not valued in this situation, but teamwork is. Begin by reviewing the analysis requirements with the user. Discuss output media and the nature of the reports. Move from there to rough drafts of the reports and go back for more discussions. You may wish to develop a prototype of the report. Continue to coordinate and make changes until everyone is satisfied. (See Figure 8-1.)

What makes a good report? That depends, of course, on its intended use. We can, however, offer some general guidelines.

FIGURE 8-1

Report. Prototype of vendor address list report for the King Books Accounts Payable System.

```
                     V E N D O R   A D D R E S S   L I S T

  2/11/87                                                           PAGE  1


          NAME                    ADDRESS                CITY       STATE  ZIP

  Addison Wesley        Jacob Way                    Reading        MA    01867
  Bantam Books          666 Fifth Ave               New York       NY    10103
  Benjamin/Cummings     2727 Sand Hill Road         Menlo Park     CA    94025
  Delacorte Press       1 Dag Hammarskjold Pl       New York       NY    10017
  Harcourt Brace        757 Third Ave               New York       NY    10017
  Harper & Row          10 East 53rd                New York       NY    10022
  Houghton Mifflin      Two Park St                 Boston         MA    02108
  Penguin Books         40 West 23rd St             New York       NY    10010
  Reston                11480 Sunset Hills Road      Reston         VA    22090
  Science Research      155 N. Wacker Drive         Chicago        IL    60606
  Simon & Schuster      1230 Avenue of Americas     New York       NY    10020
  West                  50 W. Kellogg Boulevard     St. Paul       MN    55164
```

Organization. Users should be able to see immediately how data is arranged in the report and be able to locate needed data. The most usable report is the one that is completely transparent. An experienced user should be able to pick up the report and understand the data without any special training. Organize the data so that the most important appears first and in the sequence in which the user actually uses it. Also, a consistent labeling system is essential.

Level of Detail. We can categorize reports three ways. A **detail report** lists every record of the file in use. An **exception report** lists only out-of-the-ordinary circumstances, such as overdrawn accounts or items that are out of stock. A **summary report** presents only an overview of the data, showing totals and trends. Analysts have a tendency to favor detailed information; hence, a number of users are drowning in data. Excessive length hurts usefulness. Consider shortening reports by having some of the detail data available as an optional on-demand report. This is an especially useful concept when dealing with screen output.

Effectiveness. The catchall word *effectiveness* can be used to include several measures of report usefulness. *Timeliness* should be considered first. Reports that present out-of-date information are worse than useless. Information on reports also must be *complete* and *accurate*. And, of course, *cost* is important. The cost of a report depends on the medium, volume, forms, and speed of the report. But cost factors need to be considered within the total picture. For example, it may be less expensive to print an inventory report weekly, but if that sparse schedule renders the report ineffective and results in the loss of sales, then a more expensive online system of reporting is, overall, more cost-effective. (See Figure 8-2.)

Printed Output Forms

There was a time when recipients delighted in spotting computer-produced output in the guise of a personal letter. It was not too difficult to figure out—with your address (". . . can be sent right to your home at 65 West Circle Drive . . .") slightly aslant in the middle of the different-type letter body. Now all that has changed. Computer-produced letters, and other documents, are beautifully printed in consistent type, on a variety of attractive forms. Although most printer output goes on plain lined paper, the billion dollar business forms industry thrives on offering all manner of output forms to be imprinted by the computer.

Some printers can print only on **continuous fanfold** paper, forms that come with perforated strips on the sides to fit over the printer alignment sprockets. These strips can be removed, sometimes with hardly a trace. Other printers can print on only **cut forms**, single sheets of paper. Many printers can print on either type of form.

FIGURE 8-2

Screen reports. Prototype of screen reports for the King Books Accounts Payable System.

```
            V E N D O R   S T A T U S   R E P O R T

              1. QUERY VENDOR STATUS

              2. ALL PAID VENDORS

              3. ALL UNPAID VENDORS

              4. DISCOUNTED VENDORS THIS MONTH

              5. EXIT

          READY FOR NUMBER OF YOUR CHOICE: 1
```

```
            Q U E R Y   V E N D O R   S T A T U S
      VENDOR NUMBER?  3245-94567

      Vendor is:       Yourdon Press
                       1133 Avenue of the Americas
                       New York NY 10036

      DO YOU WISH:     1. ALL DATA ON YOURDON THIS MONTH

                       2. OUTSTANDING PURCHASE ORDERS

                       3. OUTSTANDING INVOICES

                       4. SUMMARY DATA THIS MONTH

                       5. EXIT

          READY FOR NUMBER OF YOUR CHOICE: 3
```

```
            O U T S T A N D I N G   I N V O I C E S

                        YOURDON PRESS

      INVOICE      PURCHASE    DATE        DATE        INVOICE
      NUMBER       ORDER       ORDERED     DELIVERED   AMOUNT

      13-222-439   14601       7/21/87     8/12/87       57.50

      13-236-860   14954       8/10/87     8/28/87      250.00
```

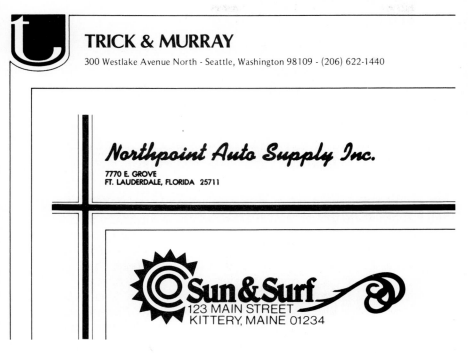

FIGURE 8-3

Preprinted output forms. Many companies use preprinted letterhead forms for their computer-produced correspondence. These can be continuous fanfold or cut forms.

Let us examine the various types of printed output forms and their appropriate uses. As an analyst, you can get free advice and even free samples from business forms companies.

Custom Preprinted Forms.   Replete with preprinted letterheads, designs, logos, headings, and descriptions, custom forms can be made to suit the fancy of the buyer, as shown in Figure 8-3. These forms come in a variety of paper types and colors. They would be used for external reports, that is, reports that go to persons outside the organization. Typical external reports include stockholder information, customer bills, and the IRS W2 form.

Laser-Printed Forms.   Some laser printers have the ability to use a *forms flash*, a template between the laser beam and the paper, to cause lines and letters to print on plain paper, giving the appearance of a custom preprinted form.

Convelopes.   Bearing descriptive product names such as Self-mailer or Speedi-mailer, "convelope" is a generic name for continuous envelope. Convelopes are envelope/reports that can be printed continuously and later separated at perforations between the envelopes as shown in Figure 8-4. Their special feature is that the form is a set of carbon-backed paper layers, with a detachable top layer that can be kept for company records. The remaining portion is a computer-addressed envelope, with the printed

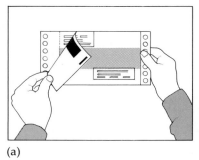
(a)

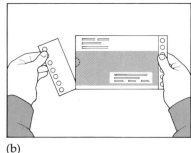
(b)

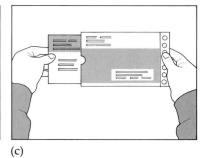
(c)

FIGURE 8-4

Convelopes. (a) First, print the message and address. Keep the outside copy for your files. (b) Customer tears off perforated stub to open. (c) Customer removes printed message.

information inside, ready to mail. The carbon backing is in place in such a way that only the name and address appear on the outer envelope layer. Student grades are often prepared this way, as are other to-be-mailed outputs such as automobile registration notices and pension checks.

Mailing Labels.   Computer-printed mailing labels, an efficient way to use files of names and addresses, are in widespread use. Note the example in Figure 8-5. The labels have a pressure-sensitive backing that permits them to be lifted easily from the print sheet to the mailing envelope. The manual part of the operation—moving the computer-printed label from the sheet to the envelope—precludes using such labels for large volume files.

Such labels are practical for small business mailings and for the periodic mailings of clubs and organizations. Mailing labels are particularly popular in elections, as politicians have become aware of the potential for computerized fund-raising from targeted audiences. The politician prepares a standard letter that can be folded, stapled, and marked with a mailing label. All this can be done by a volunteer with little training. Political campaign staffs are likely, however, to use the plain vanilla labels, purchasing them in endless continuous rolls. Businesses using labels often have them preprinted in colors, with the company name as a return address.

When a report is to be produced on a preprinted form, the computer operator must locate the correct forms and set them up on the printer. The immediate concern is to align the computer-produced output so that it appears properly on the form; it would not do to have last month's balance erroneously appear in the box prelabeled "Finance Charge." Programmers often produce several preliminary phony reports that operators can use to test output alignment.

Any kind of preprinted form will be more expensive to use than generic lined printer paper. The extra expense means that you must be able to justify the need for using the special forms. A final word of caution to the analyst: Do not be pressured into buying large quantities of preprinted forms before the system has stabilized. There may be necessary forms changes, and you do not want to be stuck with an unusable inventory.

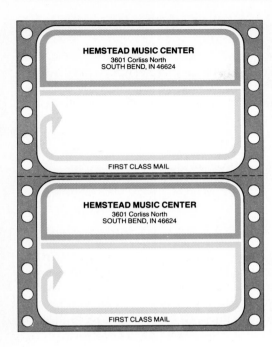

FIGURE 8-5

Mailing labels. Continuous mailing labels usually come "one up," like these, or "two up," side by side.

◇ DISTRIBUTION

Any output produced on a physical medium that is to be handled by people, notably printed reports, must be moved from the place where it is produced to the place where it will be used. This process is called **distribution**. As the analyst, it is your responsibility to plan the distribution procedures and see that they are put in place.

Microcomputer output usually is printed in the same location as the user, so there is no need for distribution procedures. The same may be true of larger computer systems that transmit output to remote printers stationed near the user organization. In other cases, however, medium and large systems have central printers grinding out reports that need to be sent to various recipients, both internally and externally. There are many ways to deliver output: delivery cart, hot truck (a truck dedicated to the speedy delivery of computer reports), company mail, U.S. mail, and special commercial mailing services such as Federal Express. There may also be occasions when output is distributed using Greyhound or Trailways package express or even taxi door-to-door service.

A word of caution about using the U.S. mail: You do not dump customer bills or any other kind of mass mailing into the local mailbox slot. Planning your mailings may be a complicated procedure. The post office has printed guidelines that you may need to study. Perhaps the best advice is to allow yourself ample time to plan ahead—the post office will not be rushed, even if you are.

In addition to deciding how output should be distributed, it is your responsibility to interact with the production personnel who will eventually keep the system running when it is in production mode. They will need detailed written instructions from you regarding who gets what, when, and how. Keep in mind that, as the analyst, you are only on the scene as the system is coming into being. Once a system is up and running, you will not be available to tell everyone how things are supposed to work—hence, advanced coordination and written procedures.

◇ USERS AND THEIR OUTPUT

Computer system output is spread far and wide. The most casual user will sometimes speak of having had "computer training" because someone showed him or her how to read a computer-printed report. It is reasonable to assume that everyone in the civilized world, including some people you want to impress, will see computer output in some form. The stakes are high. Well-planned output is important.

The pervasiveness and importance of computer output is no secret. What *is* a secret, or so it sometimes seems, is the critical importance of input to the computer system. In the next chapter we shall make a strong case for providing users with the means to present the very best input.

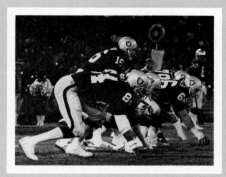

FROM THE REAL WORLD

Computers at the Superbowl

If you are a football fan lolling in your living room watching the Big Game on TV, you probably are not thinking heavy thoughts about the technology that makes it possible. But down-in-the-dirt blocking is accompanied by another bag of tricks—the skillful blending of computer and telecommunications technology. The whole thing is blessed by satellites beaming down from the sky.

At the heart of CBS electronic wizardry are three Chryon™ color graphics computers. These compact computers, which sit in mobile units outside the stadium, permit finely detailed graphics to appear on your TV set. A scene is captured with a camera, digitized, and then saved in computer storage where it can be retrieved at any time. The graphics generated join the words and pictures generated by cameras, videotape machines, microphones, and miles of cable.

You might take a look at the commercials and network plugs offered up during halftime—you will see more computer graphics. Did you know that the CBS "eye" is graphic computer output? So is—if we can switch networks here for just a moment—the NBC peacock logo. In fact, if you keep a sharp eye, you can observe that almost all commercials that do not feature in-the-flesh humans ("I've noticed my wash is brighter . . .") are actually computer graphics. Meanwhile, pay attention to the game.

—From an article in *Government Computer News*

SUMMARY

- Initial output considerations in discussions with the user include what output information is needed, how it should be presented, what format it should have, when it is needed, and what the volume will be.

- Printers can be classified as impact or nonimpact, and as solid character or dot-matrix. Solid character printers are also called letter-quality printers.

- Screen output considerations include screen size, screen shape, resolution, and color. Screen resolution is related to the density of pixels (addressable points) on the screen. Screen color is important for graphics applications.

- Windowing is the ability to display several different outputs in different locations on the screen at the same time.

- Screen output forms include reports, queries, and graphics output. The process of recreating screen graphics on paper is called a screen dump.

- Graphics software packages come in many varieties. In particular, integrated packages allow users to cull data from databases for graphics.

- Hard-copy graphics include dot-matrix graphics, plotters, and slides.

- Special output includes microfilm (or microfiche), audio output, and robotics.

- Good report design depends on organization, level of detail, and general effectiveness criteria. Level of detail can vary from detail reports to exception reports to summary reports. Report effectiveness factors are timeliness, completeness, accuracy, and cost.

- Some printers accept only continuous fanfold paper, some only cut forms, and some both. Some output is printed on custom preprinted forms, while others can be printed on laser-printed forms.

- Convelopes are special forms that are mailable envelopes, with special pressure-sensitive report paper inside, that permit the report paper to be printed while it is inside the envelope. Mailing labels can be computer-printed and then lifted from the sheet and placed on envelopes.

- Distribution is the process of moving output from the place where it is produced to the place where it is used.

KEY TERMS

Audio output	Cut forms
Convelope	Detail report
Custom preprinted forms	Distribution

Dot-matrix printer
Exception report
Graphics output
Hard-copy graphics
Impact printer
Laser-printed forms
Letter-quality printer
Mailing labels
Microfiche
Microfilm
Nonimpact printer
Pixel

Query
Resolution
Robotics
Screen dump
Screen output considerations
Screen output forms
Screen shape
Screen size
Solid character printer
Summary report
Windowing

REVIEW QUESTIONS

1. The chapter suggested that output considerations to discuss with users include what output information is needed, how it should be presented, what format it should have, when it is needed, and what the volume will be. What kinds of conclusions might be reached on these items for the following situations (make any assumptions you need to make):

   Presenting patient information during a hospital stay for surgery

   Producing information about a stolen car in response to a license plate check

   Presenting completed computer-produced tax information for a client in a form ready to mail to the IRS

   Showing stockholders, in their annual glossy brochure, statistics comparing the (rosy) profit picture this year as compared to the last five years

   Preparing detailed engineering drawing information data so it is available when needed. There is a heavy volume and the engineers need to be able to see all the data for one drawing at the same time

2. Compare the different resolutions on the screens you can find on campus or at your local computer store. Can you note obvious differences in clarity?

3. What computer-printed output arrives regularly in your mail? Count and classify it for a week. Are you sure you recognize all computer-generated letters?

4. Compare two or more windowing software products by finding articles in your campus library. What seem to be the advantages of one product over another?

5. What applications can you think of that would justify putting graphics output on a plotter rather than a dot-matrix printer?

CASE 8-1
MISSING PERSONS REPORTS

Analyst Martin Drury was a novice with a minimum of training, but he felt he could succeed by dint of hard work. Unfortunately, this was not true. His first assignment was to design some reports for a new missing persons tracking system for a regional law enforcement consortium. The senior analyst, who was in charge of the overall effort, was out of town for two weeks on another assignment. Martin wanted to have the report specifications done when he returned.

Martin began by examining the manual reports used by the system; these were something of a jumble, prepared haphazardly by overworked personnel. He felt that, to be useful to the police and other law enforcement agencies, there should be online access to missing persons data, and also weekly activity reports listing new and revised entries. Martin set about designing screen formats and a printer spacing chart for the weekly report.

The inquiry screen format contained data about a missing person: physical description, date reported, where last seen, and so forth. The weekly report contained all these details about each person plus an extra column for status comments: sightings, new leads, and so on. The formats were carefully drawn and looked professional.

Martin called a meeting of the officer task force committee assigned to the project part time. He was dismayed to discover that they paid little attention to his formats because they did not even agree with the type of reports or the data presented on them. Separate reports were needed for missing children, for example, whose records contain extensive associate-adults data. Further, after much haggling, the officers convinced him that they needed not weekly detail reports but daily summary reports.

Martin returned to the drawing board. He drew up neat versions of what he thought the reports should be. At the next meeting the committee, apparently satisfied with the direction the reports were taking, tackled the specific data on the reports and the formats. Martin tried to pin them down specifically but was referred to the duty officers who would actually be using the system on a 24-hour basis.

At this point Martin threw out most of his work and started from scratch. He made appointments with the officers who would be interacting directly with the system and learned from them what was needed. He made rough drafts and went back to these same people for a revision session. After one more round, Martin was ready to commit to a formal, attractive output format. These were approved by the task force committee. The entire process had taken 27 days.

1. Martin made some obvious mistakes. What were they?
2. Less obvious are the mistakes made by Martin's lead analyst. Perhaps you noticed that the whole planning scheme is a little loose here. What steps should Martin's boss have taken to ensure a more successful result?

CASE 8-2
A COMPUTER ON THE TELEPHONE SALES STAFF

Telemarketing is a catchword for selling by telephone. Analyst David Correa was being sent to talk to J. J. West and Associates, a telemarketing company that raised $5 million a year for nonprofit corporations. He wondered how anyone could make money in such a business, since many people considered telephone solicitations a rude annoyance. There was obviously a lot he did not know.

David found out that telemarketing fund-raising appears deceptively simple, requiring only a staff of telephone salespeople, telephones, phone numbers of prospective donors, and pledge slips. The complexity, he discovered, lies in managing a constantly changing staff, who often stay only until they can find "something better." In subsequent meetings and interviews, David found that the staff problems seemed to revolve around training.

David considered handling the training problem in the traditional way, providing classes and documentation. But the constant staff turnover meant that traditional approaches were not cost-effective; it would mean running continuous classes for two or three people. In an inspired approach, David turned to the computer to train new employees.

The software he designed used two basic components. The first was a program to tell the computer to read telephone numbers from a data file and then dial them through an automatic dialing modem. This program was activated by a menu-driven program used by the telephone salesperson. The screen output was actually a sales script that could be used with almost no formal training. The computer begins by asking, "Are you ready to make the next call?" If the salesperson keys a Y (for Yes), the next name and telephone number on the call list appear at the top of the screen and the system starts dialing the number. The first script then appears, with the introduction and first sales pitch. Based on answers received, the salesperson can call in other scripts. The system proved very effective and provided an unexpected bonus: Employees stayed longer than before.

1. This case, for a change, contains no major bumbling that we must explain away. Using the screen as an output device for training, however, is an idea whose time has come. Can you think of other training applications that might be appropriate?

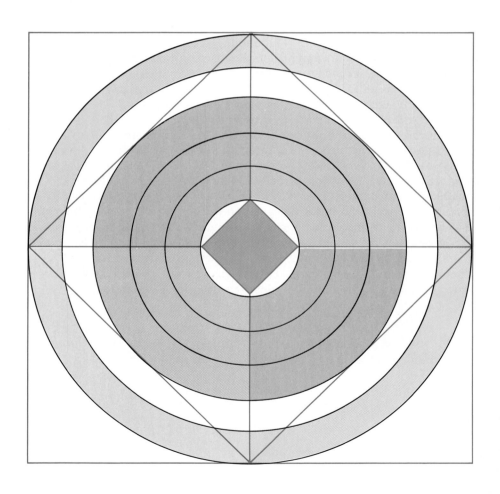

Chapter 9

Detail Design: Input

The Republican Party of Missouri was pleased to receive a check from the Monsanto Company for $12,439. It did not occur to party officials to question the rather odd amount of the check, nor did they compare it with the more modest $500 that Monsanto had contributed the year before. Monsanto eventually realized something was wrong when a vendor demanded payment due in the amount of— you guessed it—$12,439.

You have just encountered another of the famous goofs for which the computer always receives the blame. This little vignette, however, describes not a computer error but a human error. A data entry operator typed the incorrect digit that sent the money winging to the wrong recipient. Input errors are by far the most common type of error in computer systems. The analyst will not be around to personally key in the data in situations such as the one just described. What, therefore, can the analyst do to ensure that correct data is provided? When planning input procedures, the wise analyst will place primary emphasis on accuracy. This chapter will examine all aspects of input data, including accuracy.

◇ INPUT CONSIDERATIONS

You have probably seen the acronym GIGO: garbage in, garbage out. Even the computer cannot produce good output if it is not given good input. We want good input. So do users, managers, and customers. Our number one goal is accuracy. Getting the data right would seem to be a straightforward task, but there are many pitfalls on the way to the computer, almost all related to the possibility—and probability—of human error.

Humans make mistakes when they complete or read forms, when they use their fingers to key data, and when they misunderstand what is needed. Analysts make mistakes when they fail to plan adequately for such events as unreasonable data or sudden high data volume or a systems audit.

Planning is the key. You should have no illusions about eliminating human error, but a well-designed system can reduce errors significantly, and, perhaps more important, catch many of the errors that do occur. In our quest for good data, we shall begin with an overview of planning, followed by input editing techniques. Then we shall consider various input devices, and, lastly, give special attention to screen input and forms design. Let us begin with the most basic subject of input planning: capturing the data.

Data Capture

How will the data get to the computer for processing? Traditionally, data has been handwritten on some paper medium by one person and then transcribed by another person onto some computer-accessible medium such as punched cards, tape, or disk. This method is rather slow and subject to error from two sets of people.

It is common today to capture data at its source, as a by-product of the event that causes the data to be generated. This process is called **source data automation**. A familiar example is a supermarket bar-code reader that scans the Universal Product Code of the item being sold. In addition to retrieving price data, the computer generates a transaction to be used as input to the inventory management system.

Data is captured at the source for online systems and for some batch systems. An example of source data for an online system is the data supplied by a guest for an automated hotel reservation system. As the reservations clerk gets the data, he or she keys it right into the system. For an example of a batch system with source data, consider an automated factory payroll system. Instead of filling out time cards that then must be keyed in by other people, factory employees use identification-marked badges to punch into a time clock. The time clock data is captured at that source and sent to the payroll system for batch processing. Data is generated in a special data preparation step for batch systems when it is economical to do so and time is not a critical factor.

Bar-code reader. The zebra-striped codes on these supermarket items can be scanned by the bar-code reader, which will retrieve the items' prices from the computer system.

The Data: Content, Format, Volume

What will the data actually be? That is, what is the *data content*? Does the input consist of annual tree growth figures, number of books backordered, employee insurance selections, truck identification number, patient blood type? By the time you get to the input planning stage, data content is no longer a puzzle. This is because you carefully studied existing data in the analysis phase and have already designed the system output. You know what is required for output. Now you must determine what input is needed to obtain the output. Your familiarity with the system requirements makes that task fairly straightforward. You need to know exactly what data items are needed and where to get them.

The *data format* is the next consideration. You need to describe each field in detail, noting its maximum length, whether it is alphabetic or numeric, its sequence, and other details, such as the existence of punctuation marks.

Data volume is a topic often overlooked by an inexperienced analyst. Input volume is directly related to the number of data entry personnel and is one indicator of expected run time. The greater the number of data entry personnel, the greater the volume of input data they can handle. In addition to normal volumes, and even average volumes, you need to be aware of peak volumes. A peak volume refers to very high volumes, usually expected

BOX 9-1	**Button Input**

If you are staying at the Disney World Hilton, input commands to the computer are as close as the buttons on your telephone. Begin with the computer-controlled room environment: You can press one button to turn up the heat, another to turn on the air conditioning. Another button lights up if the door is not securely bolted.

But this is only the beginning. Each room phone has two lines so that two calls can be handled simultaneously: You can have a normal voice conversation on one line while transmitting data over another. Since word and data processing equipment are connected directly to the telephone, guests can use the phones for a variety of data transmissions to the computer.

---

in a given time frame. Peak volumes are particularly important to companies whose business is cyclical or seasonal. Accounting firms, for example, expect a rush of business at quarterly report time. Toy companies expect to do the bulk of their business during the Christmas season.

Data Control

To be in control of the data, you must have control over accuracy, scheduling, security, and auditing.

The case for *accuracy* was already presented with some vigor in the beginning of the chapter. It is mentioned again here because it is part of the total control package.

The *scheduling* issue means that you must make sure the data is ready for processing when it is needed. A batch system is usually run according to a specific periodic schedule—2:00 to 2:30 A.M. each Thursday, or whatever—and the data must be available at the appointed hour, whether it is waiting on tape or being sent over phone lines from the branch office. When you plan the user system guide later on, you will spell out what organization and which personnel are responsible for getting the data where it belongs on time. Systems that process data as it is generated are in a state of readiness to receive it, so the issue becomes one of being able to handle the data volume.

Security refers to the safety of the data. Security is important for processed data, especially when derived relationships, such as return on investment, could be useful to unauthorized users. But individual unprocessed data items, such as salaries or medical history, must be subject to security measures also. It is often easier to tamper with unprocessed data. This is because unprocessed data is more easily accessible to nontechnical personnel and also because the data is not yet subject to computer system controls. The whole subject of security will be considered in detail in Chapter 15.

When businesspeople think of having control of their data, they often are thinking of the *audit* function. The reports on which the data will eventually appear may be subject to auditing by both management and outside auditors, including the IRS. The system must provide a means for tracing any given transaction through the system: from input to process to files to reports. This topic will be included in Chapter 14 on project management.

◇ EDITING INPUT DATA

There is no way of making sure that all data is correct, but many errors can be detected. If data is captured at the source using an automated process, chances for accuracy are excellent. These methods include using devices such as wand readers and bar-code readers and magnetic ink character readers. These devices have one important element in common: No human transcription is required. That is, an automated device reads characters prepared by another automated device and no human has to intervene.

In many cases, however, data is keyed by people, and people do make mistakes. The first line of defense against errors is the people themselves, as their eyes scan over their work in progress. It is common to present data entry personnel with a set of keyed data for their approval before the data is actually entered into the system. For example, if an insurance adjuster is entering data about a claim, the screenful of related data he or she just keyed will be displayed, and the adjuster will be asked to approve it before the data is accepted for processing. The adjuster can make changes if necessary.

Deciding What To Edit

How much editing is enough? How much editing is too much? There have been somewhat facetious estimates that 90% of the coding is for 10% of the data. In other words, there is a great deal of code for editing bad data that occurs infrequently. Still, unchecked bad data may ruin a production run, not to mention preparing the way for inaccurate files to be used in the future. Programs must be prepared to handle whatever data comes along. But sometimes novice programmers continue to edit data once it has passed into the system. After a data item has passed initial screening edits, it need not be edited in subsequent programs. However, captured data must be subjected to rigorous editing routines in the first programs that process the data. If data is entered online, it can be edited as it is received. Errors can be brought to the attention of the data entry person, who can make immediate changes and reenter the corrected data.

Data that is input in batch mode is usually processed first by an editing program that rejects transactions with erroneous data and prints an error report. The rejected transactions are held in a suspense file until the error report is processed and corrections are made.

Types of Edits

Whether editing is done online or in a batch program, a set of fairly standard types of tests can be applied.

Sequence Check.   Batch data, such as vouchers, are often assigned numbers in order. The sequence order is checked in the edit procedure, because a break in the sequence could mean lost or improperly processed data. Another important type of sequence check is transaction date. If it is not processed in the order received, for example, an account withdrawal might be processed before the deposit transaction that covered it.

Reasonableness Check.   Is it reasonable to have a bill payment in the amount of $0.00? Is it reasonable to have a negative rate of pay, say $-$11.00 per hour? Is it reasonable to ask for $2,000 to be dispensed from a cash machine? Concerns such as these can be raised about data being entered into the system. Input data can be evaluated against some reasonable standard. Many reasonableness tests are related to range checks.

Range Check.   A range check tests data against upper and/or lower limits. For example, the number of patients in the coronary care unit can be no more than 24. Or a thread of a screw must be between $\frac{1}{32}$ and $\frac{1}{8}$ inch.

Validity Check.   Data can be established as valid by comparing it with previously established data that is already in the system in the form of tables. An example is a state abbreviation. WV (West Virginia) can be found in the table, whereas WW cannot. However, there is no guarantee that WV is correct, only that it is among the correct alternatives.

Comparison Check.   Certain data items are related to one another. For example, the federal government has job classifications labeled by grade, and each grade has a matching salary range. You could make sure that a salary was appropriate for a certain job classification grade.

Class Check.   A class check determines whether data in a field is numeric or alphabetic. Any data that will be used for calculations must be numeric, so this check is critical. If data fields are known to be always numeric or always alphabetic, then this additional check may detect an error.

Completeness Check.   Some data items must be present if the data is to be usable. The field must be checked to be sure it is non-blank. An example would be a zip code, which is required for a mailing address. From a broader perspective, some systems require that a new entry consist of a full set of transactions; a completeness check would make sure they are all present.

◇ INPUT DEVICES

There was a time when input devices were a mere blip on the systems planning horizon. An analyst's biggest decision might have been what color the punched cards would be. This is a far cry from the input device variety of today. Given a dizzying array that includes everything from bar codes to light pens to voice input, the analyst must decide which device is appropriate, based on utility (all the factors listed in the preceding section) and cost. As noted in the previous chapter, basic knowledge of computer hardware is assumed, so this discussion emphasizes uses and advantages rather than mechanics.

Key-to-Tape.   Used for low-intensity batch systems such as payroll, a key-to-tape system is economical but subject to transcription errors. However, the operator can correct errors he or she makes as they appear.

Key-to-Disk.   There are several variations to the key-to-disk method. An operator might key data via a word processor or microcomputer to a floppy disk that is used later as input to a batch operation. Phone company employees might transcribe directory update forms by keying them on an intelligent terminal that edits the transactions and places them on hard disk for later processing. Or a group of people, such as insurance claim adjusters, might be hooked up collectively to an online system that uses their transactions for instant updating and also places the transactions on disk storage for reporting and auditing purposes. In all these cases, and many more, the use of the CRT screen is of critical importance. Designing screen input is not a trivial task. This important subject is given its own section.

Point-of-Sale (POS) Terminals.   Soon the last of the old-fashioned cash registers will be committed to the Smithsonian. POS terminals in just a few years have become a standard in retail stores. The unit may be self-contained: keyboard, operator guidance panel (seen to light up, function by function, to tell the store clerk what data is needed next), a narrow printer to produce the customer cash receipt, another printer to keep the register journal on a continuous roll, and a small transaction display screen so the customer can watch the action. A **wand reader** or **bar-code scan device** may be connected to the main unit. The advantages are many: price accuracy, clerk productivity, improved customer service, and decreased costs. Many POS terminals are hooked directly to computers that use the POS-generated inventory transactions.

Wand reader. A wand reader scans the price tag for product identification and price. OCR WAND® is used by permission of Recognition Equipment, Dallas, TX 75266.

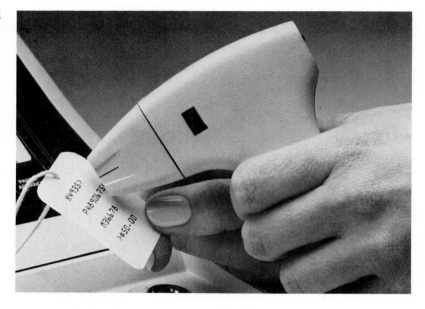

Graphics Input Devices. The **joystick**, used for rapid movement of figures on the screen, is a popular input device but has little application apart from games. A **light pen** has a light-sensitive cell that closes a photoelectric circuit when placed against the screen, causing lines to be drawn. The primary applications of the light pen are in drafting and engineering, but there are also some applications in the business environment. A **digitizer** can scan a drawing and convert the picture to digital data that can then be represented on the screen and processed by the computer. A **graphics pad** uses a stylus applied to a pressure-sensitive surface that causes a corresponding drawing on the screen.

Mouse. Although the mouse can be considered a graphics input device, its use has become so varied and so pervasive that it deserves a category by itself. This hand-held device is rolled on a flat surface, usually the table near the computer terminal. The rolling movement causes a corresponding movement by the cursor on the screen. Although it takes a little getting used to, many users like its freewheeling action. Some computers, such as the Macintosh, use the mouse exclusively and have no keyboard cursor movement.

Touch Screens. With touch screens, the touch of a finger on the screen can be detected by sensors on the edges of the screen. The sensors can pinpoint the touch location and cause corresponding action on the screen.

Mouse. The cursor on the Macintosh personal computer is controlled by a mouse. When the mouse is rolled on a flat surface, it causes corresponding cursor movement on the screen. A click of the mouse's button can cause action related to where the cursor is pointing.

Although considered user-friendly, this method needs special software and has limited practical application in business systems at this time.

Magnetic Ink Character Recognition.   Commonly referred to as MICR ("miker"), the magnetic ink character recognition input method is used by banks to machine-read data from checks. This direct data capture eliminates costly intermediate data preparation steps.

Optical Recognition.   Optical recognition methods, which use a light source to scan data and input it directly to a computer system, all have the same key advantage: They capture data at the source. **Optical mark recognition** machines sense penciled marks on paper—test scores and meter readings are common examples.

Optical character recognition (OCR) works the same way, with a scanning device (often a wand reader) recognizing digits and letters to be input. OCR is commonly used for retail price tags.

Bar-code readers operate on the same principles and are part of the daily scene in the supermarket. A scanning device inputs the product identification data from the bar code and uses it to retrieve the price from the computer system. If you are planning an application with bar codes, a major

| BOX 9-2 | **All Is Not Lost: Boeing Tools** |

A leader in the aerospace industry, the Boeing Company employs thousands of factory workers, many of whom use several different tools each day. The basic tool tracking system had been a loose checkout routine whereby clerks passed out tools as needed. Unfortunately, tools were often misplaced or lost altogether.

Boeing systems analysts then designed a new system that has proven much more effective. Each tool is marked with a bar code and each employee has a badge with an identifying bar code. As each tool is checked out, the clerk invokes a program by passing a wand reader over both the tool and employee bar codes, recording the matchup. The process is repeated in reverse when the tool is returned, undoing the relationship. The number of tools now properly accounted for is almost 100%.

consideration is getting all the objects to be scanned marked with the codes in the first place. Finally, legible **handwritten characters** can be scanned optically and input directly to the computer.

Vision Recognition.   Vision systems are still in their infancy, but some experts find them promising. A camera, which takes a picture of an object and then digitizes the image, can be aimed at a fairly large area to perform simple factory jobs, such as inspection tasks. A typical task might be to determine which metal object is flat. For this purpose the machine is faster than the human eye.

Voice Recognition.   Voice input devices convert voice input to digital code that can be accepted by the computer. These devices are particularly useful in environments where users are limited by "busy hands"; that is, they cannot easily free themselves from their tasks to type input data. An example is a butcher preparing meat and wanting to attach a computer-printed meat-quality label (prime, choice, and so on), as determined by the inspection.

◇ SCREEN DESIGN

What could be more alarming to the computer novice than the beckoning computer screen! The command the novice sees on the screen almost always includes the word ENTER. If we know nothing else, we all know that the computer is *fast*. Therefore, it must be awaiting a reply impatiently so it can get on with its speedy work. Studies have shown that user anxieties can be reduced significantly by changing the word ENTER to READY, thereby placing the computer in an apparently more subservient position. Tending to the relationship between the user and the computer, as represented by the screen, is just one of the tasks of screen design.

The immediacy of screen interaction captivates users in a way that traditional paper-shuffling never could do. The user perceives that there is no time to dawdle. The user feels a need to be alert. These are good signs for productivity but they are signs of caution to the analyst. You must design screens in such a way that the input process is clear, the methods used are non-threatening, and the data input is accurate. You must put the novice user at ease and, possibly at the same time, free the sophisticated user from unneeded instruction burdens.

Screen Design: Three Basic Approaches

When working with screen data input, you are actually working with both input and output. Although helping users to provide input and to query the system, you must also plan screen output that tells the user what to do and how to do it. Screen contents, of course, are controlled by computer programs. There are three basic approaches for presenting screens that request input from users: menu, prompt, and template.

Menu Input. The word menu is borrowed from the English language because a menu offers a range of choices to the user. In a typical menu-based system, the user is presented with a list of numbered choices and asked to make a selection by keying in the number of the choice. Other common methods of making a choice are with light pen, finger (to touch the screen), and mouse. The selection causes another menu to appear on the screen. This menu may be another in a series at the same level. Or the new menu may be the next lower menu in a hierarchy, listing choices related to the first selection. This process continues until the choice is narrowed down to one particular action, such as locating specific information (an agency's phone number and operating hours) or causing action to be taken (order more raw materials).

A menu system has clear advantages: It has the potential to list a wide range of alternative selections and it is easy for the novice to use. If used in its most fundamental form, however, it can prove tedious for experienced users. A knowledgeable user knows exactly what final menu he or she wishes to use and wants to leap directly to it rather than viewing each intermediate menu along the way. There are various ways to provide this service. A common method is to allow combined menu choices. If, for example, a user wants choice 4 from the first menu, then choice 2 from that menu, and finally choice 6, then 4.2.6 can be typed at the start. Another necessary facet of menu-driven systems is that the user must be able at any time to return to the previous menu or to the main—beginning—menu.

Automated teller machine. Although its screen is tiny, it can prompt users one question at a time.

Prompt Input.   Prompt systems come in two basic varieties. The most common is for users with various skill levels; it simply asks for specific information, one question at a time. As the questions and answers alternate, a computer/person conversation takes place. The questions are as concise as possible to save reading time—something like NEW ACCOUNT? or READY FOR HIRE DATE (MM/DD/YY). Note that it is sometimes necessary to include the expected format when asking for data.

Although simple to use, this kind of system has limitations. The process is relatively slow, and the constant screen dialogue can become tiresome. A major advantage, however, is that the programs can provide a sequence of prompts determined by user responses, which gives the system flexibility. A very simple example of a prompt system is a bank cash machine— you are your own teller and follow the prompts on the tiny screen.

Another kind of prompt is designed for experienced system users. As a programmer, you may be familiar with the kinds of prompts used by operating systems, database management systems, and other layers of system software. A simple prompt such as OK or > or : may be fine for a programmer who is familiar with a set of commands, but it is obviously not suitable for nontechnical system users.

Template Input.   Users of a template system are presented with an entire screen at a time, rather than a simple line prompt. A template has the appearance of a printed form on the screen, as you can see in Figure 9-1. Instructions are in the form of labels such as VOUCHER NUMBER and

FIGURE 9-1

Screen template. The cursor moves from item to item on the screen template.

```
                         VOUCHER INPUT

     VOUCHER NUMBER _____     VOUCHER DATE __/__/__

        VENDOR NAME _____

        ADDRESS _____

        CITY _____ STATE _____ ZIP _____

     INVOICE NUMBER _____     INVOICE AMOUNT _____.__
     INVOICE NUMBER _____     INVOICE AMOUNT _____.__
     INVOICE NUMBER _____     INVOICE AMOUNT _____.__
     INVOICE NUMBER _____     INVOICE AMOUNT _____.__
     INVOICE NUMBER _____     INVOICE AMOUNT _____.__

                                   VOUCHER AMOUNT _____.__
```

VENDOR NAME. Blank areas near the labels are filled in with data keyed in by the user. The cursor, controlled by the program, bounces from item to item. The user controls only the actual data entered, not the data requested or the order of the requests.

Template screens are often called **template pages**. Each page contains a logical grouping of data. If a traffic accident investigator were entering an accident report, for example, one page might contain the personal data of the driver, another the insurance policy coverage, and a third the accident description.

These three methods are sometimes used in combination when it suits the particular application. For example, a data entry worker in the accounts payable department might first be offered a menu with input data choices such as VENDOR, PURCHASE ORDER, INVOICE, and VOUCHER, as shown in Figure 9-2. If the INVOICE option is chosen, the next input mode could be a set of prompts for the invoice information—number, date, and so on—or perhaps a template where the cursor moves automatically from item to item until all invoice information has been entered.

Invalid Screen Entries

Since accuracy is a primary goal of data entry, you must anticipate input errors and design the system to intercept them. Perhaps the most obvious expectation is an incorrect menu entry. For example, a user is given choices 1 through 5 but presses the 7 key instead. The system should note and

FIGURE 9-2

Menu and prompt combination. The second screen scrolls up as the prompts for invoice data continue.

```
        O P E N I N G   M E N U   - I N P U T


                1. VENDOR

                2. PURCHASE ORDER

                3. INVOICE

                4. VOUCHER


        READY FOR NUMBER OF YOUR CHOICE:   3
```

```
                I N V O I C E   I N P U T

        INVOICE NUMBER: 95532-457

        INVOICE DATE: 11-13-87

        VENDOR NAME: Holt, Rinehart and Winston

        ADDRESS: 383 Madison Avenue

        CITY: New York
```

acknowledge the error in a brief, matter-of-fact, and polite message that also describes how to fix the problem. Messages should be meaningful and explanatory: they should not just note INVALID DATA, for instance. In this case INCORRECT MENU CHOICE—PLEASE ENTER SELECTION 1 THROUGH 5 would be sufficient.

Prompt or template systems usually check data for reasonableness and validity as it is entered. Postage costs in six figures would not be reasonable. Nor would alphabetic salary data be valid. Again, in such cases the user should be given a descriptive error message—not something cryptic like E59614—and an opportunity to reenter the data correctly.

It is common to permit the user to see a completed set of data before it is actually entered into the system. After a template page is completed, for example, the user is asked if the data is correct. An affirmative reply would cause the data to be entered, and the user then proceeds to the next template screen. But if the user spots an error and answers NO, then he or she will be asked which data item is in error and be given an opportunity to reenter that data right over the wrong data.

One more point about taking care of errors: Ensure that error messages are erased from the screen after the user has acted on the messages.

Screen Ergonomics

The subject of **ergonomics** includes all computer/person considerations, including the physical considerations shown in Figure 9–3. However, no consideration may be as important as the relationship between a user and the information on the screen. Well-planned screen design increases user comfort, accuracy, and productivity.

Let us begin with a check list.

- The data on the screen should be *uncrowded*. The phone company has spent thousands of dollars on studies to determine the best way to present names on the screen when a customer calls directory assistance. Their main conclusion was that the operators were speediest when the screens had lots of "white space" and were uncluttered.
- Use *blank separators* to distinguish data groups. Analysts with a programming background have a penchant for streams of asterisks, but asterisks lend themselves poorly to screens.
- Use *logical grouping* of data, presenting similar data together. As you will see, the same advice applies to forms.
- Use *frequent closure*; that is, break each task into finite subtasks and give the user feedback at reasonable intervals. Users become nervous—Am I still OK?—if left "unattended" too long. Because of this, the computer should not continue an extended task without giving the user occasional reassurance. If copying a backup diskette, for instance, one technique that provides assurance is to state in

FIGURE 9-3

Workstation ergonomics. The distances shown are for optimum comfort.

advance the time it will take (say, 2 minutes) and then print the time remaining on the screen (1:59, 1:58, and so on). Never leave a blank screen or the user may do something unfortunate, like turn off the machine!

- Use *consistent terminology* to avoid confusion. Do not, for example, use the words *erase* and *delete* interchangeably.
- Spare your user that fishbowl feeling—*limit sound effects*. Many computer systems offer various bells, beeps, and buzzers. If you employ them to highlight errors you will have very grumpy users.
- Use a *consistent error message location*. If you zone a certain portion of the screen for error messages, system users will learn to recognize their appearance.
- Use *businesslike messages*. Nothing remotely cute should appear on a screen, not even "that's better" or "thank you."

There are times when we need to make sure we have the user's attention. **Highlighting** a message makes it appear more obvious than other screen text nearby. Highlighted text can be of brighter intensity or it can be blinked on and off. (The cursor typically blinks continuously.) Another highlighting technique, particularly popular for templates, is **reverse video**, in which brightness and darkness are reversed.

Other physical approaches to directing a user's attention are boxes, arrows, and variable colors. If you use colors, stick to a few primary colors such as

---

BOX 9-3   **Ergonomic Equipment**

The facets of computer/person interaction need consideration if comfort and productivity are paramount.

- The ability to *tilt and swivel* a terminal or microcomputer becomes especially important if there are multiple users (who are presumably of different sizes) of the equipment.
- Varied users will appreciate a *detachable keyboard* that can be placed in a location to suit comfort and convenience.
- The *keyboard configuration* should make typing easy, with keys well spaced for touch typing (some microcomputer keyboards are notably poor in this area). Special function keys and a separate numeric key pad may be important. Also, the person keying should receive a tactile feedback so it is clear when contact with the keys has been made.
- A *nonglare screen* will enable users to work longer periods without eye strain.
- The *screen characters* should be easy to read and, in particular, the lowercase characters g, j, p, and y should have descenders.
- A most important consideration is the *ergonomic chair*, which has push-button movement controls and excellent back support.

---

red, green, yellow and blue. Rainbows of magenta, chartreuse, and avocado are confusing at best. Many software systems available today use the **window** concept, as discussed in Chapter 8, showing various activities or files, in certain boxed-off areas—"windows" on the screen.

◇ FORMS DESIGN

We use poorly designed forms every day. Surely it was not long ago that you puzzled over just what line to write your name on, or tried to scrunch "Minneapolis" into the half-inch space allotted. Many people who make up forms know nothing about forms design.

Some people, however, make their living as **forms specialists**. These people usually work for businesses that sell forms or for large corporations that use a lot of forms. As an analyst, you may be lucky enough to have access to their expertise. Chances are good, however, that you will be designing forms yourself. This discussion presents the rudiments of forms design.

Reverse video screen. This template screen highlights data needed by using a light color.

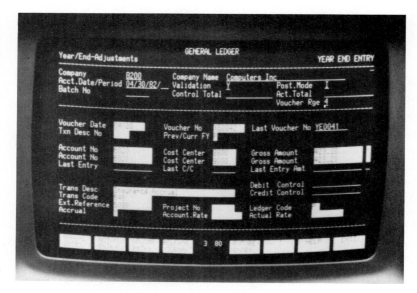

Input Forms: Why We Need Them

Although we know about voice input and other methods that do not seem to need paper accompaniment, we are a long way from the "paperless office." We still need paper forms to help us do our jobs. A form is a *carrier of data*—specifically, data to be input to the system. The type of form can be as diverse as the data it carries, such as invoice figures, student registration data, or factory tooling requirements.

Forms also are used as *authority for action*: to pay an employee overtime, purchase feed grain, increase a boutique's inventory supply. Lastly, a form is a *visible record* of data input and authority given.

Designing a Good Form

Perhaps the form that gets the most bad publicity is the income tax form you must complete every year. Despite continued IRS efforts to present a simplified form, taxpayer grousing persists. IRS analysts consider three general goals when designing a form. The form should be (1) simple to complete, (2) designed so that its use is error-free, and (3) easy to read and transcribe. If a form passes these not-so-simple tests, frustration and costs are reduced. Consider the following aspects of forms design, keeping these goals in mind.

Identification.   Each form needs a clear and descriptive title to communicate its purpose. If the form is for external use (a bill for customers, for example) then the identification may be enhanced with the company logo.

Order and Grouping of Data.   Printed requests for data should be in some logical order, related to the need for the data. On a job application, for example, it would certainly be distracting—not to mention peculiar—to be asked for your name somewhere in the middle of the second page. Grouping like data together is called **zoning**: All identification data (name, address, phone, social security number) normally would be in close proximity on a form, while past employment data would be grouped together in another location. Note the zoning in the form in Figure 9–4.

Form Size.   Choose standard sizes. Drawing a form that ends up being $4\frac{1}{2}$ by $12\frac{3}{5}$ inches will cause the user untold problems with file drawers, binders, and notebooks. Some common sizes are 3×5, 5×7, 8×10, $8\frac{1}{2} \times 11$, and 11×14.

Spacing.   Lack of ample writing space is the most consistent complaint about forms. You need to consider what instrument and what kind of user will fill out the form; a typist in an office has different needs from a meter reader in the field.

Paper Stock.   The type of paper used will be determined by how the form is used. Generally, the rougher the use, the heavier the paper. How often is the form used? How gently—or not gently—is it handled? Is it going to be folded? Will it be reused? Will it be used outdoors?

Color.   Color is often a consideration for multiple-part forms, so that users can tell at a glance who gets which copy. (Yellow to bank, blue to customer, pink to file, and so on.)

Instructions.   Instructions are perhaps the most important consideration. Instructions must be concise and unambiguous. The word *date*, for example, tells very little. Today's date? Birth date? Month-day-year? And so on.

As you can see, designing a form is not a simple matter. It is complicated further by the many types of forms to consider.

Rules and Captions

Rules are the lines that divide a form into various sections and columns. **Captions** are the brief instructions used on a form: a word or words that tell users what to write in the space provided. The relationship between rules and captions is important, one that makes the difference between clarity and confusion. No prose description could illustrate this as well as a picture, so note the examples in Figure 9–5. Do you have a vote for the most straightforward, can't-miss form?

MOORE

COMPUTER FORMS AND SUPPLIES
CATALOG ORDER FORM
P.O. BOX 20, WHEELING, IL 60090

① BILL TO

☐ Check here if name and/or address are incorrect, and enter corrections

601671

Linera Lucas
3601 Meridian North
Dedham MA 02026

Your Phone Number _____

② SHIP TO (if different from address above)

Name _____
Title _____
Company _____
Address _____
City _____ State _____ Zip _____
Person Ordering _____
Title _____ Date _____ ( ) _____ Business Phone
Signature _____

③ SHIPPING   Please check one:

☐ Regular Delivery (Usually UPS, except for heavy, bulky items.)
☐ UPS 2nd Day Air Service
☐ Emergency Next-Day Delivery (PLEASE CALL IN YOUR ORDER **1-800-323-6230**.)
For Customers in Alaska, Hawaii and Puerto Rico: UPS does not serve your area entirely. Please check one of the boxes below for shipping preference.
☐ 1st Class Parcel Post   ☐ 3rd Class Parcel Post

④ NEW ACCOUNT INFORMATION

To help us expedite your order, please provide your DUN & BRADSTREET number (if you have one):
If you've bought from the Moore Catalog before, enter your
Moore Customer No. _____
Bank Name: _____
Account Number (if known) _____
City _____ State _____ Zip _____

⑤ CATALOG NUMBER	QUANTITY	ITEM DESCRIPTION	UNIT PRICE (each, doz., etc.)	TOTAL AMOUNT

ORDER IMPRINTED ITEMS BELOW

⑥ CATALOG NUMBER	QUANTITY	ITEM DESCRIPTION	UNIT PRICE (each, doz., etc.)	TOTAL AMOUNT

Type Style ☐ Garamond ☐ Alphatura Book ☐ Alphatura Medium ☐ Kaufman Bold ☐ Logo ☐ Code ☐ No Logo

Type Style ☐ Garamond ☐ Alphatura Book ☐ Alphatura Medium ☐ Kaufman Bold ☐ Logo ☐ Code ☐ No Logo

Type Style ☐ Garamond ☐ Alphatura Book ☐ Alphatura Medium ☐ Kaufman Bold ☐ Logo ☐ Code ☐ No Logo

⑦ IMPRINT INFORMATION: See the forms ordering page for complete details and specifications regarding the ordering of imprinted forms and checks. **PLEASE PRINT CLEARLY**

Your Firm Name _____
Street or Mailing Address _____
City, State, Zip Code _____
Phone Number _____

☐ Check this box if you want your custom imprinting done in brown ink for a flat charge of $15.00 more. (If not specified, imprinting will be black ink.)

MINIMUM ORDER $15.00

Sales Tax _____
Total $ _____

⑧ PAYMENT

(CHECK ONE)
☐ Check or Money Order enclosed
☐ Open Account—new customers, see left
☐ Company Purchase Order (required on orders over $1,000)

Purchase Order No. _____

Authorized Signature _____ Title _____
☐ Visa ☐ MasterCard ☐ American Express
Account Number _____
American Express account number ends here
Signature _____
Expiration Date _____

FIGURE 9-4   **Zoned form.** Each separate type of data on this order form is in a separate numbered zone.

BOX 9-4	Paper Weights

The weight of a paper is determined by the weight of a ream, that is, a sheaf of 500 17 × 22 sheets of paper. The greater the weight of the ream, the heavier and more durable the paper. Here are some common types of paper by weight.

Weight	Paper	Use
9 lb	Onionskin	Flimsy: filing only
10–20 lb	Bond	Medium: letters, reports
24–32 lb	Ledger	Medium plus: permanent records
60–72 lb	Index cards	Heavy: repeated handling
90–140 lb	Card stock	Heaviest: durable; rough handling

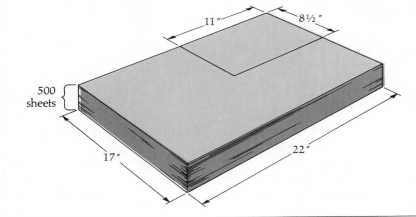

Multiple Forms

Forms that require multiple copies need special attention. The first issue is whether to use carbon paper or NCR paper. Carbon paper, whether interleaved or attached to the back of the copies, is messy and bulky. NCR (for *no carbon required*, not National Cash Register) is a chemically treated paper that allows lower copies to receive data by exerting pressure on the top copy. It is clean and easy to use. The disadvantages are that it is more expensive than carbon paper and that some people are allergic to the chemicals in the paper.

Multiple forms can be used in a complete **unit set** with tear-off copies or in **split carbon sets** (Figure 9–6) with partial carbon to duplicate only some of the data.

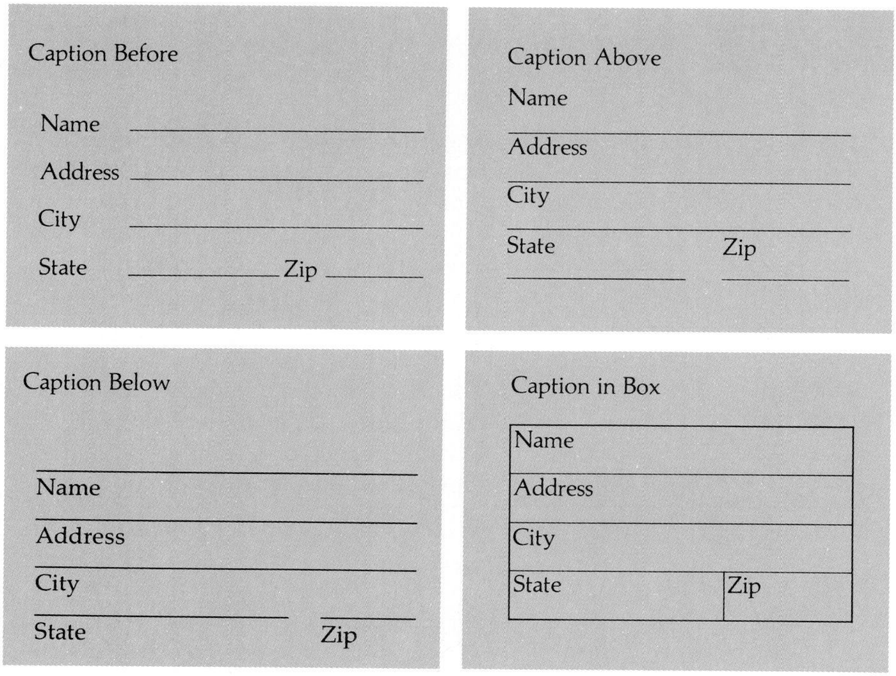

FIGURE 9-5

Form captions.
Which set of captions
gives the clearest
instructions?

Caption Before

Name _____

Address _____

City _____

State _____ Zip _____

Caption Above

Name

Address
_____

City
_____

State _____ Zip _____

Caption Below

_____
Name

_____
Address

_____
City

_____
State   Zip

Caption in Box

Name	
Address	
City	
State	Zip

Forms Management

Analyst Kathy Rockefeller was inundated with user change requests and felt the requests needed to be cataloged and prioritized. To help get organized, she used her straightedge to prepare a form with column headings such as date received, priority assigned, and expected work hours required. This hand-drawn effort was not particularly formal, but Kathy did not expect anyone but herself and a user to see it. Kathy was surprised—and annoyed—when reproduction personnel refused to make copies of the form because it had no official sanction. Why should anyone care what she used at her own desk?

Note the dilemma, so familiar in our society: trying to balance the rights and needs of the individual with those of the larger organization. In this case we are not talking about taking a vote, however. The rules of the organization prevail. There is a good reason for this: If forms are allowed to be developed without any controls, they soon will sprout like mushrooms all over the company.

Many organizations have established some version (a person or a group of people) of a **forms management group**. The forms management group, also called the forms control group, monitors forms from their inception—assigning a form name and number—and keeps records of all forms in use. Although this may seem like more unwelcome red tape, there are powerful

FIGURE 9-6

Split carbon set.
Note that the center portion of the first page will not copy to the succeeding pages because there is no carbon backing there.

reasons why an organization should establish forms control procedures. These include the following:

- *Avoid redundant forms.* Another analyst is likely to have the same problem Kathy did and solve it in a similar way.
- *Eliminate obsolete forms.* Paperwork has a way of staying around indefinitely unless someone makes a conscious effort to reduce it.
- *Maintain adequate forms stock.* Someone must monitor form stock levels and reorder on a timely basis.
- *Control costs.* Forms control personnel encourage the use of standard sizes and weights of paper, and discourage elaborate preprinted, multicolor forms if not needed. Further, costs can be controlled through large-volume ordering.

An unauthorized form is known as a **bootleg form**. The proliferation of unauthorized forms in an organization is usually a sign of inadequate management controls in this and other areas.

◇ DATA CODING

Consider an inventory file for a sporting goods franchise. Each record in the file represents one particular product: a tennis racket, soccer ball, golf shoes, or whatever. The fields within the records can easily be imagined—item number, item name, vendor, quantity on hand, last order date, and so on. We can also imagine what kinds of questions a company executive might ask: How do orders of water skis this year compare with orders of water skis last year at this time? Which franchise had the lowest average clothing inventory last year? How do inventories compare among the various stores? To be able to answer these and other questions, a little advanced planning is required. You will prepare the item number—the record key—in such a way that each part of it is meaningful.

FIGURE 9-7

FIGURE 9-7

Coding examples.
Three different types
of codes are illus-
trated in different
applications.

Code	Item	Code	Item
20	Sailboat—Columbia	40	Aluminum hull—Johnson
21	Sailboat—Coronado	41	Aluminum hull—Evinrude
22	Sailboat—Lightning		⋮
23	Sailboat—Santana	50	Inflatable—Achilles
⋮			
30	Cruiser—Chriscraft	51	Inflatable—Spencer
31	Cruiser—Grumman		⋮

Block sequence codes assign a block of numbers to a characteristic of the item being coded.

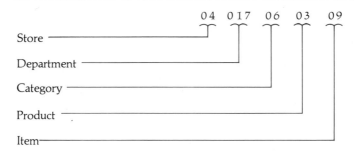

Group classification codes are used to identify an item using meaningful categories. In this example, the code reads store 4 (Jackson), department 17 (court sports), category 6 (equipment), product 3 (tennis racket), item 9 (Prince deluxe).

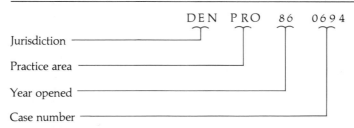

Mnemonic codes are derived from the description of the item to be coded. This is an example of a code for a legal case file system. The first part of the code refers to jurisdiction, indicating city of county. The second part of the code is major practice (probate), RES (real estate), or TAX.

The point is this: Data that is input to a computer system eventually needs to be retrieved. If it is important to retrieve the data by category, such as type of product or department, then it is useful to plan an identifier code that is not just a unique key but has built-in meaning related to the data itself.

Data coding is a common practice. For the sporting goods just described, a typical code would include numbers representing store, department, and category of product (clothing, shoes, tennis equipment, and so on). Although you may not have noticed that they had significance, the digits and letters that make up your zip code, social security number, driver's license identification, and magazine subscription number are all in some code. These identifiers can be decoded by a person, or a computer, who understands their composition.

There are many ways to plan a code. The subject is quite complex. In fact, entire books have been written on coding. We shall let the examples in Figure 9-7 of block sequence codes, group classification codes, and mnemonic codes suffice.

There is one more important point to make about codes. They are sometimes introduced to reduce the time required for data entry. For example, 1 could be keyed to represent first shift, 2 for second shift, and 3 for third shift. Similarly, the months of the year could be represented by the numbers 1 through 12. And so forth.

◇ USERS AND THEIR INPUT

Users interact with a system through its input and output. They prepare the input and use its output. There is more than a subtle difference between the two activities, however. Users examine, peruse, and react to output. They may even complain about the content or appearance. Even though they are indirectly involved, the output actually was produced by the computer.

Input, on the other hand, is produced directly by the user. Any flaws in the input can be laid at their doorstep. It is your responsibility to prepare an environment where users can move surefootedly through the input process. Help them be successful.

Ergonomists Take on the Traders

To Bankers Trust, the trading room in Manhattan's downtown financial district was a place where $10 billion in transactions could occur in a day. To the ergonomists who remodeled it last year, however, it represented a vast challenge. They had to design new facilities to meet the demands of 243 fast-thinking traders who deal in everything from state and local government bonds to foreign currencies.

Management wanted the job done with dispatch. Tom Ringkamp, then the bank's senior vice president in charge of real estate, was given a mere six months from planning to completion. The room was finished on schedule with the help of a firm called Interior Facilities Associates of New York. The furniture was turned out by a woodworking subcontractor, Spec' Built of New Jersey, which had produced rooms for such investment firms as E.F. Hutton, Salomon Brothers, Goldman Sachs, and Lehman Brothers.

Getting the traders to cooperate required coaxing. "Change was anathema to them," says Ringkamp. "They were making lots of money in the old room." To help persuade them—and to find out what they wanted and needed—architect Gerd Althofer placed mock-ups in the corridor outside the old trading room.

What the traders got was a room that should remain state-of-the-ergonomic-art for at least several years, and be flexible enough to take future changes. Trimly tailored efficiency replaces the jammed quarters of the old room, with its make-shift tables of plywood laid on file cabinets and visual display terminals (VDTs) plunked on top. Each trader has up to four VDTs. All of a trader's new terminals respond to a single keyboard, while before each had its own keyboard, taking up most of the tabletop. To fight reflections on the VDTs, the lighting is dimmer than in most offices, and is primarily reflected off the many-vaulted ceiling. Acoustically the room is muffled, but not too much. "A lot of shouting is good and usual for a trading center," says Ringkamp. Telephone jacks were installed, right- and left-handed. Spec' Built produced all the trading stations, prewired, in just 90 days.

Fire protection was important, not primarily to protect the equipment (which is insured and can be replaced), but to ensure continuity of operation. A bank of backup generators is stationed on the floor below to provide power in an outage. "The building can go down," says Ringkamp, "but this room cannot." It doesn't pay to lose a $10-billion day.

—from *Fortune*

SUMMARY

- Data input errors cannot be eliminated, but a well-designed system can reduce errors and catch many of the errors that do occur.

- Getting the data to the computer for processing is called data capture. Traditionally, data has been handwritten and then transcribed to a computer-accessible medium. Today it is more common to capture data as a by-product of the event that causes the data to be generated, a process called source data automation.

- Input planning also concerns data content (what data is needed), data format (a description of the data), and data volume (how much data).

- Data control involves accuracy (correctness of the data), scheduling (having the data ready for processing on schedule), security (safety of the data), and auditing (tracing the data through the system).

- If data is entered online, it can be edited as it is received, with errors brought to the attention of the data entry person. Data that is input in batch mode is usually processed first by an editing program that rejects transactions with erroneous data and prints an error report.

- Types of edits include checks for sequence, reasonableness, range, validity, comparison, class, and completeness.

- Input devices come in a wide variety. Key-to-tape devices are used for low-intensity batch systems. Key-to-disk systems have users keying data to disk for later use or users sharing a system that uses their keyed data for immediate online updating.

- Point-of-sale (POS) terminals are cash registers with automated features, including the ability to capture sales data. A wand reader or bar-code device may be used to scan the product being sold and input its identification data to the system.

- Graphics input devices include the joystick for rapidly moving figures on the screen, the light pen for drawing lines on the screen, the digitizer for converting a picture to digitized data to be processed by the computer, and a graphics pad with a pressure-sensitive surface on which a stylus is applied, for causing a corresponding drawing to appear on the screen. The mouse is rolled by hand on a flat surface, causing corresponding cursor movement on the screen.

- Touch screens use the touch of a finger on the screen to detect and act on desired screen selection.

- Magnetic ink character recognition (MICR) is used by banks to machine-read data from checks.

- Optical recognition methods use a light source to scan data and input it to the computer. Optical mark recognition machines sense penciled marks on paper. Optical character recognition (OCR) devices scan and recognize character data. Bar-code readers scan product identification codes and use that data to store or retrieve data in the related computer system. Legible handwritten characters can be scanned optically and input to the computer.

- Vision recognition systems use a camera to take a picture of an object, digitize it, and feed the input to the computer. Voice recognition systems convert voice input to digital code that can be accepted by the computer.

- There are three basic screen design approaches: menu, prompt, and template.

- Menu input, an easy option for beginners, gives a user a series of choices. The menus may be hierarchical.

- Prompt input comes in two varieties. For users with more basic skill levels, the prompt is in the form of simple questions that lead to a computer/person dialogue. Although sometimes tedious, an advantage of this type of system is the flexibility of a series of prompts based on user responses. Another type of prompt is the symbol prompt recognized by experienced technical personnel for system software.

- Template input has the appearance of a printed form on the screen. When inputting data via template, the cursor moves from item to item, directing input. A template page contains a logical grouping of data.

- Analysts must anticipate invalid screen entries and acknowledge the error with a polite message that also describes how to fix the problem. Also, it is common to permit the user to see a completed set of data before it is actually entered into the system, giving the user the option of making corrections at this time.

- Ergonomics includes all computer/person considerations. Screen ergonomics include an uncrowded screen, blank separators, logical grouping of data, frequent closure to give feedback at reasonable intervals, consistent terminology, limited sound effects, a consistent error message location, and businesslike messages.

- To get the user's attention on the screen, we may employ highlighting to make a message stand out, or reverse video, in which brightness and darkness are reversed. Also, many software systems use a window concept to show certain activities or files in boxed-off areas.

- The expertise of forms specialists is designing forms. They work for large corporations or companies that sell forms.

- An input form is a carrier of data, has authority for action, and is a visible record of data input and authority given.

- Aspects of good form design are identification of the form, the appropriate order and grouping of data, standard form size, ample spacing, suitable paper stock, suitable use of color, and good instructions.

- Rules are the lines that divide a form into various sections. Captions are the brief instructions on the form.

- Multiple forms can be used in a complete unit set with tear-off copies, or in split carbon sets with partial carbon to duplicate only some of the data.

- Many organizations have established a forms management group. There are a number of reasons for establishing forms control procedures: to avoid redundant forms, eliminate obsolete forms, maintain adequate forms stock, and control costs. An unauthorized form is known as a bootleg form.

- If data that is input to a system needs to be retrieved according to categories, such as type of product or department, then the data record key needs to be prepared using data coding. Data codes act as a unique key that also serves as an identifier that is directly related to the meaning of the data itself.

KEY TERMS

Auditing
Authority for action
Bar-code reader
Blank separators
Bootleg form
Captions
Carrier of data
Class check
Closure
Comparison check
Completeness check
Data capture
Data coding
Data content
Data control
Data format
Data volume
Digitizer
Ergonomics
Form size
Forms management group
Forms specialists
Graphics input
Graphics pad
Handwritten characters
Highlighting

Input devices
Input form
Instructions
Invalid screen entries
Joystick
Key-to-disk device
Key-to tape device
Light pen
Logical grouping of data
Magnetic ink character
  recognition (MICR)
Menu input
Mouse
Multiple forms
Optical character recognition
  (OCR)
Optical mark recognition
Optical recognition
Order and grouping of data
Point-of-sale (POS) terminal
Prompt input
Range check
Reasonableness check
Reverse video
Rules
Screen design

Screen ergonomics	Unit set
Sequence check	Validity check
Sound effects	Visible record
Spacing	Vision recognition
Split carbon sets	Voice recognition
Template input	Wand reader
Template page	Window
Touch screen	

REVIEW QUESTIONS

1. Charge customers sometimes have merchandise incorrectly charged to their accounts or perhaps mistakenly are not charged for some item. Which types of input devices mentioned in the chapter might lend themselves to such an error? How could the error occur?

2. What applications in your daily environment use some form of source data automation? List and describe.

3. Consider planning records for a personnel file. In a general way, describe the data content, data format, and data volume.

4. Which input devices are likely to be used with microcomputers? Stop by your local microcomputer store and ask which devices are the best sellers.

5. Which screen design method would you choose for a new charge account for a pharmacy? For a stock inquiry program? For a patient medical evaluation? Justify.

6. If you were designing screen input for a user to prepare tax returns, what kind of considerations would you make for screen ergonomics?

7. Survey the forms you use at school for activities such as registration, financial aid, library use, and graduation application. Rate them on the aspects of good form design: standard size, ample spacing, and so on.

CASE 9–1
ONLINE CARD CATALOG

How would you like to go to your college library, step up to a terminal, and give the computer a few hints about what book you want? You do not even have to have the title right—just a title word or two will send the system searching for the book and then flash its exact library location on the screen. Even a vague subject matter search will yield several choices on the screen. Sound good? Read on.

This is exactly the kind of system that analyst Carole Kirkpatrick had in mind when she began working with the librarians at Coe University. As the systems team worked together on this project, their requirements list grew until they felt they had a coherent plan. The approach was two-pronged: (1) Give librarians improved control, and (2) give easy access to students.

The issue of input was straightforward for the librarians: they could easily be trained, and could refer to documentation and each other for assistance. Easy access for students was trickier. The input methods for both the librarians and the students were to be a series of prompts.

When the system was ready for student testing, Carole put together very clear instruction sheets on pasteboard. She positioned the sheets next to a mockup display in a prominent location in the library and invited students to try it out. She was proud of what the system could do and surprised to see that students showed little interest. They would glance at the directions and move on. Finally, Carole enticed a group of students to review the system and make comments, which they did. They basically admitted that they did not want to read the long sheets of directions.

Down came the pasteboard signs. Carole took a new tack: online instructions. Each terminal not in use had a full screen sign:

```
PUSH
ANY
BUTTON
```

Students who did so were then taken through the system step by step, with careful on-screen instructions for using the prompt input system. This time the mockups were so crowded—and noisy—that it was decided to move the card catalog to another location.

1. Success at last. Can we fault Carole for the aborted first attempt or is that just part of the process?
2. Even though a goal was to have the students be self-sufficient, we know that librarians are always on the premises. Do you think it is possible to design a system so that novice users can access a computer system with no assistance whatever?

CASE 9–2
GETTING THE PATIENT RECORDS STRAIGHT

The Community Health Clinic (CHC), a new private organization serving people with immediate medical needs, had been well-accepted in the community. When the clinic began, Director Ralph Juarez hired a software consulting company to install a computer system to handle the clinic's medical and financial needs. As part of a continuing growth pattern, he planned to automate the clerical operations, beginning with the patient intake procedure and the update of services received.

Analyst Martha Hunt was assigned this task by the consulting company. After reviewing the documentation for the existing clinic systems, she first interviewed Juarez and then his assistant, nurse Gayle McCormick. After extended discussions, they were able to agree on an overall menu system. The menu would quickly switch to a template screen format to input original patient data such as name, social security number, address, and medical history. Other parts of the menu would be used to update the patient file with symptoms given and treatment received.

These plans were documented, approved, and implemented. Medical and clerical personnel seemed to adjust well to the new system but, alas, things began to unravel within two months. First in a trickle, then in a flood, patient complaints started to come in. In follow-up visits, patients discovered that their records were incorrect. There were also complaints of being billed for the wrong services or services not received. Some patient bills were returned because of incorrect addresses. Some patients (eventually) mentioned that they had not been billed at all.

Juarez launched an investigation, concluding that users were introducing errors into the system. Their defense was that no one, including the computer, had told them that anything was wrong until it was too late, and that even the errors they realized they had made could not be fixed.

1. Who is to blame for this state of affairs?
2. What data could have been checked for accuracy?
3. What method could be used to allow users to make changes to data before it enters the system?

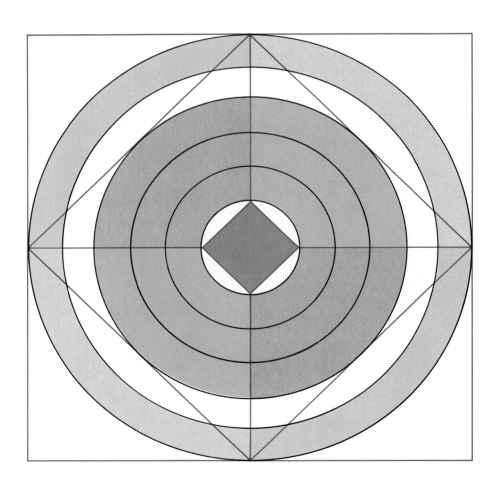

Chapter 10

Detail Design: Processing

As student analysts operating in a classroom environment, Ed Gorski, Terry Ho, Joe Smith, and Peggy Weideman were organized as a team. Their task was to design a new inventory system for a chain of hardware stores. After considering various alternatives, they felt they could justify a system that used point-of-sale terminals, capturing inventory data as a by-product of the sales transaction.

From this preliminary design concept, they proceeded to more detailed design issues. They needed to be able to represent their design on paper in ways that could be understood by others, especially the programmers who would be coding from the design specifications. They chose a combination of tools to represent the processing: Warnier–Orr diagrams, pseudocode, and a few decision tables for specific complicated processes. They made these selections because these tools fit the logic of their planned system and because they felt these representations would be easy to maintain.

The merits of these and other detail design tools will be examined in this chapter. But first, consider where you are in the systems life cycle.

◇ THE TRANSITION FROM PRELIMINARY TO DETAIL DESIGN

Review for a moment what you have done in the systems life cycle. After you established the nature, scope, and objectives in the preliminary investigation phase, you moved on to the analysis phase to learn about the system and establish its requirements. Next came the preliminary design, in which you planned an overall design that was cost-effective and acceptable to the client.

From there you entered the detail design phase, planning first the output and then the input. Now we come to the third part of the detail design phase, the design of detail processing. This step is the one that is of greatest importance to programmers. If the specifications you plan in this phase are clear and straightforward, the development phase may be rewardingly smooth. Now it is time to say exactly what is to be done, and how.

◇ DETAIL PROCESSING CONSIDERATIONS

The preliminary design is finished, so you know the overall design plan. Further, you have completed detail design of the system outputs and inputs. Now, specifically, how do you plan the processing to get the outputs from the inputs? A typical data processing program—a small subset of the system—may read a record, process that data in some way, and then write a record, continuing this way in an iterative fashion. But the total set of possible operations far exceed this one example. The single input-process-output routine just described might be much more complicated.

Suppose, for instance, that the input record just mentioned is actually an inquiry from a user at a terminal for a hotel reservation. An overview of this system is shown in Figure 10-1. Some editing of the input data is probably appropriate—Is the date numeric? and so forth. This will be followed by reading a reservation master file or, more likely, a database of reservation information. Several comparisons will have to be made regarding the availability of resources: dates requested, rates, room capacity and location, and possibly complications such as connecting rooms or seasonal rate discounts. At some point, customer data will be input, the files will be updated, and reports generated on room status and the like. If the reservation is for someone planning to attend a convention, there may be a separate convention file to be queried and updated, with related transactions generated.

Eventually some decisions will be made about the content of the output, such as yes/no accommodations, alternative dates or accommodations, estimated costs, and the like. In addition to a reply to the user, the program will probably write a transaction that will cause a reservation confirmation to be printed. Such a system would doubtless produce several reports based

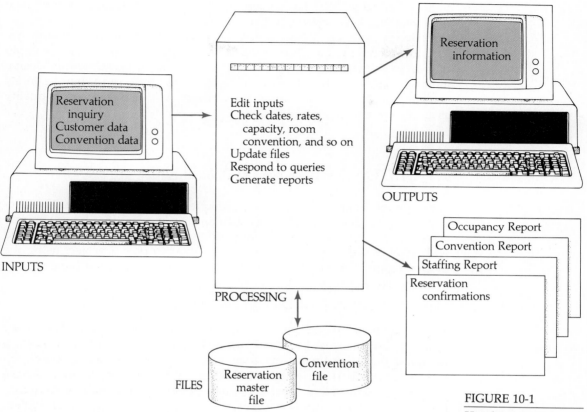

INPUTS

Reservation
inquiry
Customer data
Convention data

Edit inputs
Check dates, rates,
   capacity, room
   convention, and so on
Update files
Respond to queries
Generate reports

PROCESSING

Reservation
information

OUTPUTS

Occupancy Report
Convention Report
Staffing Report
Reservation
   confirmations

FILES   Reservation
         master
         file

         Convention
         file

FIGURE 10-1

**Hotel reservation
system.** Overview
diagram of inputs,
processes, and out-
puts for a hotel reser-
vation system.

on anticipated staffing and supply needs, daily occupancy, accounting records, and other information that would be useful to management. And
so on.

Although we just brushed over this example, it is not difficult to see
how complicated the design of such a system can be. Consider some of the
common activities of this and other system programs: read, compare, edit,
move, sort, calculate, retrieve, update, summarize, classify, store, and write.
In planning such a system you could prepare the design graphically using
one of the design methodologies we shall discuss shortly. Or you might
prefer to turn to a classic detail design tool such as a systems flowchart.
You may augment these design schematics with the tools discussed in a
later section on detail processing design tools.

All the activities just listed, from read to write and everything in between,
are related to processing data. The program receives data input, uses previously processed data from files, updates the files, and writes output in
the form of new information. Although you have already planned the detail
design of the output and input, you must still plan the files and databases

that are related to the detail processing. Files and databases are a significant topic that we shall examine in Chapter 11.

In many systems, including the hotel example just offered, online system considerations are an important part of the system and add complications to the design. To sum it up then, we shall proceed from design methodologies to other detail processing design tools, and then to online considerations. We shall complete the chapter with a discussion of quality assurance. First, the methodologies.

◇ SYSTEM DESIGN METHODOLOGIES OVERVIEW

A **design methodology** is a set of techniques based on a concept. The originator of a methodology has a particular idea in mind that he or she perceives as a useful way to get things done. This idea is then embellished with a set of tools that will enable an adopter to plan a project using the methodology. These tools are used to express and document the design.

Historical Setting for Design Methodologies

The introduction of design methodologies flourished in the 1970s. They came on the heels of the most turbulent period in computer history, a decade of disappointments that included shattered budgets and some projects that were, literally, years behind schedule. Some problem projects have become so famous that they have gone down in the annals of computer history. An early failure was an aborted major airlines reservation system that left shaken managers wondering if computerized reservations systems were even possible. The long-awaited and bug-ridden IBM 360® operating system was so traumatic for project manager Frederick Brooks that he wrote a book about it *(The Mythical Man-Month)* to help others avoid his tribulations. The New York State online betting system project caused sickening headlines in the trade press for months. (The most memorable scene was the programmers sleeping on the floor in the office Christmas Eve in a desperate attempt not to fall further behind in their schedules.) The time was ripe for a little help.

Which Methodology?

Now to the design methodologies themselves. There were many of them. They were usually introduced in the trade press, often with some fanfare. The methodologies frequently were supported by books and formal course offerings. For a substantial fee, a company could bring a design methodology expert on-site to teach resident analysts and programmers the new approach. In fact, the methodologies that succeeded took hold in industry and only later made their way into textbooks and college courses.

Gradually, it became clear, however, that there was no panacea. No one methodology emerged as the chosen or established winner. (For a time managers thought there might be a single answer, much as structured programming had become the universal elixir for improved programmer productivity.) The net result is that today no single methodology has emerged that would be an asset in every design problem. Some methodologies, however, are particularly appropriate for certain types of design problems.

The design methodology field has narrowed in two ways. The first is that some methodologies were not promoted or accepted enough to get off the ground. These have faded into obscurity. Second, the few that remain are not always accepted in total. In fact, some tools from different methodologies have found their way into the body of generally accepted analysis and design tools. Collectively, the new and old tools now form the set of tools from which many analysts draw.

Despite all of this, these two methodologies do deserve special attention: Structured Design and the Logical Construction of Programs. We shall examine them in terms of their underlying concepts, techniques, and tools.

STRUCTURED DESIGN METHODOLOGY

Structured design is based on concepts originated by Larry Constantine and expanded by his colleagues Edward Yourdon and Glenford Myers. The structured design methodology has been widely promoted and has gained considerable acceptance in the computer industry. **Structured design** is based on the concept of data flow. The cornerstone of structured design is a top-down modular approach. The discussion focuses first on the structure chart as a top-down tool and then on the concept of modularity.

Structure Chart

A key concept of structured design is that design should begin with a high-level picture of the system. This high-level picture identifies major functions that act on the data, as you can see in Figure 10-2. The major functions are the initial component parts of the structure chart. Each major component is then broken down into subcomponents that are, in turn, broken down still further until sufficiently detailed components are obtained. This is considered a **top-down** approach to design. The drawing that shows these components and their relationships is called a **structure chart**. Since the components are pictured in hierarchical form, the drawing is also known as a **hierarchy chart**.

A structure chart has been compared to a business organization chart. Imagine such a chart with only two levels. The manager of an urban bank, for example, has each of these employees reporting directly to him: 23 tellers, 6 commercial loan officers, 3 consumer loan officers, 4 new account

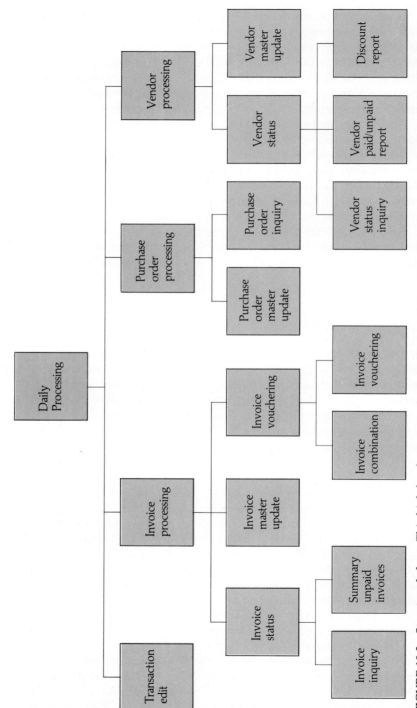

FIGURE 10-2   **Structured chart.** This high-level structure chart shows the relationship of programs for daily processing operations.

officers, 6 investment counselors, 5 secretaries, 2 marketing representatives, 3 credit card advisors, 2 bank cash machine officers, 1 foreign exchange representative, and 2 janitors. This manager probably dreads going to work because there are just too many people and too many details that need attention. The obvious solution to the problem is to establish intermediate management positions, with only the top people reporting directly to the bank manager. And, of course, these formal business relationships could be represented on a hierarchical chart. The point is that anything large is more manageable if it is partitioned and organized. That includes systems.

The way in which components of a structure chart are related is an important concept in structured design. But before we discuss these important relationships, note the limitations on structure charts. A structure chart does not show the sequence in which system components will be executed; execution does not necesarily move from left to right as shown on the chart. Also, the structure chart does not indicate what logic conditions cause what activity. This missing information can be supplied in other ways. A common explanatory tool is pseudocode, which can describe logic in English-like fashion. Pseudocode is presented in a later section of this chapter. Meanwhile, the picture—the structure chart—tells the story.

Modularity

The way that a system is divided into various pieces has a significant effect on its structure. High-level components of the structure chart are programs; lower level components are called modules. Once converted to programmed form, a **module** is a set of logically related statements that perform a specific function.

Large complex programs were once viewed with awe. This awe mingled respect for the program complexity with sympathy for the programmers who had to develop and maintain them. Now computer professionals recognize that the way to efficient development and maintenance is to break the programs in a system into manageable pieces—that is, modules.

We have already noted in our structure chart discussion that structured design involves organizing the pieces of a system in a hierarchical way. There are also two module-related objectives of structured design. The first is that the modules be small and easy to manage. The second is that the modules be relatively independent of each other. After we examine overall module concepts, we shall look at the independence issues, coupling and cohesion.

Module concepts.   Some key structured concepts are embodied in the goals for constructing modules. Note the use of the word *goals* instead of *rules*. Rules are often rigid. Thus, since these desirable module characteristics are not always attainable, they are called goals instead. The first goal is that a module must have a **single entry** and a **single exit**. This is relatively straightforward and provides a standard that makes future program perusal much simpler. Next, a module should have **limited data access**; that is, it should access only the data it needs to perform the module's function. Limited data access, which we will emphasize again in the coupling discussion, means that any future changes to data will impact a minimum number of modules. Finally, a module should have minimum coupling and maximum cohesion.

Coupling.   The relationship between modules is called coupling. This relationship should be as weak as possible. Consider a programmer who maintains a set of large programs written by someone else. He or she might say something like this: "Sometimes I fix a bug in a program and find out later that the fix caused a problem somewhere else in the program." The program probably does not have weak coupling.

Coupling is a concept that is often difficult to implement in the purest sense, because sometimes modules are undeniably related. Nevertheless, the less related they are, the easier future modifications will be. Ideally, a change to one module should not affect other modules. However, if you must revise a module that feeds six others, then you have a potential of seven revisions instead of one. Worse, you may revise the original module without recognizing its impact on the others.

One factor that influences the strength of coupling is the complexity of the interface between two modules. Of course, every module must have one interface, in order to be connected to the system. But what kind of connection is it? If one module calls another, then their relationship strengthens with each type of data that is passed. The more data item types, the more highly coupled the modules. But since loose coupling is desirable, the calling module should send only the data needed. For example, the calling module should not send a complete record, as a matter of convenience, when only a certain field is needed by the called module. Notice that this means access has been limited to the data required.

A worst-case coupling would be the alteration of a variable that is used by other modules. That is, the data is not passed module to module, but the net effect is that the altered data was distributed to all modules that use that variable.

Cohesion.   A module has maximum cohesion when it has a single well-defined function. Consider what knowing the function of a module might do for the aforementioned maintenance programmer, who might also say, "When a bug shows up in one of these programs, it takes forever to find it." Finding a bug in a 20,000-line program involves a lot of detective work. But, if we could zero in on the correct 20-line module, by function, the task would be considerably reduced.

Breaking a system into functional programs, and programs into functional modules, is not always an easy task. The key word here is *functional*. Consider this worst-case scenario: A programmer knows that the program should be in manageable chunks, so he or she makes a new module every 60 lines of code. This, of course, leads to a jumble of functions in every module—reading, calculating, writing, and so forth. A cohesive module, however, performs only one function. Examples of a single function are editing a transaction, or calculating discount on an order, or comparing current inventory level with the reorder level.

A module that is highly functional is said to have **modular strength**. To maximize cohesion, no extraneous functions are permitted in the module. In particular, input/output functions should be confined to their own modules. If input/output activity is required in another module, that module should invoke the appropriate input/output module.

Coupling and cohesion are related concepts. The greater the cohesion in each module, the weaker the coupling between modules. The goals of strong cohesion and weak coupling lead to independent modules, modules that are easy to develop and maintain.

◇ LOGICAL CONSTRUCTION OF PROGRAMS METHODOLOGY

The Logical Construction of Programs methodology was originated by Jean-Dominique Warnier (pronounced warn-yea) in France and Kenneth Orr in the United States. The methodology is commonly referred to as the Warnier–Orr method. The Warnier–Orr methodology is widely accepted in France and has a substantial following elsewhere, including the United States.

Warnier–Orr Concepts

The key concept of the **Logical Construction of Programs (LCP)** methodology is that the design is based on the structure of the data. The graphic illustration of this—that is, the drawings used to depict LCP design—are called Warnier–Orr diagrams. At first glance, a Warnier–Orr diagram might seem like a structure chart turned on its side. But there are differences. As you can see in Figure 10-3, a Warnier–Orr diagram begins with the result and then works backward, right to left, through the activities that produce the result. Also unlike structure charts, Warnier–Orr diagrams include information about logic flow.

Some analysts prefer Warnier–Orr diagrams because, in addition to being easy to draw and understand, they are easy to modify. Ease of modification becomes an important graphics considerations after you have redrawn entire charts for modest changes.

Drawing Warnier–Orr Diagrams

To draw Warnier–Orr diagrams, you must first identify major functions. Then, keeping in mind that LCP operates on the principle of data structures,

FIGURE 10-3

Warnier–Orr diagram. Braces define the subsets: Payment can be cash, check, or credit card. Check and credit card are further divided into subsets.

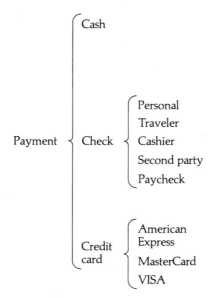

identify the data structures that are associated with these functions. Begin with the output data on the left and work backwards, using braces to separate each subset, to determine processes and inputs that produce the derived output.

Figure 10-4 illustates notational techniques that relate to logic flow. The use of a plus sign in a circle shows that adjacent modules are mutually exclusive; only one will be chosen at a time. The numbers in parentheses indicate the number of iterations of a certain routine. The use of (0,1) means that a process can occur either zero (if not chosen) or one (if chosen) times.

The Warnier–Orr method is well suited to problems where the input and output are hierarchical and the relationship between them is fairly straightforward. Warnier–Orr diagrams also illustrate iterative patterns very well.

◇ THE ANALYST AS PROPONENT OF A METHODOLOGY

A point that needs to be made in any discussion of methodologies is that no methodology can be a total answer to anything. Designing is problem solving, a very fundamental human task. No plan, no book, can replace human judgment. Someone has to think, someone has to decide. A set of tools can enhance the process but in the end the design is in the hands of the analyst.

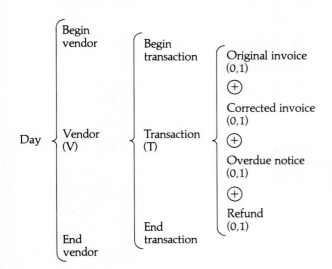

FIGURE 10-4

Warnier–Orr diagram. Processing transactions for vendors. There are *T* transactions from each of *V* vendors. Any given transaction will be one of the four types indicated on the right.

Still, as an analyst, you can most certainly use a design methodology. Few computer environments make equal use of different methodologies. That is, you probably will not be equally proficient and comfortable in several methodologies, as you may have been in several programming languages. Also, many data processing installations have mandatory standards to be used by everyone. However, as we have noted before, it is not always necessary to use a methodology in its entirety, and many analysts use a collection of tools from different sources. In the end, the advice is personal: If it works for you, and the company you work for, use it.

OTHER DETAIL PROCESSING DESIGN TOOLS

The following systems design tools are not promoted as part of a methodology. That is, they may stand alone or in combination with other tools. Not all of them will be needed or used in each new system design. They are offered as options for use—as is any set of tools.

Data Flow Diagrams

Data flow diagrams were first introduced in Chapter 6 as an analysis tool. They are particularly suited to describing the existing system in a graphic way. Once drawn, they can be understood quite readily by users.

Data flow diagrams can also be used for design, as you can see in Figure 10-5. Recall, however, that DFDs do not show control information, no whys or whens—only that certain data is transformed by a certain process, at

FIGURE 10-5

Data flow diagram.
The new system has some manual procedures that are similar to the old system. But the initial purchase order and invoice processing, shown here, places these records on the accounts payable database, where they are available for automated inquiries, vouchering, and reporting.

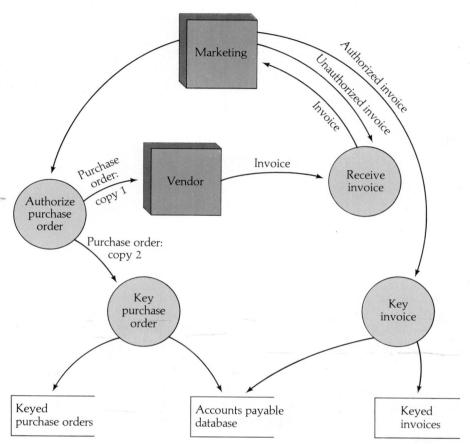

least some of the time. Data flow diagrams do not show data structure or hierarchy either. But they can be useful for expressing the first cut of the design of the new system. When used for design, initial data flow diagrams are often converted to some other design tool, such as a structure chart or system flowchart.

System Flowcharts =summarize DFD's ⟹ 1 box = entire program

The most enduring system design tool remains the system flowchart. A **system flowchart** is a drawing that shows the flow of control through the system. It represents the program processes and the files that are related to those processes. The system flowchart also shows the order in which processes are run if they are dependent on one another. For example, a system flowchart will illustrate that a program to edit batched transactions will be run before the program that uses the edited transactions to update the master file.

Before we describe the use of system flowcharts in more detail, note the difference between system flowcharts and data flow diagrams. Data flow diagrams depict the flow and transformation of data. System flowcharts stress the flow of control from one process to another. Data files from a design data flow diagram translate to data files on a system flowchart. In addition, several processes (bubbles) on a data flow diagram may be combined into a single program on a system flowchart.

The system flowchart is drawn using symbols approved by the American National Standards Institute (ANSI), as noted in Figure 10-6. These standardized symbol shapes represent certain actions and storage media. Note that a system flowchart is not the same as a logic flowchart. Some programmers are introduced to the logic flowchart to plan the detailed logic of programs. A system flowchart does not show any logic detail, only the "big picture" of the system. In fact, a single box represents an entire program.

Few formal rules exist for system flowcharts, but a few practices have become commonplace and accepted. These suggestions will make your system flowcharts more readable.

Process/File Relationships.   Some organizations draw their system flowcharts to connect processes (programs) to each other directly with arrows. That is, processes are not connected via a file. Recall that system flowcharts emphasize process flow, not data flow. For example, if one process produces a file that is used as input for the next process, you do not draw lines only from the first process to the file to the second process. This might only cause confusion if that file were subsequently eliminated from the system. An alternate approach is to connect the processes directly and have additional arrows for the relationships of the file with the processes.

Although most people are taught that logic flowcharts go top to bottom and left to right, system flowcharts usually go straight down, and in the center of the page. Files are on either side of the processes. Generally, input files are on the left and output files are on the right. If a file is output from one process and then used as input to a subsequent process, it is usually placed on the right, with appropriate arrows. This works fairly well if the part of the system represented on a page is small and has only a few files. If there are many files, however, this method becomes too complicated; at no time are crossed lines permitted. An alternative is to place all files strictly to the left for input and to the right for output. If a single file is first output from one process and then input to another, it is drawn twice, to the right of the first process and then to the left of the later process. As the file is output, it is labeled with the destination step. As the same file is input to another step, it is labeled with the step of origin. Figure 10-7 shows process/file relationships in three different ways.

Time Relationships.   A single drawing of processes should contain only the activities that follow each other in a time line. For example, a batch process that begins with the manual keying of data transactions, followed

FIGURE 10-6

ANSI system flow-charting symbols.

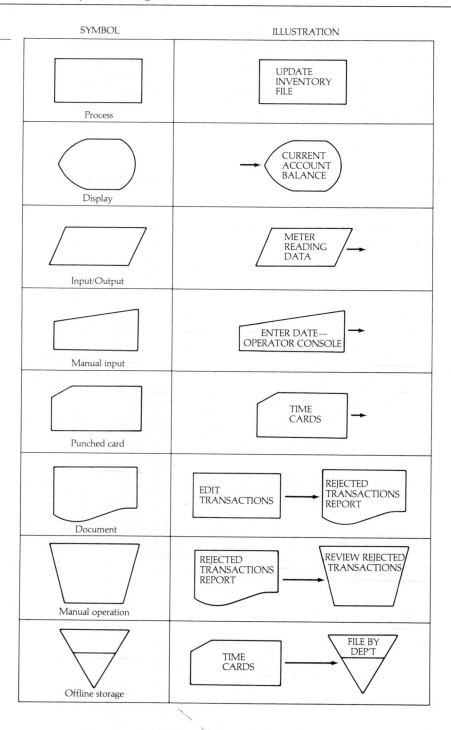

SYMBOL   ILLUSTRATION

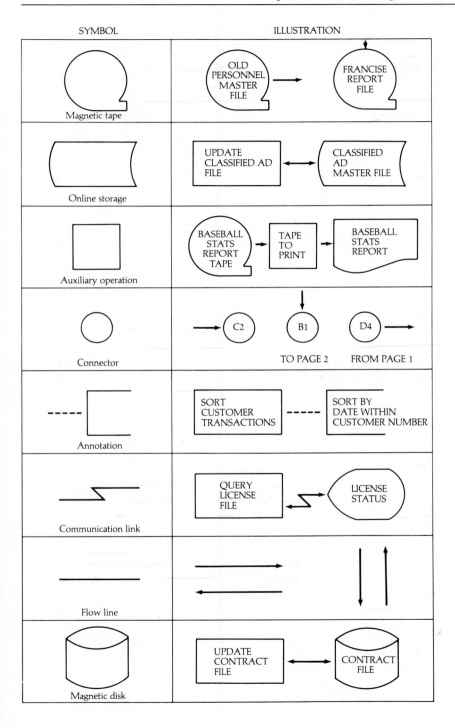

Symbol	
Magnetic tape	OLD PERSONNEL MASTER FILE → FRANCISE REPORT FILE
Online storage	UPDATE CLASSIFIED AD FILE ↔ CLASSIFIED AD MASTER FILE
Auxiliary operation	BASEBALL STATS REPORT TAPE → TAPE TO PRINT → BASEBALL STATS REPORT
Connector	→ C2 B1 D4 → TO PAGE 2 FROM PAGE 1
Annotation	SORT CUSTOMER TRANSACTIONS ----- SORT BY DATE WITHIN CUSTOMER NUMBER
Communication link	QUERY LICENSE FILE ↔ LICENSE STATUS
Flow line	→ ← ↓ ↑
Magnetic disk	UPDATE CONTRACT FILE ↔ CONTRACT FILE

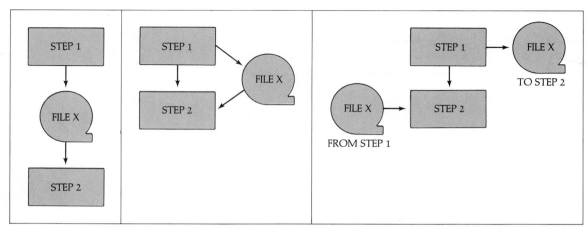

FIGURE 10-7

Process/file relationships. Three different ways to show process/file relationships.

by programs that, in one job, edit, update, and report, could be placed on a single system flowchart sheet. But, a system that updates a master file in a real-time environment would not be on the same sheet with the processes that draw scheduled periodic batch reports from that file. The updating process is going on constantly during business hours, but the reports are produced only occasionally, and probably during the middle of the night. Analysts are tempted to put processes like these together on the same page because they use the same files.

It is appropriate at this point to reiterate that these rules are not hard and fast. Analysts can and do make exceptions. An example would be placing activities on the same drawing that are not synchronized timewise, but then compensating for this by carefully labeling the time differences. For instance, words like *Weekly* or *Online Update* could be written near the related processes.

Although logic flowcharts have fallen somewhat out of favor, system flowcharts remain in the mainstream of systems design. The reason for the diminished use of logic flowcharts is the difficulty of keeping them current. Each program change, sometimes rather small, theoretically means a new drawing. In fact, programmers have long joked that the logic flowchart is drawn after the program is all checked out. Since system flowcharts are not directly related to the program logic, they are less affected by program changes and thus subject to fewer revisions. (See Figures 10-8 and 10-9.)

Nassi–Shneiderman Charts

As a tool, Nassi–Shneiderman charts fall somewhere between process design and development. This technique is named for its originators, Isaac Nassi and Ben Shneiderman. A **Nassi–Shneiderman chart** is a drawing that uses

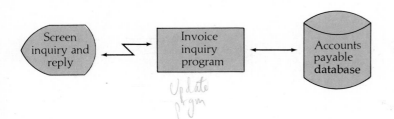

FIGURE 10-8

System flowchart.
Typical system flow-
chart for part of an
online system.

rectangles and triangles to depict process logic, as shown in Figure 10-10. These charts were originally introduced to enforce the three basic logic control structures of structured programming: sequence, selection, and iteration. Because Nassi–Shneiderman charts can show only these basic control structures, they are sometimes referred to as **structured flowcharts**.

Nassi–Shneiderman charts are very easy to read and follow. Note the example in Figure 10-11. They show the logic in a clear manner. Their construction, however, is quite complicated. Worse, their form leaves no room for adjustments of any kind. Any change means redrawing the chart. These restrictions make Nassi–Shneiderman charts less than desirable for volatile applications. However, automated tools do exist to draw and redraw Nassi–Shneiderman charts.

Nassi–Shneiderman charts are often supplemented by pseudocode.

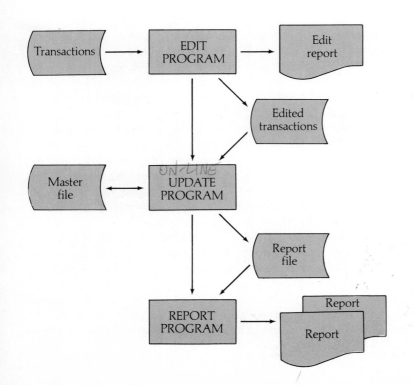

FIGURE 10-9

System flowchart.
Typical system flow-
chart for a batch
system.

FIGURE 10-10

Nassi–Shneiderman symbols. General shapes. Sizes and ratio of height to width may vary.

SEQUENCE

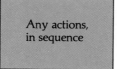

ITERATION

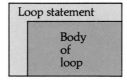

SELECTION

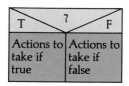

Pseudocode

The computer industry has embraced pseudocode in recent years because it has distinct advantages. **Pseudocode** is an English-like way of expressing design that is more precise than the English language but less precise than a programming language. The key advantage of pseudocode is that design ideas and logic can be expressed without worry as to the formal syntax of a programming language.

Pseudocode is used in different ways by different people. Pseudocode can be used quite early in detail design processing as a tool to outline processes in a general way. These early descriptions list files read and written, and reference the major activities of the process. At this point some analysts may turn the pseudocode over to the programmers for more detail work. In other words, the detailed pseudocode of program logic will be prepared by the programmer in the development phase.

Other analysts, however, work in a different kind of environment. Due to company policy or programmer experience level or analyst preference, it may be the analyst who prepares detailed pseudocode and turns it over to the programmer for translation to a programming language. It can be argued that well-planned, detailed pseudocode can go a long way toward keeping programmers on the correct path and on schedule in the development phase. It can also be argued, however, that the analyst has stepped over the line into programmer territory.

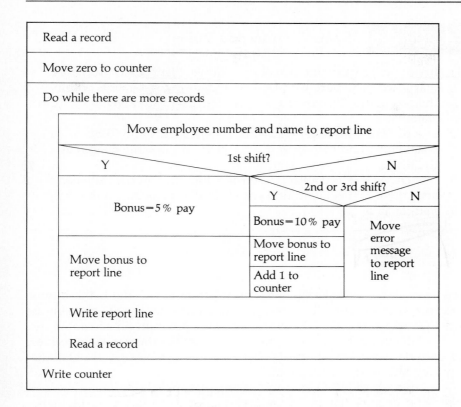

FIGURE 10-11

Nassi–Shneiderman example. Let us say that company management decides to distribute bonuses to employees. Assume a file already exists that contains employee data and a computation of regular pay for this pay period. First-shift people are to be awarded 5% of their pay as a bonus, while the other employees get a 10% bonus. Management also wants a count of the employees not on first shift.

Another popular use of pseudocode is as a supplement to other design tools. Structure charts, as we noted earlier, contain no sequence or logic instructions. This information can be supplied by pseudocode.

Pseudocode is easy to maintain. Since it is text material, pseudocode can be written with a text editor or a word processor. A change to the pseudocode is simplicity itself compared to changing a drawing. This factor alone is a powerful influence on choice of processing design tool.

HIPO

HIPO stands for **hierarchy plus input-process-output**. HIPO consists of a set of diagrams that describes system functions from a general level to a detailed level. Developed by IBM, HIPO is used for both design and documentation. As Figure 10-12 shows, it is a visual tool.

HIPO supports top-down development. The first of the three types of diagrams is called a "visual table of contents," or VTOC. The VTOC presents a structure chart along with a short description and a legend explaining what the arrows on the overview and detail documents mean. The second

BOX 10-1	Some Pseudocode Rules

1. Break the program into small procedures. Name each procedure. Start each procedure with BEGIN. Indent the statements within the procedure. Terminate each procedure with an END statement.
2. Use IF-THEN-ELSE for decisions. Indent THEN and ELSE statements. End each IF with an ENDIF, in the same margin as the IF.
3. Use DOWHILE or DOUNTIL for iteration. Indent the statements after the DO statement. End each DO with ENDDO, in the same margin as the DO.

document, the overview diagram, shows, from left to right, the inputs, processes, and outputs. The overview diagram—there will probably be more than one—describes the processes of the programs in general terms. The third diagram, the detail diagram, also describes the inputs, processes, and outputs, but in much greater detail. There can be several detail diagrams for each overview diagram.

The strength of HIPO as a tool is that it supports different levels of design and documentation. Initial design efforts can focus on the higher level drawings. These high-level drawings can be understood by managers and users. A programmer needing more detailed information can use the detail diagrams. The HIPO visual package has become a popular tool for designing systems and programs within systems.

The design processing techniques discussed so far can apply, generally speaking, to any type of system. Systems that process data online, however, require special attention in several areas. These online issues are so pervasive that they deserve a separate section.

◇ ONLINE SYSTEMS OVERVIEW

Most systems being developed today are online systems. Computers are being used for all kinds of applications where direct communication with the computer is essential. Most of these systems operate in real-time, reacting to input data and giving users immediate information. Banks, retail stores, and the airlines are typical users of online, real-time systems. These systems have special considerations over and above those needed for batch

BOX 10-2	Pseudocode Example

```
Start program
Open file
Read a record
Move zero to counter
DOWHILE there are more records
        Move employee number and name to report line
        IF first shift THEN
            bonus = regular pay × .05
            move bonus to report line
        ELSE
            IF second shift or third shift THEN
                bonus = regular pay × .10
                move bonus to report line
                add 1 to counter
            ELSE
                move error message to report line
            ENDIF
        ENDIF
        Write report line
        Read a record
ENDDO
Write counter
Close file
End program
```

This pseudocode describes the logic for the employee bonus problem that was illustrated in the Nassi–Shneiderman chart in Figure 10-11.

systems, which can be run in the middle of the night without direct user intervention.

Users need to have their terminals connected to the computer. Some people want their microcomputers to be connected to each other. Big computers, medium computers, and small computers can be connected in various configurations. Everyone seems to want to communicate and to use their computers to do the communicating. Since the communicating is often over a wide geographical area, data communications become a system consideration. We shall first examine data communications and then consider special factors related to online systems.

FIGURE 10-12

A HIPO package.
The HIPO package consists of a visual table of contents, one or more overview diagrams, and (usually) several detail diagrams.

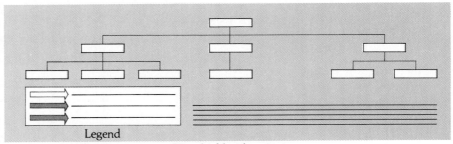

Visual table of contents

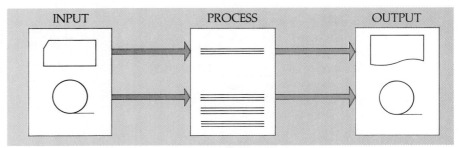

Overview diagram

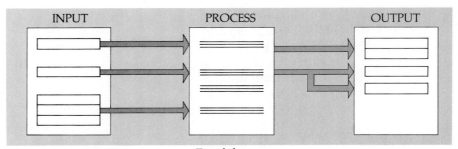

Detail diagram

DATA COMMUNICATIONS

It is possible for an online system to have all its components in one contained physical area, with the components connected to each other by direct wiring. Such a system is said to be **hard-wired**. It is more likely, however, that components of the system will be geographically dispersed, in ways that suit the organization being served. Many businesses and government agencies, for example, have a central headquarters office and many remote branch offices, all of which need computing power (see Figure 10-13). Since these systems cannot be hard-wired, they are connected through some type of communication channel.

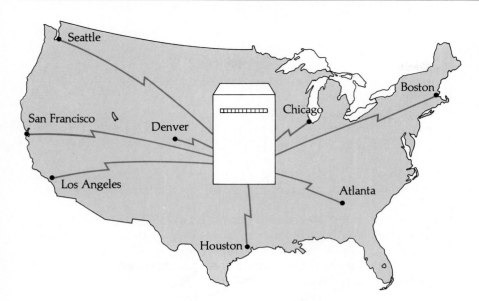

FIGURE 10-13

Online system. Users in different geographical locations can access computing power over data communication lines.

Data Transmission

Data communications means using computers to send data over communication channels. Most long-distance communications between computer devices occurs over telephone transmission lines. The problem is that most telephone lines are designed to carry continuous **analog signals** rather than the pulses of **digital signals** from computing devices (Figure 10-14). Thus, when transmitting over telephone lines, it is necessary to convert data from digital to analog signals, and then back to digital code again at the receiving end. Conversion from digital to analog is called **modulation**, while the reverse is called **demodulation**. Devices that handle these functions are called **modems** (Figure 10-15). The word *modem* is an abbreviated version of *mod*ulate/*dem*odulate. Some organizations use private digital lines that need no signal conversion.

Technical details of data transmission, such as types of modulation, types of transmission lines, and bit rate per second are beyond the scope of this book. This information, as well as various service options, is available from the vendors and data communications experts with whom you coordinate. However, you do need general knowledge of distributed systems, so let us proceed with that topic.

Distributed Systems

Some online systems use a single centralized computer for all processing, hooking up terminals from remote locations. Accessing centralized computing power from remote locations is called **teleprocessing**. The first airline

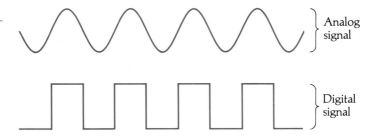

FIGURE 10-14

Analog and digital signals. Most long-distance communication between computer devices is done over telephone transmission lines, which normally use analog signals. Computer equipment, however, uses digital signals.

reservation system, developed by American Airlines, was a teleprocessing system, with hundreds of reservation terminals communicating with a central computer.

A more common system configuration is one that includes processing at several locations. A **distributed data processing** system is a decentralized data communications system that includes processing in remote locations. All locations, called **nodes**, in a distributed system can capture and process data. The processing, of course, is done by computers. The remote locations often have microcomputers, but may have minicomputers or even mainframes. For example, a fast-food chain could have microcomputers in each franchise location. Local processing can track supplies, monitor customer sales, and produce reports for the on-site manager. That same computer can send data to the headquarters computer for financial reports and for reports combining data from the other franchises.

Line configurations between a computer and its terminals can be point-to-point or multidrop. A **point-to-point line configuration** is one that connects each terminal directly to the computer (Figure 10-16). A **multidrop line configuration** connects several terminals to one line, which is then connected to the computer; that is, several terminals share the same line (Figure 10-17). Terminals connected to a computer in the same physical location can be connected point-to-point. If a teleprocessing system, however, connected six terminals in Tulsa to the main computer in Houston, it would make no sense to send data over six separate lines. Instead, a multidrop system could use a single line to send the data for all the terminals.

Distributed data processing systems can be designed in a variety of network configurations. Common network configurations are star, ring,

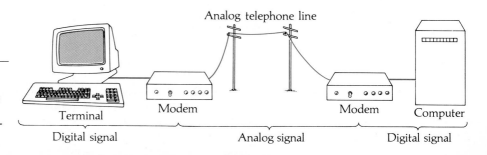

FIGURE 10-15

Converting digital and analog signals. Modems are devices that convert digital signals to analog signals and vice versa.

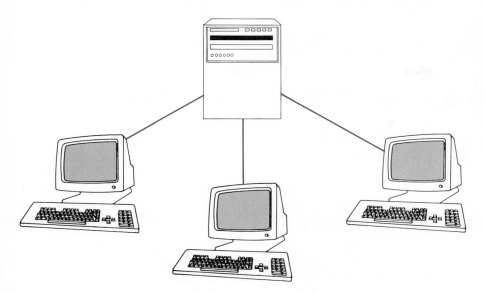

FIGURE 10-16

Point-to-point line configuration. Each peripheral device, such as these terminals, is connected directly to the computer.

and hierarchical. A **star network** has a central computer that is connected to several subordinate computers at remote locations. The central computer is called the **host** computer (Figure 10-18). Typically, the local sites use their own computer for most of their processing but turn to the host to submit summary information or to access sophisticated databases. A variation on the star network is the multistar network, as shown in Figure 10-19. In a **multistar network**, several host computers are connected, but each has its own star network of computers. For example, a major bank forms a star network with its own branch office banks. But this bank also is connected with other major banking institutions, which have star networks of their own.

A **ring network** is a circle of connected computers with no host computer. The computers in the ring transmit data to one another in a pre-planned circular fashion. As Figure 10-20 shows, an appropriate application for a ring network is an organization of several businesses or agencies that need to share information.

A **hierarchical network**, as you can see in Figure 10-21, looks something like the star network, with local computers attached to a central host computer. However, in the hierarchical network, these computers may also play host to smaller computers. For example, the central computer may be a mainframe, and the next layer may be minicomputers that access the mainframe database to assist the lower level microcomputers.

A popular and growing type of network is a local area network. A **local area network** (LAN) is a medium for sharing hardware and information. Since the computers in a local area network are usually microcomputers, this topic will be discussed in Chapter 16.

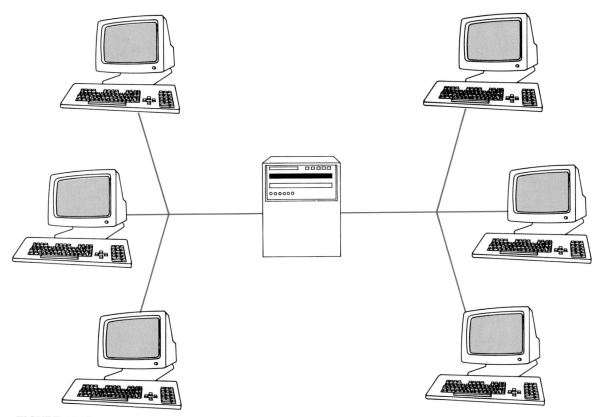

FIGURE 10-17

Multidrop line configuration. Several devices can be connected to the computer through one line.

There are many advantages of distributed systems. Workers at the local site have easy access to computing power and fast response time. Furthermore, they can tailor applications to their own needs. Local processing also reduces the load on the host computer, which is often overburdened. The disadvantages of distributed data processing are the software and memory limitations of the smaller machines, lack of on-site expertise at the local level, and compatibility problems. The last can be significant.

The Compatibility Problem

Few organizations today have only one computer, and those that have more than one may have a compatiblity problem. In many cases, the hardware and communications software are different and incompatible. Although there is talk in the industry of standardization, this is not yet reality. Nonstandardization is a key problem. Multivendor installations complicate the compatibility problem. The vendors themselves are sometimes not anxious to assist in making their hardware compatible with other hardware you may have, since it is counterproductive to their purposes. They would prefer,

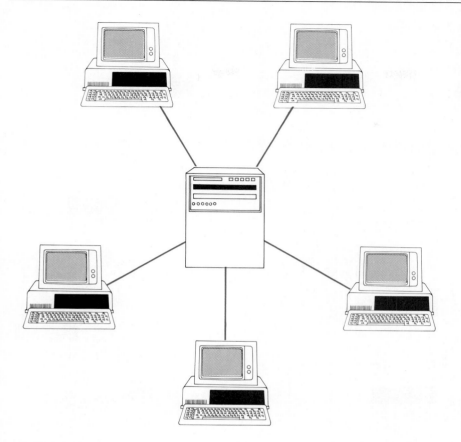

FIGURE 10-18

Star network. Micro-
computers are con-
nected to a minicom-
puter, which is the
network server.

of course, that you be locked in to their system. An alternative is the black
box syndrome, a euphemism for custom-made compatible connections. Such
connections take considerable time in meetings with vendors and representatives from the phone company or other carrier.

No one expects you, as the analyst, to possess technical expertise on
this important but peripheral topic. But you need to be aware of the problems, recognize your need for assistance, and plan your schedule to accommodate those needs.

◇ PROCESSING CONSIDERATIONS FOR ONLINE SYSTEMS

The special considerations for online systems fall into these categories: online
programming, online files, online system controls, and online system performance.

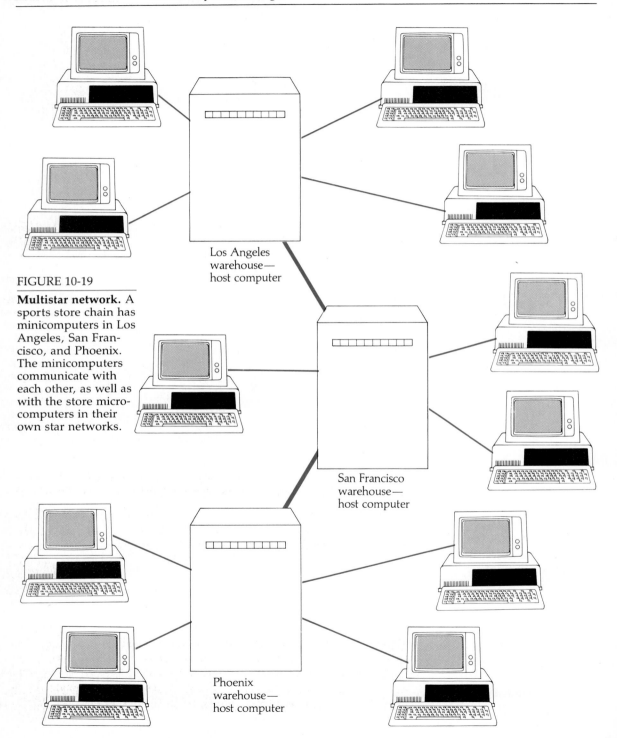

FIGURE 10-19

Multistar network. A sports store chain has minicomputers in Los Angeles, San Francisco, and Phoenix. The minicomputers communicate with each other, as well as with the store microcomputers in their own star networks.

Los Angeles warehouse— host computer

San Francisco warehouse— host computer

Phoenix warehouse— host computer

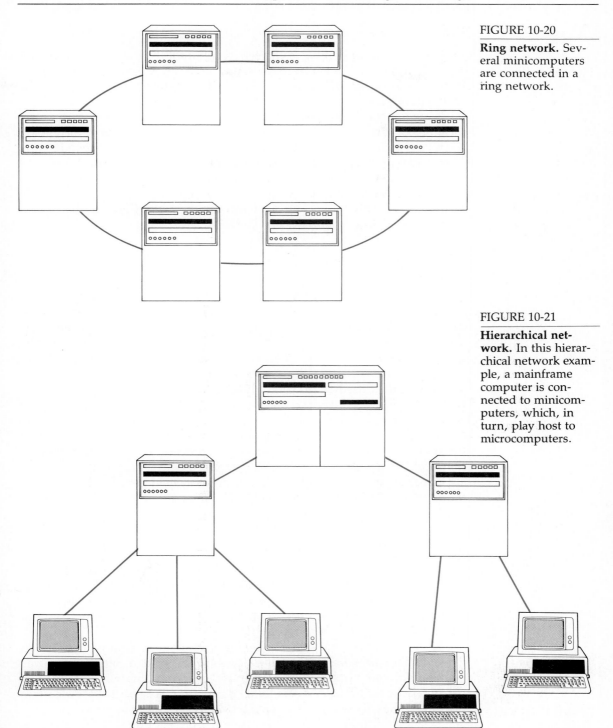

FIGURE 10-20

Ring network. Several minicomputers are connected in a ring network.

FIGURE 10-21

Hierarchical network. In this hierarchical network example, a mainframe computer is connected to minicomputers, which, in turn, play host to microcomputers.

Online Programming

Online programming involves screen generation, the display of output on a terminal screen. The software that must be in place for screen generation is beyond the scope of this book, but it is worthwhile to mention this topic in a general way.

Screen generation programs must be able to accept input from different locations on the screen. This involves controlling the location of the cursor, a process that can be quite complicated. Cursor control demands extra knowledge and extra coding. Also, cursor control varies with the type of terminals used. Some terminals are designed specifically to handle screen generation; the terminals that are not so designed present further coding complications. This entire subject is hardware-dependent, and can be learned best in the environment that uses the hardware.

Online Files

Files used in online systems must be on disk, rather than on tape. There are two reasons for this: speed and direct access. File organization must be direct or some compromise method that includes direct access as one of the access modes. Many systems, however, are database-oriented, not file-oriented. We shall distinguish between file and database processing in Chapter 11.

Transactions used to update online files must be validated as they are received. You would want to make sure, for example, that an airline agent sending a reservation transaction has keyed in a flight number that exists.

A system may be suitable for deferred online processsing if instant updating is not essential. **Deferred online processing** means that transactions can be accepted in real time but processed later. An example is a supermarket point-of-sale system, whose system-generated inventory transactions can be processed in a batch at a later—deferred—time.

Online System Controls

The online control mechanism that is most obvious to users is **identification** and **authorization**. Each user must be identified to the system, usually through an account number that can be verified against a stored table of legitimate account numbers. In addition to identification, each user must present authorization, usually in the form of a known password.

Passwords on online systems are often multilevel. That is, in addition to the initial password that permits access to the system, other passwords may be required to permit specific functions. For example, a user may be permitted access to the system but need a second password to view certain privileged files. Or a user may be allowed to see the main menu but may need further authorization to select the file update option.

Another approach is to permit initial access to the system with different types of passwords, essentially "owner" and "nonowner" passwords. Then each file can be labeled with the activities permitted for owners and non-owners. For example, normally an owner can do anything, including changing the passwords, but a nonowner typically is limited to retrieval only.

In an online system, it is conceivable that transactions may be lost. Therefore, it is necessary that the sender receive confirmation that the transaction was received. For audit purposes, all online transactions must be **time-stamped**, that is, the transaction generation time, as computed by the computer's internal clock, automatically is placed as a field on the transaction. Each transaction also must be labeled with its source, that is, the identification of the sender. This identification includes both the terminal and the identification of the person.

If transactions update files as they are generated, then a safety-net approach is to keep an **image file**, a file of before and after images of the master file record updated. Also, copies of the transactions are kept in **journal files**. The image and journal files can be used to reconstruct the file activity if the master is lost or improperly updated. For this reason, these files are sometimes called **recovery files**.

Online System Performance

When a system is being used by people waiting for answers, and possibly with customers waiting for service, performance becomes critical. Each aspect of an online system's performance must be regularly monitored.

Online response time is an obvious factor to be measured. **Response time** is the time that elapses from when a user request is made at a terminal and the reply is received. When you are waiting for a response, five seconds is too long. If the system is bogged down with heavy activity, the delays can become intolerable. Responding to a chorus of complaints is not enough. The system must be monitored regularly. Disposition of a problem can be anything from adding more memory to scattering usage to times other than peak periods. If the solution is technical, you may need to deal with online systems specialists to determine the best course.

An issue directly related to online response time is volume. Performance criteria to watch are average volume, peak volume, and expected volume growth. Software packages are available to monitor these statistics. This information will provide guidelines for future changes in the system configuration, so that slow response time can be anticipated and avoided.

◇ QUALITY ASSURANCE PLANNING

Throughout the development of any system, whether online or not, a quality product must be planned and assured. The concept of quality assurance has been developed to meet that need.

| BOX 10-3 | **Quality Assurance: Now We'll Listen to You** |

Every manager wants systems of high quality. Setting up an organization called the Quality Assurance Group, or some similar name, is a step in the right direction. But more is required for success—that is, for a successful working relationship with project personnel. The key factor is the way the group is perceived by those who interact with it. The following factors can have an important influence on positive perceptions of the quality assurance group:

- The first factor is *organizational hierarchy*. Or, to put it another way, is the quality assurance group located high enough in the organizational scheme of things to indicate management support? People tend to form negative perceptions when they are asked to do something that they see as a burden to themselves but sense that the company is not committed.
- The quality assurance group *staff* is the second key factor. Managers tend to keep their best people in the slots where the action is, shunting more secondary people to staff positions. But the quality group must be staffed with relatively senior people and, most especially, people who have a history of success and the clear respect of their peers.
- Another aspect of the quality assurance group that shapes perceptions is the *training program*. To provide a solid foundation on which solid positive perceptions can be based, an extensive training program must be put in place.
- The *processes* by which the group helps project personnel build quality products must not aggravate the resistance that normally occurs when people feel that others are assessing their work.
- The last issue is *joint responsibility*. It should be made clear that, even though the project manager is the person who manages the effort, the quality assurance group is jointly responsible with the project manager to make sure that all quality issues are addressed.

Much has been written in the computer industry trade press about quality assurance, and people are writing entire books about it. It is not a new topic, but the current emphasis is new. Although the discipline called quality assurance is relatively young, software problems have plagued the industry for too long.

The history of software has been marked by late schedules and budget overruns. Even worse, the completed project often has failed to meet user specifications and needs. In fact, some programs have seemed failure-prone, breaking down with alarming regularity.

But help is on the way. **Quality assurance** is a discipline that includes a set of tools and techniques to ensure that software is reliable and meets certain expectations. These techniques include setting software standards, planning controls, establishing testing techniques, searching for software defects, and others.

How is quality assurance infused into the systems process? The methods are variable and sometimes random, depending on who happens to know what, who is on the project at a given time, and what kind of political clout is carried by someone espousing quality assurance. Many companies take a formal approach, setting up a quality assurance group and giving these people a charter to establish and enforce standards on anyone creating software. Results, unfortunately, often fall short of expectations. The reasons for this are many, but they usually are related to the lack of company commitment and the natural resistance to outsiders passing judgments on one's work.

◇ DETAIL PROCESSING WRAPUP

The question that students most often ask when they come to the detail design stage is "How do I know what to do?" Some are looking for a formula, a predictable step-by-step design process. There is no formula. In fact, if there were such a formula, analysts would not be in demand. Systems could be designed and developed routinely by less qualified personnel.

But analysts are in demand. Analysts are needed for their judgment. Every system is different. This book and others like it can offer only suggestions and guidelines. It is up to the analyst at the scene to apply techniques that are appropriate to the system at hand.

So the question still remains, "How do I know what to do?" You read, you study the system, you discuss with your teammates, you ask a lot of questions, you decide what fits this particular system. Gradually, some assurance will develop from the confusion. With each new system, your expertise and confidence will grow. It takes time.

FROM THE REAL WORLD

Name, Job Title, Blood Type

Even when the role of personnel departments was more narrowly defined, the work of collecting and collating employee data was a time-consuming task. When mainframe computer power became available to corporations, one of the first tasks assigned to the machines was the keeping of personnel records. Personnel software was written many times over in the early days of business computer use, and managers turned their attention to other matters.

But now new roles, along with new personnel software, are being carved out for personnel professionals. A manager of a cosmetics firm, for example, insists on recording the blood type of each employee. He reasons that, if there were ever a community disaster, they would be all set up to cross-match donors. But this interesting fact is but a footnote on the personnel scene. New systems have been developed to respond to one or more of these needs:

Employee services. Whether honoring employees for longevity or alerting them to in-house job opportunities, the process must begin with data collection. Personnel systems are long past the standard fare of biographical and historical data. Additional data on employees now includes items such as employee counseling and quality of work life.

Combining payroll and personnel. In many cases, such as new hires or changes in employee status, the data that payroll staffers key into the system is also entered into personnel records. It makes no sense to maintain two clerical teams to record the same facts. Software in use today often combines the two functions.

Cost tracking. Since labor is a key cost ingredient, it makes sense to use personnel records to help track costs. Typical built-in tracking includes the costs of hires and transfers, benefit liabilities, labor cost projections, and tabulations of department costs by job classification. In fact, personnel records are now used more for cost decisions than for rudimentary tallying.

All of these can be used to provide management with better visibility of the business and quicker access to information about the personnel who support it.

SUMMARY

- Possible processing operations include read, compare, edit, move, sort, calculate, retrieve, update, summarize, classify, store, and write.

- A design methodology is a set of techniques based on a concept. Design methodologies were introduced in the 1970s because of an industry history of projects that were overbudget and behind schedule. They were introduced through the trade press and classes for industry. It has become clear that no one methodology is an asset in every design problem, but some methodologies are appropriate for some types of design problems.

- Structured design methodology is based on the concept of data flow, and takes a top-down modular approach. Structured design focuses on the structure chart, or hierarchy chart, and modularity.

- A module is a set of logically related statements that perform a specific function. A module must have a single entry and a single exit. A module should have limited data access—that is, access only to the data it needs to perform its function.

- The relationship between modules, which should be weak, is called coupling. A module has maximum cohesion when it has a single, well-defined function. A module that is highly functional is said to have modular strength.

- The Logical Construction of Programs (LCP) methodology is based on the concept of data structure. A graphic illustration of the LCP method is called a Warnier–Orr diagram. Warnier–Orr diagrams begin with the output data and work backward, right to left, to determine processes and input data that produce the output.

- An analyst may have a choice of methodologies, or may have to conform to an organization standard, or may use a combination of techniques, depending on the application.

- Data flow diagrams (DFDs), used as an analysis tool, can also be used as a design tool. DFDs are sometimes used as the first design tool and then converted to another form, such as structure chart or system flowchart.

- A system flowchart is a drawing that shows the flow of control through a system. It represents the program processes and the files used by those processes. System flowcharts are drawn with symbols approved by the American National Standards Institute (ANSI).

- Generally speaking, system flowcharts are drawn top to bottom, with input files on the left and output files on the right, are connected process-to-process, and are drawn with other processes in the same time frame.

- Nassi–Shneiderman charts (sometimes referred to as structured flowcharts) use rectangles and triangles to enforce using the three control structures—sequence, selection, and iteration—in the expression of the design.

- Pseudocode is an English-like way of expressing design that is more precise than English but less precise than a programming language. Pseudocode can be maintained easily on a text editor or word processor.

- HIPO (hierarchy plus input-process-output) consists of a set of diagrams that describes system functions from a general level to a detailed level.

- An online system whose components are physically wired together is said to be hard-wired. Systems whose components are geographically dispersed are connected through some type of communication channel.

- Data communications means using computers to send data over communication channels, usually telephone lines. Systems use modems to convert computer system digital signals to the analog signals required by telephone lines (modulation), and vice-versa (demodulation).

- Accessing centralized computing power from a remote location is called teleprocessing. A distributed data processing system is a decentralized data communications system that includes processing in remote locations. Each processing location in the system is called a node.

- A point-to-point line configuration connects each terminal directly to the computer. A multidrop line configuration connects several terminals to one line, which is then connected to the computer.

- A star network has a central computer, called the host, that is connected to several subordinate computers at remote locations. A multistar network connects two or more star networks through their host computers. A ring network is a circle of connected computers with no host computer. A hierarchical network connects a central host to lower level computers that may themselves be host for smaller computers. A local area network (LAN) is a medium for sharing hardware, usually microcomputers, and information.

- The advantages of distributed data processing systems are easy access, fast response, tailored applications, and reduced load on the host. The disadvantages of a distributed system are software and memory limitations, lack of on-site expertise, and compatibility problems.

- In a distributed system, hardware and communications software may be different and incompatible, especially if there are many vendors. Custom-made connections may be necessary.

- Online programming involves screen generation, which means extra knowledge and extra coding. Online files must be on disk. Transactions to update online files must be validated as received. Deferred online processing means that transactions can be accepted in real time but processed later.

- Online system controls include identification, probably an account number, and authorization, usually a password. Each transaction to an online system must be time-stamped with the transaction's generation time, and labeled with the identification of the sender. If transactions update files as they are generated, an image file of before and after master file records is kept. Also, copies of the transactions are kept in journal files. These files are called recovery files. Online performance must be monitored regularly. In particular, response time, the time elapsed between user request and system reply, must be minimized.

- Quality assurance is a discipline that includes a set of tools and techniques to make certain that software is reliable and meets certain standards. These techniques include setting software standards, planning controls, establishing testing techniques, and searching for software defects.

KEY TERMS

Analog signal	Module
Cohesion	Modular strength
Coupling	Modulation
Data communications	Multidrop line configuration
Data flow diagram (DFD)	Multistar network
Deferred online processing	Nassi–Shneiderman chart
Demodulation	Point-to-point line
Design methodology	configuration
Digital signal	Pseudocode
Distributed data processing	Quality assurance
Hard-wired	Recovery file
Hierarchical network	Response time
Hierarchy chart	Ring network
HIPO (hierarchy plus input-	Single entry
process-output)	Single exit
Host	Star network
Image file	Structure chart
Journal file	Structured design
Local area network (LAN)	System flowchart
Logical Construction of	Teleprocessing
Programs (LCP)	Time-stamped
Modem	Top-down approach

REVIEW QUESTIONS

1. Describe the industry background that led to the development of design methodologies.

2. Why has no single methodology become a standard for design, as structured programming has become an unofficial industry standard for programming?

3. Compare the two methodologies described in the chapter. What people originated them? What are their underlying concepts? Describe their documentation tools.

4. How do coupling and cohesion complement each other?

5. What is meant by modular strength?

6. Describe the notation for Warnier–Orr diagrams.

7. Draw or describe the ANSI symbols for the following: display, document, annotation, process, magnetic disk, offline storage.

8. What is the alternate name for Nassi–Shneiderman charts? What is their key characteristic? Draw a Nassi–Shneiderman chart to find and print the average current charge balance for all department store credit customers whose zip code is 98008 or 98006.

9. What special advantage does pseudocode have? Under what circumstances might it be used, and by whom? What other design tools might pseudocode supplement?

10. What does HIPO stand for? Describe the three types of HIPO diagrams.

11. Which of the design tools described in the chapter seem easiest to maintain? Which seem the most difficult?

12. Do you have a preference for one or more of these design tools? For what reasons?

13. Differentiate between a teleprocessing system and a distributed data processing system. Give a business example of each.

14. Personal computers use modems and communications software to hook up to networks of other personal computer users. Check out modems and software at your local computer store. What networking services are available, and what kinds of costs are involved? What services do they offer?

15. For the summer tourist season, two hotel chains have agreed to share reservation information and steer customers to each other through their computer systems, which includes a computer in each hotel. The Harley Hotel chain headquarters office and mainframe computer is in Denver, with hotels in San Francisco, San Antonio, Dallas, Houston, Seattle, Sacramento, Los Angeles, Minneapolis, Salt Lake City, and Kansas City. The Camalla Inn chain has headquarters and its main computer in Chicago, with hotels in St. Louis, Tulsa, Phoenix, Tucson, Portland, San Francisco, Dallas, and Oklahoma City. What kind of a network configuration do they form? Sketch a drawing.

CASE 10-1
MAKING THE NEW FROM THE OLD

In the systems analysis class at Goodrich College, the classes were divided into teams for project work. They were presented with a situation to which they could apply the tools and techniques of the systems life cycle. The team of Henry Fremont, Ellen Wallingford, Chuck Corliss, and Paul Ravenna worked on a mail order inventory system.

In the analysis phase, they used data flow diagrams as their principal tool to describe the existing system. The DFDs worked well to follow the paper flow from order entry to order processing to warehouse to shipping. As they studied the existing system, they noted several problems. For example, it was common for orders sent to the warehouse to be returned to order processing if items ordered were out of stock. The paper flow system, all manual, was slow and unreliable.

The student team wanted to design a new online system that would reduce out-of-stock situations and reduce paper flow. They planned to place terminals in each of the departments so that each of the user groups—order entry, processing, warehouse, and shipping—could query and update the inventory file. The team felt that, since the basic functions were still the same, the same data flow diagrams that were used for analysis could be used as part of the design description. Their instructor intervened, observing that the same diagrams would not work for the two purposes.

1. There is definitely a problem here. Why will the original diagrams not work for the design?
2. What have the team members misunderstood about the use of data flow diagrams?
3. Can new data flow diagrams be used for the design? Would some other design tool be preferable for this application? Discuss.

CASE 10-2
CATS AND DOGS IN DETAIL

The Meridian Veterinary Clinic is a large enterprise with branches throughout the Los Angeles area. The clinic hired Burke Enterprises, a service bureau, to introduce a computer system to manage the animal records. Jim Wiggin and Ed Flores were the analysts assigned to the project. They studied the existing system in depth and were amazed to discover the amount of detail kept on animal patients.

Among other things, the veterinarians had information on each animal's date of birth, lineage, markings, shots, medications, allergies, surgery history, dates of stay, source of funding, owner, and much more.

Jim and Ed proposed a preliminary design plan for a host minicomputer and network of supporting microcomputers, together with query capabilities and batch reports. They described the preliminary design in detailed narrative form. In the detail design phase, they continued to be very specific and detailed, choosing design tools that reflected their approach. The heart of their documentation was an extensive set of Nassi–Shneiderman charts. These were supported by several decision tables. Jim and Ed felt that the programmers could code quite readily from these specifications.

As it turned out, no one could use their specifications very well. The users could not follow the documentation and neither could the programmers.

1. What is the problem here? Is different documentation needed? Additional documentation? Discuss.
2. Why do you suppose the programmers could not work from these detailed specifications?

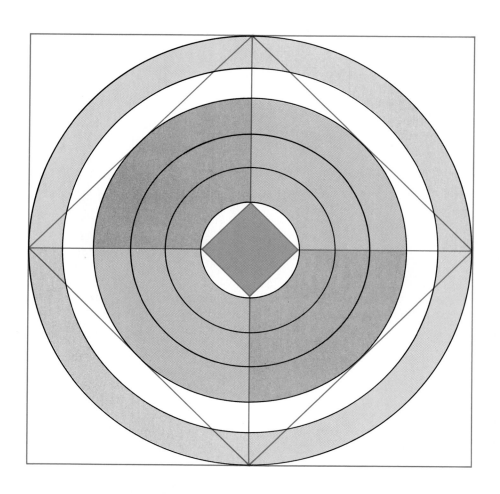

Chapter 11

Detail Design: File Processing and Databases

The student analyst team of Jill Jefferson, Al Seneca, and Todd Pike was assigned the task of designing a system to track member billing at a country club. As they gathered background information, the three noted a repeated theme: that many members were busy executives who sometimes wanted instant answers about their accounts at the club. Managers therefore wanted access to individual member records from terminals in their offices. The team planned a direct file organization for the member master file.

Many other tasks captured the team's attention in the next few weeks. When they turned in their project document, their instructor noted the direct master file but also the daily reports, in member order, that were drawn from the file. This discrepancy was brought to the attention of the students, who first considered sorting the file every day into member number order for reporting, but abandoned that approach as economically unsound. They finally decided to consider another type of file organization or even a database.

File decisions are among the most confusing for beginning analysts. In this chapter we will explore some of the problems associated with file and database planning.

◇ SECONDARY STORAGE

Before we begin the discussion of file processing, let us review secondary storage devices. Our discussion will focus on magnetic tape and magnetic disk media. We shall briefly consider physical characteristics, data representation, fixed- and variable-length records, blocking, and access motions.

Magnetic Tape

Standard **magnetic tape** is $\frac{1}{2}$ inch wide and comes on a $10\frac{1}{2}$ inch diameter reel with 2,400 feet of tape. Small cassette tapes are available for some microcomputers. Magnetic tape has an iron oxide coating that can be magnetized. Data is stored as extremely small magnetized spots. The amount of data on a tape is expressed in terms of **density**, which is the number of characters per inch (cpi) or bytes per inch (bpi) that can be stored on a tape. Although some tapes can store as many as 6,250 bpi, a common density is 1,600 bpi.

One character of data can be represented on a cross section of tape. As you can see in Figure 11-1, a set of like positions in cross section is called a **track** or **channel**. Most tapes use nine binary digits to represent a character, one per channel. Each location has either a magnetized spot, to represent the 1 bit, or no magnetization, to represent the 0 bit. A common data representation code is **Extended Binary Coded Decimal Information Code (EBCDIC)**, which uses eight binary digits to represent a character. The ninth position is used as a parity bit to check the correctness of the data.

Records may be stored as either fixed length or variable length. The term **fixed length** means that all records on a file are the same length—that is, have the same number of characters. If the records on a file have different numbers of characters, they are **variable-length** records. Variable-length records are appropriate for an application that has records with a field that appears a variable number of times, such as one per customer.

A **logical record** is one written by the applications program. It is common practice to group several logical records together in one **physical record**, or **block**, to be written to the physical storage device. This process is called **blocking**, and the number of logical records in a physical record is called the **blocking factor**. The space between blocks, as noted in Figure 11-2, is called an **interblock gap**, or **IBG**. The advantages of blocking are that space is saved on the physical device and there are faster read/write times.

Magnetic Disk

A **magnetic disk** is a platter coated with ferrous oxide, which can be magnetized. A set of disks is called a **disk pack**. Each disk surface has concentric tracks, on which data can be written, in the form of magnetized spots.

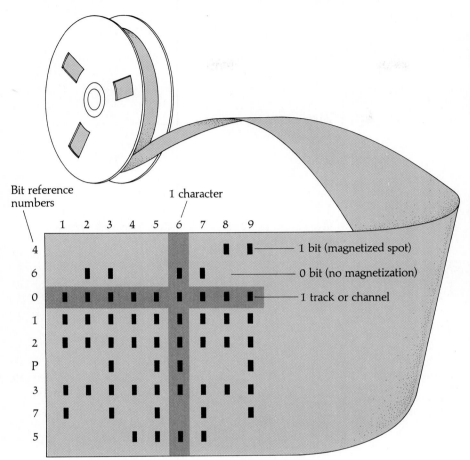

Bit reference numbers

1 character

4 — 1 bit (magnetized spot)

6 — 0 bit (no magnetization)

0 — 1 track or channel

1
2
P
3
7
5

FIGURE 11-1

How data is represented on magnetic tape. This shows how the numbers 1 through 9 are represented, using combinations of 1 bits and 0 bits. For each character there are eight bits. The ninth bit—represented here under the bit reference numbers by the letter P—is a parity bit. In the odd parity system illustrated here, each byte contains an odd number of 1 bits. In an odd parity system, an even number of 1 bits indicates that something is wrong with the data. Note that the parity-bit track appears in the middle of the tape.

Disks are stacked in a pack, but not like a stack of stereo records. (See Figure 11-3.) On a disk pack, there must be space between disks, so that the disk device access arms can get between the disks to read or write data.

Data is usually organized on disks vertically, by cylinders. A **cylinder** is a set of tracks accessible by one position of the access arms. Disk data is read or written with four **access motions**: (1) **seek time**, during which the access arms move into position on the correct cylinder, (2) **head switching** to activate a particular read/write head over a track in the cylinder, (3) **rotational delay** until the correct record on the track rotates into position, and (4) **data transfer**, the movement of data to or from the disk.

After this basic material, we can examine how to plan files to use on secondary storage devices.

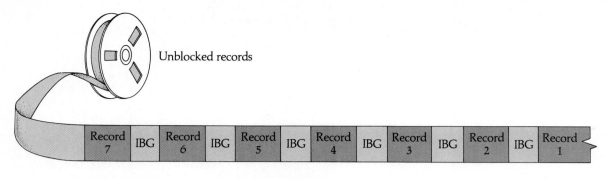

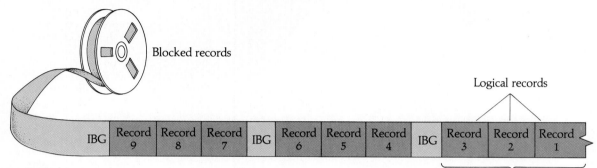

FIGURE 11-2

Unblocked and blocked records. In the unblocked records, each block is a physical and also a logical record. Each physical record (block) is separated from the next by an IBG. In the blocked records, three logical records are grouped into one physical record. This saves space, because of fewer IBGs, and increases processing speed.

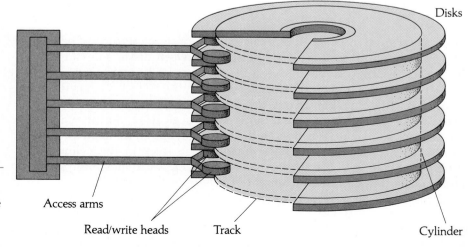

FIGURE 11-3

Magnetic disk. The read/write heads on the access arms move between the disks to read and write data.

FILE PROCESSING

The subject of data files is a complicated one, and one that is a key part of detailed processing. The way data files are planned is directly related to program logic. **File processing** is a generic term that refers to the planning of separate data files to match program logic. File processing is the traditional way to use data files, and is still appropriate in many systems. An alternative is database processing, which we shall discuss in a later section. We begin by considering the types of data files commonly used in systems.

Types of Files

Several different types of files are used in a system: master files, transaction files, table files. In addition to these general types, there may be a number of files holding data temporarily, for sorting or transporting or reporting purposes. A **master file** holds data records of a semipermanent nature; the file is updated as the status of the data changes. A **transaction file** contains records with new data for updating existing master files. A **table file** contains reference data that is relatively static.

As an example of file types, the state of Florida keeps a master file of all cars that have been reported stolen. Transactions add or delete cars from the file, or make changes to existing records. A reference table file lists license plate colors by state.

Let us turn now to the basic questions that need to be asked when planning data files.

Planning Data Files

An early consideration in planning data files is the file content. The records, and fields within each record, can be planned on a record layout chart. This chart defines the order of data fields within the record, and includes information such as field lengths and numeric and/or alphabetic status. Figure 11-4 shows a typical record layout form.

What medium shall be used for file storage? What organization and access method are suitable for a given file? There are several file characteristics to be considered, as noted in the following list. Each of these may affect file design decisions.

RECORD LAYOUT FORM

INVOICE NUMBER	INVOICE DATE			VENDOR NUMBER	PURCHASE ORDER NUMBER	DUE DATE			AMOUNT DUE
	MO	DA	YR			MO	DA	YR	
9 9 9 9 9 9 9 9	9 9	9 9	9 9	9 9 9 9 9 9 9 9 9	9 9 9 9 9	9 9	9 9	9 9	9 9 9 9 9 9 9
1 2 3 4 5 6 7 8	9 10	11 12	13 14	15 16 17 18 19 20 21 22 23	24 25 26 27 28	29 30	31 32	33 34	35 36 37 38 39 40 41

FIGURE 11-4

Record layout form. Blank record forms can be marked to plan data records.

Access.   The way the file will be used may dictate the medium and access method. You would not consider, for example, a sequential tape file for an airlines reservation system because individual records are not accessible. We shall discuss file organization and access at some length.

Volume.   The capacity of the medium must be directly related to the volume of the data. You need to plan ahead so the file does fit on the chosen medium.

Volatility.   A volatile file is one that has frequent additions and deletions of records, such as a file of classified advertisements for a newspaper. As we shall see, volatility has a direct effect on choice of file organization.

Activity.   An active file is a file that is used often or perhaps constantly. An active file may be used as a reference file. A good example is the file of phone numbers used by directory assistance. Also, the existing records of some active files may be updated constantly during processing. An inventory system will constantly update records as orders are filled and quantities reduced. Each update is referred to as a **hit**, and thus the percentage of records updated in a given time frame is called the **hit rate**. An active file may or may not be a volatile file.

Cost.   As always, cost is an important consideration. But the issue is not just the purchase costs of the tapes or disks or drives. The total cost includes the operating costs of the running system.

These file characteristics are related to the choice of file organization and access, and this fact will be so noted in the discussion in the next section. But the main factor affecting selection of file organization is the proposed use of the file.

File Organization and Access

Planning the organization and use of data files is often a source of confusion to novice analysts, even if they understand file organization and access methods. The problem is usually related to usage of the files. If the analyst is thinking in terms of an online system that queries and updates files one transaction at a time, then direct access of records is required. But the novice analyst may not realize right away that these same files will be used to produce reports, an activity that assumes a certain record sequence. There appears to be a conflict: The same file is needed in both direct and sequential modes.

Sometimes the opposite problem occurs. The novice analyst will be thinking in report mode and plan sequential files, only to realize at a later time that some kind of direct access is needed, too. There are several ways to resolve these problems. No one file organization method is ideal for every situation. In this section we shall discuss file organization methods and the trade-off in their use.

As a storage medium, magnetic tape is less satisfactory than disk. Its usage is minimal because files can be organized only in sequential form. But tapes are inexpensive and suitable for backup files. Tapes also are a convenient way to transport files.

The great majority of files are kept on magnetic disk. Access to disk files can be sequential or direct, or perhaps both, depending on the file organization. Magnetic disk is extremely dense, providing sufficient capacity for large-volume files. Diskettes, used with microcomputers, word processors, and some minicomputers, also can be accessed directly. Although the capacity of diskettes is limited, they are inexpensive and conveniently portable.

Files can be organized in a variety of ways. Access to files depends on the file organization method. A file organized sequentially can be accessed only sequentially, and a file organized directly can be accessed only directly. Other organization methods, however, notably indexed file organization, permit either sequential or direct access. Let us look at these three file organization methods more closely.

Sequential File Organization.   A sequential file is one whose records are in order in some sequence—that is, by key. A **key** is a field chosen to order the records. It makes sense, for example, to use social security number as the key for records in a personnel file or to use part number as the key for an inventory file. If a master file is to be kept in sequential order, then the transactions to update that file must also be in sequential order. This means that the transactions must be sorted before they can be used to update the master. Sequential file organization is very efficient for files that are highly volatile and/or highly active. Either situation, or both, are easy to process. But the files have to be updated by a batch process. It is not reasonable to submit a single transaction to update a sequential file because the program would have to read all the records in the file that preceded the transaction key. This would be unacceptably inefficient.

Applications that are suitable for sequential batch processing are systems that are able to gather data and accept output in a relatively leisurely time frame, as opposed to systems whose customer service depends on instant results. Typical sequential batch applications are payroll, personnel, and general ledger.

Direct File Organization.   A direct file is one whose records are stored in such a way that access to any one of them is direct; that is, the record can be read or written without reading other records in the file first. A direct file must be on a direct access device, usually disk.

Records in a direct file are not placed one after another, as in sequential organization. Instead, a file area is defined on the disk and records may be placed in any one of the available record slots within that area. Any slot will do. But once a location for the record is chosen, you must be able to find that record again. In other words, there must be a way for predicting where a given record can be found. That method is called **hashing** or, sometimes, **randomizing**. Hashing schemes vary, but all perform some sort of arithmetic tranformation on the record key, which produces a disk address for the record. The hashing formula will first be applied to the key when the record is written. Then, at some future time, the same hashing formula on the same key should produce the same disk address, and the record can be located there.

The hashing process is not entirely straightforward. The hashing formula may produce the same disk address for two different keys. These identical addresses are called **synonyms**. There are various ways around this problem. One simple solution is **linear probing**, a process of checking disk address locations near the hashed address. That is, a record being written is placed near the original hashed address, and a record being read is searched for in the same way. Figure 11-5, although intentionally simple, illustrates the point.

Synonyms can appear for one or more reasons. The keys used, for example, may be too similar. The first three digits of a social security number, for example, refer to geographical region. It is common to strip the first

FIGURE 11-5

Example of direct processing. Assume there are 13 slots (0 through 12) available on the file. Dividing the key number 661, J. Behnke's employee number, by the prime number 13 yields a remainder of 11. Thus, 11 is the address for key 661. However, for the key 618, dividing by the prime 13 yields the remainder 7, a synonym. Since 7 has already been used (by the key 137) the address for key 618 becomes, by probing, the next number, that is, 8. Note incidentally, that keys need not appear in any particular order. (The 13 record locations available are, of course, too few to hold a normal file; a small number was used to keep the example simple.)

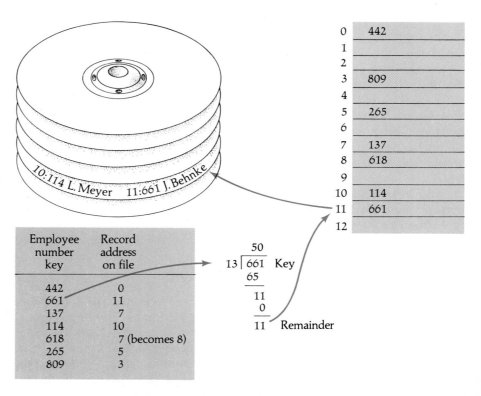

Employee number key	Record address on file
442	0
661	11
137	7
114	10
618	7 (becomes 8)
265	5
809	3

$$\begin{array}{r} 50 \\ 13\overline{)661} \text{ Key} \\ \underline{65} \\ 11 \\ \underline{0} \\ 11 \text{ Remainder} \end{array}$$

0	442
1	
2	
3	809
4	
5	265
6	
7	137
8	618
9	
10	114
11	661
12	

three digits of a social security number if the data records are for people who come from the same place. Another reason may be that the hashing algorithm chosen is not suitable for the data. A third reason may be that the file is too full. The more crowded the file, the greater the potential for synonyms and the more time spent probing. Excessive probing increases response time, the factor most closely guarded in an online system. For this reason, it is recommended that no more than 80% of the file be filled at any given moment. Thus the trade-off: unused disk space for acceptable response time.

A file that is very active—that has constant inquiries or updates to existing records—could function well in direct mode. A file that is volatile, especially if there are many record additions, will need attention because the file will eventually exceed the recommended 80% and need to be rewritten to a larger file location.

Applications that need direct access are those that use transaction processing, that is, those that submit one transaction at a time to the system in no predictable order. Many transaction processing systems need data processed in real time, so response time is important. When fast response time is important, direct files often are used. Bank tellers and airline agents, for example, want to be able to submit a single transaction to a direct file and receive an immediate response.

Indexed File Organization. A compromise between sequential and direct file organization is indexed file organization. With this method, the file is originally written in order by key, in the same way a sequential file is written. The file is accompanied, however, by an index file. The index file is really a table, and contains only two items per row: a record key and some indicator of the record's location on the disk. That indicator may be the disk address of the record or perhaps just the sector or track where the record is located. The index file is sorted in order by key. This file organization allows the indexed file to be accessed either sequentially or directly.

An indexed file can be accessed sequentially in one of two ways. The file can be read from the beginning in the same way a sequential file is read. Or, the reading can begin at any given file record and progress sequentially from there to the end of the file. To access the indexed file directly, the index is used. The key points to the disk address, which can be used to directly access the record needed. See Figure 11-6 for an illustration of indexed processing.

Indexed file organization may meet a need but it has some distinct drawbacks. You need to consider all factors carefully. The first problem is that record additions to an indexed file are not placed in the file in sequential order. They are placed randomly, and this is compensated in the index table. It is still possible to read the records sequentially, but doing so becomes burdensome if there are many additions. Also, deleted records are not really removed from the file, just flagged as deleted. If there are many additions

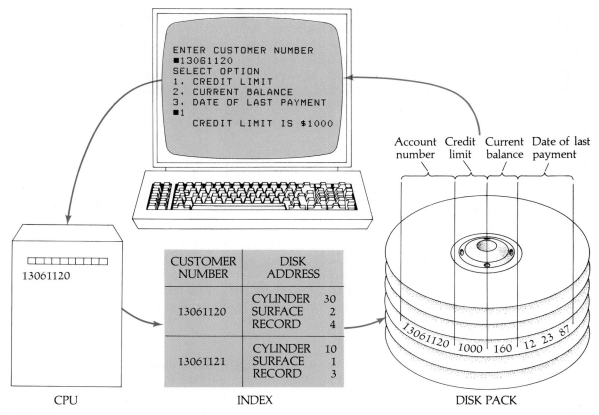

ENTER CUSTOMER NUMBER
■13061120
SELECT OPTION
1. CREDIT LIMIT
2. CURRENT BALANCE
3. DATE OF LAST PAYMENT
■1

    CREDIT LIMIT IS $1000

Account number   Credit limit   Current balance   Date of last payment

13061120

CUSTOMER NUMBER	DISK ADDRESS	
13061120	CYLINDER	30
	SURFACE	2
	RECORD	4
13061121	CYLINDER	10
	SURFACE	1
	RECORD	3

13061120   1000   160   12 23 87

CPU   INDEX   DISK PACK

FIGURE 11-6

Indexed processing example. At a retail store, a terminal operator can make an inquiry about a customer by typing in the customer number on a terminal. The index then directs the computer to the particular disk address, which indicates the customer information.

and deletions, the file will have to be reorganized. The net result is that an indexed file organization is not suitable for a volatile file.

But there are problems with highly active files as well. Direct access through an indexed file can be costly if the file is highly active. The problem is the overhead of hunting for the index first. This is just not as fast as hashing.

Consider carefully whether you really need the file in sequential order. If, for example, the file is not large, and the need for sequential reports is infrequent, you might do better with a direct file that is sorted when ordered reports are needed.

Like all compromises, indexed file organization is less than perfect for any of its functions. But it is an acceptable alternative if both sequential and direct access are needed.

As an analyst, you do not need to know or understand all the details of how these file organization methods work. Although programmers must have enough knowledge to request the desired file organization, the tech-

BOX 11-1	Volatile and/or Active Files

Examples of files that are volatile or active or both.

	VOLATILE	NONVOLATILE
ACTIVE	New contributor records often are added to a political campaign file. The records are accessed frequently to determine donation status.	Airline flight records are accessed frequently as reservations are processed. But flights are added and deleted infrequently.
NONACTIVE	Records are continuously added to the shipping file as mail orders are received, and deleted as items are shipped. No access of individual records.	Company with stable work force has few additions or deletions to the personnel file, nor is there much need to access individual records.

nical details are handled by the operating system. You do, however, need to know the consequences of your file organization selection on system performance.

File Backup and Recovery

We must assume the worst about files. It is too dangerous to assume anything less. We must assume that files could be accidentally written over, stolen, or destroyed. Therefore, copies of files must be made on a regular basis, frequently enough so that the backup files (the copies) could be used to reconstruct the originals if they are somehow lost. As we note in Chapter 15 on security, a set of backup copies is usually removed from the premises for safekeeping.

Batch systems usually operate on a **generation** basis, and prior versions of master files are kept for backup. Retention of backup files is costly and their number should be considered with care. The number of generations retained depends on the frequency of the updates and the importance of the data. Two prior master files may be sufficient for a personnel file updated weekly. A general ledger system updated daily, however, would need at least a week's worth of backup tapes because several updates could take place before an error is noticed. As a compromise, all backups could be kept for a week and then one per week could be kept for the preceding few months.

Backup plans for online files are critical. The copies must be taken at frequent intervals, in order to minimize loss of update transaction activity. Many banks take copies of customer account files on an hourly basis and then remove a set of copies from the premises. In fact, entrepreneurs have developed businesses based on the transference and protection of backup files. Copies are also kept of the transaction files. Backup schemes adopted can vary considerably, depending on the system.

We already noted that file processing is the traditional approach to developing systems. Another major approach, however, and one commonly favored today, is to plan to access data through database processing. This method represents a significant departure from file processing, and affects both access and processing.

◇ DATABASE PROCESSING

No user will come to you and say, "I think what this system really needs is a database." Instead, the database concept is presented as a solution to a series of user needs. As an analyst working with the user to define requirements, you may hear users state needs in such a way that you realize that a database system is an appropriate solution.

Consider this example. Analyst Nancy Menth was working with Chris Hemstead of Star Electronics, a growing manufacturing firm with two dozen sales representatives and several hundred regular customers. Star wanted an automated system that would track sales and provide up-to-date information on customers, sales representatives, and orders. As Nancy and Chris discussed the sales process at Star, Chris presented several situations in which he would need information in certain ways:

1. A list of all customers and customer addresses for a certain sales representative
2. The phone and address of the sales representative for a certain order
3. A display of the part description, quantity-on-hand, and bin number for a part that has been ordered

As Nancy considered these needs, she recognized that planning to place this data in traditional separate files—customer, sales representative, order, part—would lead to considerable data duplication. She also realized that the interaction needed among the files would cause a lot of processing overhead. For example, the first request would seem to indicate that all customers and customer addresses would have to be listed in the sales representatives files or else that, after reading the sales representative record, the related customer file would have to be read. Nancy saw that each of the three requests somehow came back to these types of choices, none of them desirable. But these are only the file processing choices. Database processing provides a better solution, without the disadvantages.

Nancy planned to put all the customer, sales representative, order, and part data on the same database. Each data element would appear only once. The data could be accessed using a software package called a database management system.

The example just described highlights the key reason for database processing: the complicated relationships among data. Since users often need information in a way that relies on such relationships, the database concept has become a critical one. We begin with an overview of database management systems and then move to a more detailed description of database models. Finally, we shall look at database activity on microcomputers.

An Overview of Database Management Systems

A **database** is a collection of data that is organized to minimize redundancy and maximize access to authorized persons. A company will usually have several databases, by general topic. A manufacturing company, for example, may have one database for the manufacturing process (engineering design, assembly process, quality control), another for financial applications (general ledger, accounts payable, accounts receivable), and still another for people records (personnel, payroll).

From a *physical* viewpoint, a database is organized into direct or indexed files. From a *logical* viewpoint, however, a database can be accessed in a number of ways by a number of users for a variety of purposes. In other words, the way the data is used logically is not limited to the way the data is stored physically.

Programmers and users cannot access a database directly. A **database management system (DBMS)** is a set of programs that allows users to access the database in a logical way, to suit a particular application. In other words, the DBMS shields the user from the technical details of managing the physical database. The logical organization of the data in the database is called the **schema**. (See Figure 11-7.) The schema describes the data elements and the logical relationships among them. The logical organization of the data for a particular user application is called a **subschema**, that is, a subset of the whole database.

Many database management system software packages are available on the market today. Since a DBMS is expensive to develop, you will doubtless use one of these, rather than develop your own. A DBMS package often includes access methods that can be used by programmers directly through programs, and also query language capability. A **query language** is an English-like interactive language used to access a database directly, that is, without

BOX 11-2   Advantages of Database Processing

Reduced Redundancy. Data placed in separate files usually leads to considerable redundancy. To take just one example, a bank customer's name and address could appear in files related to checking, savings, loans, personal line of credit, and a bank charge card. On a database, the customer name and address would appear just once, but be accessible to each of the bank processes. We say "reduced" redundancy because sometimes data is configured with some acceptable redundancy.

Data Integrity. Using the bank example again, you can see that separate files can lead to data that is inconsistent. Suppose, for example, that the checking account files are updated with personal change information—your address, for example—on a daily basis, but the charge card file is updated with such data only twice a week. For a period of time, two different files may have two different addresses, one correct and one not. Data integrity has been lost. Two reports may be printed from the two files, with the conflicting data. The net result is customer and employee confusion. However, once data on a database is updated, it is the same for all users of the database.

Data Independence. In file processing systems, data files are planned for a specific application and are then dependent on that application; that is, they are tied to the logic of the program that processes the data. The data may be related to other data in ways that are not connected to that particular program logic, but the data is locked into that logic nonetheless. Databases free the data, making it independent and thus accessible to a variety of applications.

Shared Data. In a database system, all data is potentially available to all users of the database. (There may be security limitations, however.) In particular, if one user has input that brings certain data up-to-date, then that same fresh data is available to all users of the database.

Fast Response to User Requests. Since database data is not locked in to the logic of certain applications, it is more readily available for new applications. This factor alone has had a tremendous impact on the computer industry. Once frustrated users, who knew the data was "in there somewhere," can now use data that was not easily accessible under the limitations of file processing systems.

Centralized Security. From a security standpoint, it seems a little scary to put all your data on one database. A knowledgeable person can access and damage one database more easily than several scattered files. However, the opposite side of that argument is that it is easier to protect data—and the access to it—when it is all in one place. Database security is often quite sophisticated, including passwords related to individual files, records, and data items.

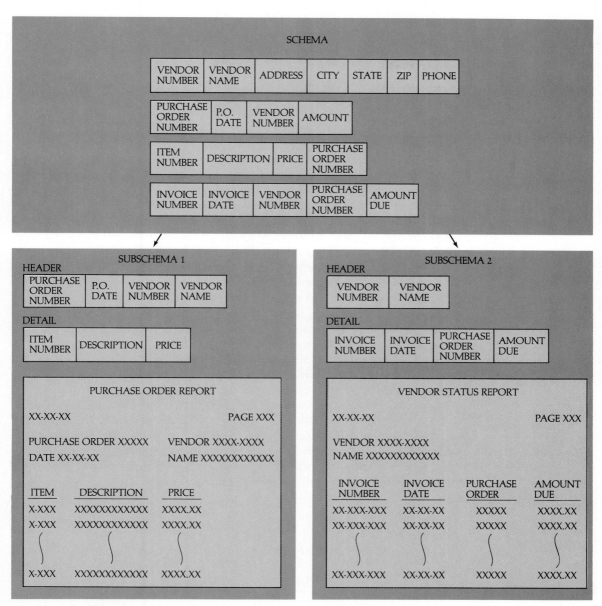

FIGURE 11-7 **Schemas and subschemas.** A database schema can be viewed differently by different users, each of whom uses a subset of the data as a subschema.

going through an application program. (See Figure 11-8.) Although some training is required, query languages are particularly helpful for nontechnical users. Another important tool that is often part of a DBMS package is a data dictionary.

Data Dictionary Revisited

The concept of a data dictionary was introduced in Chapter 6, which discussed the analysis phase of the systems life cycle. In the analysis phase, however, the data dictionary can be informal, possibly a simple file kept with a text editor or word processing program. In most organizations, however, the data dictionary used with the new system will be kept in a formal way. If the new system includes the use of a database, then the database management system probably includes a related data dictionary package, and analysts and programmers will use it accordingly.

A database management system may include software to help you develop the data dictionary. Some DBMS data dictionary software displays a screen template to guide you in entering information about each data dictionary element—its name, size, usage, and so on. Some DBMS software prepares the skeleton data dictionary from your source code. You need to become familiar with the DBMS used in the organization.

Formal organization standards are established for using the data dictionary and for determining data dictionary names. Data names receive formal approval from a central authorization group. Typically, the analyst making up a data name never used before fills out a dictionary element definition form. Information on this form includes the element name, synonyms, element definition, the use or function of the data element, data element size and description, origin, collection media, editing requirements, confidentiality restrictions, and authorization. Figure 11-9 shows a sample form for submitting a new database element name.

FIGURE 11-8

Accessing the database. Users can invoke query languages or programs to access the database through the database management system (DBMS).

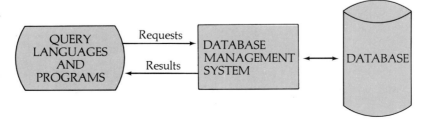

FIGURE 11-9

**Element diction-
ary.** Sample form
used to submit a
database element
name.

ELEMENT DICTIONARY

SYSTEM _____

ELEMENT IDENTIFICATION & DEFINITION

DEFINITION &
AUTHORIZATION

| 1 | Elem. I.D. | 2 | Element Cobol Name |
| 3 | File I.D. | 4 | File Cobol Name |

5 | Element Standard Name

6 | Synonyms

7 | Element Definition

8 | Main Use or Function of This Data Element

9 | Other Uses or Additional Comments

10 | Associated Elements (Elem. I.D. & Cobol Name)

11	Elem. Size (Max)	Alpha	12	Element Picture (Cobol)	13	Element Usage (Cobol)
		Numeric				
		Alpha/Num.				

AUTHORIZATION

14

Responsible for
Collection _____ _____
                        Signature   Date

Responsible for
Maintenance _____ _____
                        Signature   Date

◇ DATABASE MODELING

A database management system is based on some logical data modeling concept. These models represent data relationships in the real world. Three database models, shown in Figure 11-10, are hierarchy, network, and relational.

Hierarchy Model

A **hierarchy model** is one in which the data elements are logically associated in a parent–child relationship. Hierarchical relationships are common in real life: manager–employees, customer–sales, landlord–tenants. An element in a hierarchy database can have only one parent. Data stored in lower levels of the hierarchy can be accessed only through the parent. Note the King Books hierarchy example in Figure 11-11.

Hierarchy

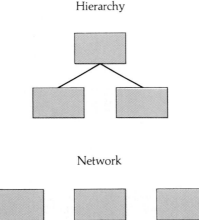

Network

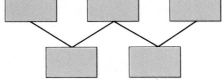

Relational

FIGURE 11-10

Database models. Schematics of three database models.

A hierarchy model for King Books

FIGURE 11-11

Hierarchy model.
Model and example
of hierarchy database
for King Books.

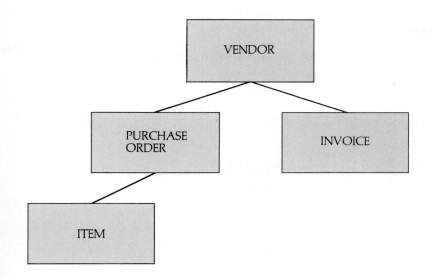

An example of a hierarchical data relationship.

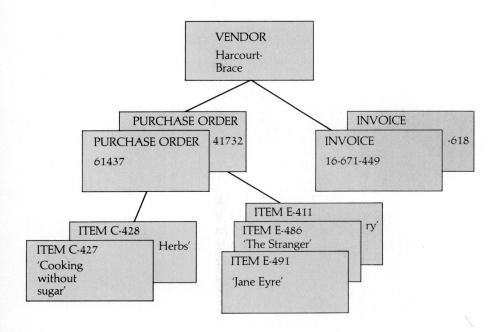

Network Model

A **network model** is similar to a hierarchy model, except that a data element can have more than one parent. For example, if we consider the hierarchical relationships class–student and major–student, then student has parents class and major. For another example, see the King Books illustration in Figure 11-12.

A network model for King Books

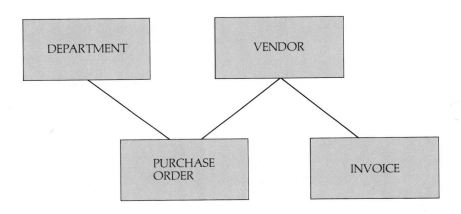

An example of a network data relationship.

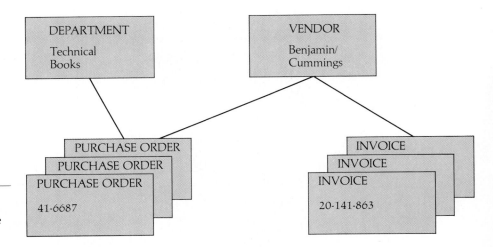

FIGURE 11-12

Network model.
Model and example
of network database
for King Books.

Relational Model

A **relational model** organizes data logically in tables, using a fixed number of elements as columns, and a variable number of entries as rows. A table is called a **relation**. Each column of the relation is called an **attribute**, and each row is called a **tuple**. Relations can be reduced or combined in various ways to answer questions about the data. The processes that manipulate relations are called **relational operations**.

Three relational operations are select, project, and join. A **select relational operation** forms a new relation by selecting certain tuples from a given relation. A **project relational operation** selects certain attributes from a relation, eliminating any duplicate tuples that may then exist in the new projected relation. A **join relational operation** is used to combine two relations, based on common attributes, into a new relation. Note the examples of each of these operations in Figure 11-13.

Using models like these, database systems became workhorses on mainframe computers. But databases have now become widely known to the public through the proliferation of database management systems for microcomputers.

DATABASES WITH MICROCOMPUTER SYSTEMS

Databases were originally conceived for use by computer professionals on mainframe computers. Although the introduction of query languages promoted direct user interaction, the assumption of professional assistance remained. Databases for microcomputers operate on a different premise entirely, one that assumes that trained users will be quite independent.

In other words, database management systems for microcomputers usually are much more user-friendly. Early database management systems were concerned primarily with getting the data on file and giving it back. For microcomputers, however, it has been estimated that 80% of the DBMS software is for the purpose of managing the user interface. Because of this attention, users are able to learn to use a microcomputer DBMS rather quickly, both to establish the data on the database and to use it for business purposes.

Those purposes—the applications of a DBMS—fall into three general categories.

- *Decision support systems.* A user may establish and query a database, asking informational questions that aid the decision-making process. Users can set up what-if situations and compare outcomes.
- *Electronic filing cabinet.* A user can keep track of all kinds of records, from what items to ship to the trade show to who should be invited to the annual Christmas party.

A relational example for King Books.

ITEM NUMBER	ITEM DESCRIPTION	ITEM PRICE
A-135	PIRANDELLO'S PLAYS	5.95
B-271	42ND PARALLEL	11.95
T-301	DATABASE MANAGEMENT	26.95
T-411	TELECOMMUNICATIONS	25.95
T-948	COMPUTER SECURITY	19.95
W-569	TRUE SPY STORIES	8.95

PURCHASE ORDER	ITEM NUMBER
41311	W-569
51648	B-271
61472	T-301
61472	T-411
61472	T-948
74819	A-135

PROJECT

Project only attributes item number and item price.

SELECT

Select only tuples for purchase order 61472.

ITEM NUMBER	ITEM PRICE
A-135	5.95
B-271	11.95
T-301	26.95
T-411	25.95
T-948	19.95
W-569	8.95

PURCHASE ORDER	ITEM NUMBER
61472	T-301
61472	T-411
61472	T-948

JOIN

Join matching item numbers.

FIGURE 11-13

Relational model. Example of relational operations on King Books database. Assume existing relations for items and purchase orders. In this example all three relational operations are used to create a new relation that supplies the information to the query: DISPLAY ALL ITEMS AND PRICES FOR PURCHASE ORDER 61472.

PURCHASE ORDER	ITEM NUMBER	ITEM PRICE
61472	T-301	26.95
61472	T-411	25.95
61472	T-948	19.95

- *Transaction processing.* Users can process new order entries or make charges to accounts.

Many database packages are available to accomplish these tasks on a microcomputer. Integrated software packages combine a DBMS with word processing, spreadsheet, and graphics processing.

Consider how a database package could be used. You begin by keying the name of the file. Then you define a record in the file, field by field. A typical definition format is: field name, data type (character or numeric), length, and number of decimal places. Then you give a command to begin entering data records. Some database management systems then print a skeleton of the record, listing all the field names in a column. You then type in the actual data next to the corresponding field name. Repeat for as many records as you like.

Once data is keyed in to the database, you can extract it in a variety of ways through a series of simple but powerful commands. Study Figure 11-14, which shows a sample file definition, data, and commands for a file of data about college buildings. In addition to these retrieval commands, most command sets include edit, delete, report, and calculate.

Overall, microcomputers represent a strong growth area for databases.

◇ DETAIL DESIGN DOCUMENT

The detail design has progressed from outputs to inputs to processing to files. Now you need to consider how to put it all together in the detail design document.

The **detail design document** will be the heftiest document you have prepared thus far. It will also be the most technical. This will be the first document containing many sections that are of only passing interest to your user management. Remember that you already carefully spelled out the design in user terminology in the preliminary design document.

As before, some parts of prior documents will be repeated in the detail design document. You may choose to make the detail design document an outgrowth of the preliminary design document. That is, you can simply add sections related to detail design to the preliminary design document and change its name to detail design document.

The following list suggests the contents of the detail design document. This list includes material covered in this chapter and the two prior chapters, all detail design topics—output, input, and processing. These will be described with narratives and supported with charts, graphs, illustrations, and formats, as appropriate. Also included in this document are ergonomic and security considerations, even though security is discussed separately in Chapter 15. The actual contents, as always, may vary, depending on the nature of the project and your choice of design tools.

- *Preliminary design documentation.* Include (from prior documentation) title page, project overview, project definition, requirements, design overview, benefits, and the general description of the new system.
- *Output considerations.* Media, content, volume, report and screen formats, output forms, and distribution procedures.

FIGURE 11-14

Using a microcomputer relational database. In the top box, the record definition is keyed, one field per line, and given consecutive field numbers 001, 002, and so forth. In the second box, the defined field names (NAME, TYPE, and so forth) appear on the screen and the appropriate field data is keyed. The third box shows the results of some typical commands to the database.

```
Record Definition for College Buildings


      FIELD        NAME, TYPE, WIDTH

      001          Name,c,20
      002          Type,c,10
      003          Year,n,4
      004          Handicap,c,1
      005          Busroute,c,1
      006          Parking,c,1
```

```
Sample Data for College Buildings

RECORD 00001               RECORD 00002               RECORD 00003
NAME      Cane Hall        NAME      Padelford        NAME      Sutton
TYPE      Classroom        TYPE      Classroom        TYPE      Library
YEAR      1943             YEAR      1958             YEAR      1937
HANDICAP  Y                HANDICAP  Y                HANDICAP  Y
BUSROUTE  N                BUSROUTE  Y                BUSROUTE  N
PARKING   N                PARKING   Y                PARKING   Y

RECORD 00004               RECORD 00005               RECORD 00006
NAME      Morley Hall      NAME      Carlisle         NAME      Union
TYPE      Theatre          TYPE      Classroom        TYPE      Student
YEAR      1951             YEAR      1967             YEAR      1958
HANDICAP  Y                HANDICAP  N                HANDICAP  Y
BUSROUTE  Y                BUSROUTE  N                BUSROUTE  Y
PARKING   Y                PARKING   N                PARKING   N
```

```
Sample Commands on Data for College Buildings


Display all for year < "1950"
00001 Cane Hall     Classroom   1943   Y N N
00003 Sutton        Library     1937   Y N N

Display all for handicap = "Y" and busroute = "Y"
00002 Padelford     Classroom   1958   Y Y Y
00004 Morley Hall   Theatre     1951   Y Y Y
00006 Union         Student     1958   Y Y N
```

- *Input considerations.* Data capture method, media, content, format, volume, input forms, screen design, and data coding.
- *Processing considerations.* Include logic developed using any detail processing design tools used:

    Design methodology tools

    System flowcharts

    Data flow diagrams

    Pseudocode

    Nassi–Shneiderman charts

    Decision tables

- *Data files and databases.* Record layouts, file organization and access, database descriptions, data dictionary.
- *Data communications.* Networking configurations, description of communications hardware and software.
- *Ergonomic considerations.* Any hardware or software ergonomic considerations made for this project.
- *Security considerations.* Any internal or external security measures taken for this project.
- *Costs, schedules, personnel.*

You will present all interested parties with copies of the detail design document well in advance of the detail design review. For a major new system, there should probably be two weeks between the time they receive the document and the time of the review. Minimum elapsed time between these two events should be a week.

◇ DETAIL DESIGN REVIEW

The **detail design review** is unlike others you may have given after the preliminary investigation, analysis, and preliminary design activities. The detail design review is the first review that has a technical orientation. For the first time, you will not be placing primary emphasis on user viewpoints, but on the technology required to do the job.

The audience for the detail design review is different, too. Although there will certainly be user representatives present, they are more likely to be from the ranks of the people with whom you have worked in some detail. In other words, attendees will be the people who can understand most of what is going on. Generally speaking, this probably precludes a contingent from upper management, since they seldom are familiar with the inner workings of the system. But you should expect first-line supervisors from the user organization to attend.

Representatives from your own organization, whether the data processing department or an outside firm, will usually include any analysts and programmers who are involved so far, and possibly programmers who will be involved in the upcoming development phase. Your own management will be there also.

As you plan the presentation, you need to assume that all attendees have read, or at least perused, the detail design document. It is not possible to cover the document in detail in the presentation. The document serves as a reference point for discussion and you should be prepared to answer any questions pertaining to its content.

The opening of the presentation should cover the origins and definition of the project. These points can be made rather briefly, since most of the audience has heard this several times by now. Then you describe the key points of design processing, showing, in particular, process functions, process relationships to one another and to files, file oranizations and database descriptions, and the expected inputs and outputs.

Plan to use half your allotted time for questions. If there are areas where you have relied on the expertise of others, you may want to have the people who can answer questions on those topics with you at the presentation.

◇ MOVING ON FROM DESIGN

Chapters 7 to 11 have been directed to the design of the system. Preliminary design was followed by detail design activities for output, input, processing, and files. Although it takes less time to describe, the upcoming development phase takes a significant portion of the project schedule. We are finally reaching the phase that you, as a programmer, probably know best. We are ready to tackle the coding and testing—the development phase.

FROM THE REAL WORLD

Climbing High with Databases

Three college fraternity brothers loved the outdoors. They spent their free time climbing, camping, or hiking. After college, Mike Price went to work for an accounting firm. Lowell Good was hired by a large mail-order catalog house as a management trainee. Doug Weber realized a lifelong dream: He opened an outdoor sports equipment store in the Pacific Northwest.

Weber's knowledge of the outdoors and his knack for retailing made the store an almost instant success. Weber saw potential for expanding the store by including a mail-order service, but he did not have the capital or the expertise. He approached his old fraternity brothers with a proposition that they pool their skills to build a successful mail-order business, Pacific Mountain Sports.

Price, the accountant, presented a business plan to a local bank, which extended the partners a $50,000 line of credit. Good, the mail-order specialist, built a mail-order list from addresses on the handwritten sales receipts. Good's list soon included other data, such as which customers were interested in what categories of products. If Weber found an excellent source of climbing ropes, for example, and wanted to run a promotional sale, Good could target the customers who had bought mountaineering equipment.

The mailing list grew to over 5,000 index cards. The partners wanted to automate the list, not just to print labels but to manage it. For example, they wanted to be able to find all customers who were cross-country skiers and then print out labels to send them appropriate fliers. The partners, consulting with ComputerLand, bought a microcomputer, a printer, and database software.

They used the database management system to store, manipulate, and retrieve their customer data. They stored customer name and address data, along with a descriptor code, representing categories such as climber, boater, or camper. They also attached a mailing label to the back of each order form. When an order came in, they looked at the label to see how a buyer was classified and noted which equipment moved best by category. They used this information to update their specialty catalogs.

Each time an order came in, the customer record was updated. If a customer did not purchase anything for a year, he or she was sent a postcard asking if catalogs should continue. This helped cut down on the catalogs that generated no sales.

Catalog sales were a resounding success. Pacific Mountain Sports grew at about 200% a year for two years. At this point, they outgrew their microcomputer and moved up to a minicomputer system.

—From an article in *Business Computer Systems*

SUMMARY

- Standard magnetic tape is $\frac{1}{2}$ inch wide, comes on a $10\frac{1}{2}$ inch diameter reel, is 2,400 feet long, and is coated with iron oxide that can be magnetized. Density is the number of characters per inch. A character of data fits on a 9-bit cross section of tape. Bits in like positions comprise a track or channel on the tape. A common data representation code is Extended Binary Coded Decimal Interchange Code (EBCDIC).

- Records may be fixed length or variable length. A logical record is one written by an applications program. Logical records can be grouped together in a physical record, or block, in a process called blocking. The blocking factor is the number of logical records in a physical record. The space between blocks is called an interblock gap (IBG).

- A magnetic disk is a platter coated with ferrous oxide, which can be magnetized. A set of disks is a disk pack. Each disk has concentric tracks on which data can be written. Data is organized vertically, by cylinders. Data can be read or written with four access motions: (1) seek time, (2) head switching, (3) rotational delay, and (4) data transfer.

- File processing refers to planning separate data files to match progam logic. A master file holds data records of a semipermanent nature; the file is updated as the status of the data changes. A transaction file contains records with new data that is used to update existing master files. A table file contains reference data that is relatively static.

- File characteristics include method of file access, volume of data on the file, file volatility (frequency of change), file activity (frequency of use), and cost. A hit is an update to a file and the hit rate is the percentage of records updated in a given time frame.

- A sequential file is one whose records are in order in some sequence by key. A key is a field chosen to order the records. Transactions to update a sequential file must also be in sequential order.

- A direct file is one whose records are stored in such a way that access to any one of them is direct; that is, the record can be read or written without reading other records in the file first. A direct file must be on a direct access device, usually disk.

- The process of finding a location for a record on disk, or locating a record already written, is called hashing or randomizing. If hashing produces identical disk addresses for two different records, these addresses are called synonyms. One answer to the synonym problem is linear probing, the process of searching nearby locations for a place to put the synonym. Synonyms may exist because record keys are too similar, or because the hashing algorithm may not be suitable, or because the file is too full.

- An indexed file is written in sequential order but has an index so that the file can be accessed either sequentially or directly through the index. Indexed organization is not suitable for a volatile file.

- Backup files must be made on a regular basis so that they can be used to reconstruct files that are lost, stolen, or destroyed. Batch systems use files on a generation basis, and prior versions of master files are kept for backup.

- A database is a collection of data that is organized to minimize redundancy and maximize access. A key reason for the database concept is the complicated relationships among data.

- From a physical viewpoint, a database is organized into files. From a logical viewpoint, a database can be accessed in a number of ways by a number of users for a variety of purposes.

- A database management system (DBMS) is a set of programs that allows users to access the database in a logical way, to suit a particular application. The logical organization of the data in the database is called the schema. The logical organization of the data for a particular user application is called a subschema.

- A query language is an English-like interactive language used to access a database directly without going through an application program.

- The database management system probably includes a data dictionary software package. Formal organization standards are established for using the data dictionary and determining data dictionary names.

- A database management system is based on some logical data modeling concept that represents data relationships in the real world. A hierarchy model is one in which the data elements are logically associated in a parent–child relationship. A network model is similar to a hierarchy model, except that a data element can have more than one parent.

- A relational model organizes data logically in tables. A table is called a relation. Each column of the relation is called an attribute and each row of the relation is called a tuple. The processes that manipulate relations are called relational operations. The select relational operation selects certain tuples from a relation. A project relational operation selects certain attributes from a relation, eliminating any duplicate tuples in the new relation. A join relational operation is used to combine two relations based on common attributes.

- The business purposes of databases on microcomputers are as decision support systems, as electronic cabinets, and for transaction processing. Many database packages are available to accomplish these tasks on a microcomputer.

- The detail design document may include information on preliminary design, output, input, processing, data files and databases, data communications, ergonomics, security, costs, schedules, and personnel.

- The detail design review is the first review that has a technical orientation. The audience will be people from your own and the user organization who have worked on the system in detail and understand what is going on.

KEY TERMS

Access motions
Attribute
Backup file
Block
Blocking
Blocking factor
Channel
Cylinder
Data transfer
Database
Database management
    system (DBMS)
Density
Detail design document
Detail design review
Direct file
Disk pack
Extended Binary Coded
    Decimal Interchange Code
    (EBCDIC)
File access
File activity
File characteristics
File processing
File volatility
File volume
Fixed length
Hashing
Head switching
Hierarchy model
Hit

Hit rate
Indexed file
Interblock gap (IBG)
Join relational operation
Key
Linear probing
Logical record
Magnetic disk
Magnetic tape
Master file
Network model
Physical record
Project relational operation
Query language
Randomizing
Relation
Relational model
Relational operations
Rotational delay
Schema
Seek time
Select relational operation
Sequential file
Subschema
Synonyms
Table file
Track
Transaction file
Tuple
Variable length

REVIEW QUESTIONS

1. Give an example of a master file, transaction file, and table file that might be associated with a department store credit system.

2. What is the difference between a file that is volatile and a file that is active? Give examples of each.

3. What is the difference between file organization and file access?

4. Using addresses 0 through 10 and a hashing scheme of dividing by prime number 11, find addresses for records with keys 14, 6, 52, 57, 76, 24, 60, and 47. Use linear probing when necessary.

5. What is the main advantage of indexed file organization?

6. Differentiate between file processing and database processing.

7. Describe the advantages of database processing for a college database that has data used for registration, grading, transcripts, and graduation requirements systems.

8. Describe the differences among the three database models. Can you apply the hierarchy and network models to data about students, instructors, and classes?

9. Check your school computer lab and/or the local computer software store for microcomputer databases. What seem to be the most popular brand names? Try tutorials that go with them.

CASE 11-1
THE BIDDING GAME

Mike Barrow, owner of Barrow Construction, walked a fine line. He wanted to provide a quality product and a known level of service in an environment that was full of unknowns: weather, material shortages, inflation, strikes, and cost estimates. One mistake, one unprofitable job, and he could be out of business.

Mike felt it all came down to the bidding game. He knew that if a bid was off by even a thousand dollars, he could lose a million dollar job, but he was not anxious to underbid and lose money either. To help him make the right bidding decisions, Mike turned to the computer.

Mike hired computer consultant Peg Jovanovich. Mike felt that the bottom line was tracking costs. Peg planned a system that would track labor and material costs that would be used as the basis for future bids. Transactions were accumulated on a daily basis, and processed sequentially by work order or purchase order number. The data was processed nightly at Peg's service bureau and the related reports were routed to Mike each morning.

At first Mike used the report data as a sort of historical reference when he prepared new bids. Later he purchased a microcomputer and became acquainted with spreadsheets. He keyed the report data and other figures to his spreadsheet program. Mike began to see that he needed data in several categories, such as full time or temporary labor, labor by job site, equipment or building materials, and so forth. As he began to recognize the true power of what he could do with a spreadsheet, he realized that his data was not set up in the most useful way. He called Peg back for further discussions.

1. The original files were established in sequential order. Does that seem appropriate for the system as first described? Should Peg have dug a little deeper into the system requirements?
2. What might be an appropriate course now for organizing the data?

CASE 11-2
TULIPS AND ROSES ONLINE

Analyst Lyle Schneider was hired to design an order entry system for Swanson Nursery, a northwest mail-order nursery specializing in bulbs and roses. A toll-free phone number had increased business fourfold in the last year. Owner Dian Wells listed customer service as her top priority. She wanted a system that would respond quickly to customer orders. Lyle designed an online order entry system that allowed the 23 operators to key orders to the system as they got the data from customers on the phone.

Lyle planned to have the order records on a direct file, entered and accessed by order number. About a month after the system was implemented, the operators started complaining that the computer response time was slowing down. Since it took longer to place a customer order, customers had to wait a longer time on the phone. This quickly led to busy signals for other customers and the consequent loss of business.

1. What is probably wrong with the system? There are several possible causes of the problem. What are they?
2. How can the problem be resolved?

PART 4

FINAL PHASES OF THE SYSTEMS LIFE CYCLE

The remaining phases of the systems life cycle are systems development and systems implementation. Novice analysts already may be familiar with the development phase because it is related to programming and testing. The implementation phase must be planned carefully so that all elements of the system are in place before conversion time. If the systems life cycle phases have progressed satisfactorily, the new system should be completed on time, within budget, and to the satisfaction of the users.

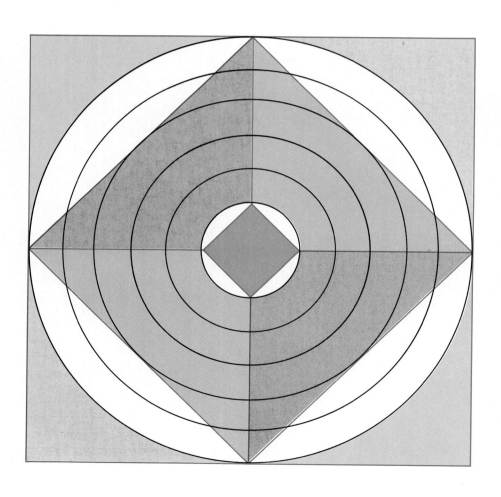

Chapter 12

Systems Development

After Michael Dole got his degree in German, he spent a lot of time trying to find a teaching job. He discovered that turnover in that field was very slow and that few people were approaching retirement. There were few jobs available. He decided to consider the computer field, which seemed wide open. He talked to friends and did some research in the library. He concluded that his best bet was to go to Southwind Community College, which had a special certificate program in data processing. In three quarters, he planned to be a computer programmer.

Michael met several people at Southwind who had varied backgrounds but similar ideas. In fact, they all thought they would become COBOL programmers in a large shop on a mainframe computer. As they progressed through the coursework, however, Michael and the other students began to realize that the computer industry was multifaceted. In addition to many computer career paths, there was an amazing variety of ways a person could work as a programmer.

Michael eventually went to work for an airline. His initial work was related to maintaining COBOL programs, but he found he had to learn several software tools just to understand, change, and test the programs. In this chapter you will study the programming processes with which you may already be familiar. But the chapter also examines some of the software tools you are likely to find in industry. Before we get to the programmer's role, let us see how the analyst fits into the picture.

◇ THE ANALYST ROLE IN THE DEVELOPMENT PHASE

The analyst role in the development phase varies according to organization policy, management style, the size and nature of the project, and analyst preference. Consider organization policy. In many organizations, for example, it is official policy that analysts do not participate in the programming process, except in an advisory capacity. The analyst role is one of liaison, coordinating between the programmers and other interested parties, especially users.

Some organizations give everyone the title programmer/analyst and everyone does some of both. However, the experienced people are the ones who are most likely to take analyst roles. But even a seasoned veteran can switch to programming to modify code during system testing or at some other crucial time. This kind of role change would be appropriate in such an organization. A third type of organization is the very small organization, in which just a few people play all the roles, including operating the computer. Obviously, flexibility is the rule in this situation.

If there is no strict organization policy, then analyst participation can vary according to circumstances. Some managers want the analyst to be very involved, especially if the programmers are inexperienced. This type of involvement can include activities such as monitoring the programmers' work to see that design specifications are properly coded, or that structured programming standards are embraced, or that all modules are tested.

The analyst role can also vary by the size and nature of the project. An analyst may handle many small projects at the same time but be only minimally involved in the details. If the project is one that is really a major revision to an existing system, then both the programmers and the users are experienced and there is less inclination for the analyst to become intimately involved at the programming level.

If the new system consists of microcomputers and some packaged software, the analyst may handle the entire project, from beginning to end. From the first talks with the users to the installation of the hardware (yes, it is usually the analyst who physically hooks it all together) to the training on the new software, it is possible that programmers may not be involved at all.

Finally, analyst preference may play some part in determining the level of participation in the project. Some analysts prefer to keep a hand on the product. If some coding activity fits manager temperament and analyst skills, this may be the way to proceed. Whatever the analyst's role in the development phase, however, some person or persons must begin the program development process.

◇ PROGRAM DEVELOPMENT

Janice Edwards, a novice programmer, was asked to consider how long it would take to develop a sales analysis program to identify consumer buying trends for a national cosmetics company. Janice was familiar with the subject matter and had discussed the analysis and design specifications already prepared by the analyst. The assignment seemed fairly routine, so Janice made her estimate based on time to code, compile, and test the program. The project leader thought her estimate was too low and questioned Janice about the basis of her estimate. Janice had left several things out.

Programmers do see the basic programming process as coding, compiling, and testing. But any experienced programmer knows that there are many other related activities, all of them time-consuming. First are the design activities. These are followed by coding, compiling, planning test data, testing, and documentation—not necessarily in that order. In the same time frame, a programmer typically interacts with others in various ways. There will be scheduling activities and design reviews, both of which will be discussed in detail in Chapter 14 on project management. Programmers may also attend coordination meetings with management and the project leader, and possibly meet with specialists who have program concerns related

to auditing or security. There may even be training in new hardware or software. Chapter 14 also notes that estimates are likely to be more accurate when activities are broken down into fine detail.

Refining the Design

The design may be complete when it is turned over to the programmer. No further design activity is then necessary. That is, the programmer may receive detailed specifications from which it is possible to begin coding immediately. But this is unlikely for two reasons. The first is that analysts, like everyone else, have limits on their time, and it is plausible to leave some of the detailed specifications to experienced programmers. The other reason is that many programmers resent being used as just "coders," and prefer to participate in the thinking and planning activities.

If some design remains to be done, then the programmer will need time to communicate with the analyst and perhaps directly with the user. Then he or she continues the activities that were described in Chapters 10 and 11: charts or pseudocode or record layouts or whatever remains to be done before coding can begin. Note this point for the sake of clarity. Design is design. Programming is programming. Just because the programmer does these activities does not make them programming. They are still design activities.

Coding

It has been a common practice for programmers to write programs on coding sheets and then key them to computer-sensible form. Students may still operate this way but, for many programmers, the time of this approach has passed. Instead, an experienced programmer today often takes an established program with some of the properties of the new program and revises it. This saves coding, keying, and testing time.

Programmers may have a say in the choice of language, but this decision is often installation-dependent. Many installations have a policy of coding in a certain language. Deviations from that standard are usually not permitted. The main reason for such a policy is uniformity, which will result in time and money savings. It is more efficient to have all programmers use the same language because this reduces the time required to learn the new language and makes it easier to move from one program to another. In spite of this, however, there are installations that support several languages, usually because they have diverse applications that are suited to different languages. Some installations use a different language on their microcomputers than on their mainframes. Also, many installations are encouraging the use of nonprocedural fourth-generation languages to increase productivity.

Sometimes large projects can benefit by streamlining certain parts of the project. For example, early coding of standard routines that can be placed in a disk library and used by everyone would be a significant time saver. A typical example is an error routine that accepts transactions and an error message code and prints a line on an error report. This routine can be called from a number of programs. Or consider a credit-granting routine that processes several different types of transactions. Different routines would process the different transactions, but all could call a prewritten routine to save the transaction and the before and after image of the updated master file.

Top-down or Bottom-up

Coding is not necessarily a linear activity. That is, programmers do not need to code system programs end to end and then begin testing. Some activities can be done in parallel. Some modules can be tested while others are still being coded.

Program coding can be done from the top down or the bottom up, or in some combination of the two. **Top-down coding** is based on the principle of coding the high-level modules first and leaving the lower level modules called in skeleton form, to be filled in later. The lower modules are only a shell, with an entry and an exit. In other words, as the higher module is being coded, references are made to lower modules as if they are coded and available. But, in fact, a call to that still-incomplete module will result in an empty action.

Bottom-up coding begins with some complete lower level modules, while the high-level driver modules are merely skeletons that call the lower modules. The point of both these approaches is to allow testing to begin on some of the modules while others are still being coded.

The advantages of top-down coding are many. The top-down approach usually tests the most important modules first. Also, top-down coding allows users to see a preliminary version of the system sooner than they do in a bottom-up approach. Once the major high-level modules are coded and tested, a few lower level modules that give some sort of output can be coded and results produced. If, however, coding starts with lower routines, that is, from the bottom up, few tangible results are available to show users.

It has even been said that top-down testing allows deadline problems to be handled more easily. That is, if the system is going to be late, at least there is something to show the user. The system, for example, may not accept all transaction format types, may not have all the edits in place, may produce only some of the required output—all of these are lower level activities. Nevertheless, the system will be capable of accepting some input, doing some processing, and producing some output—all of which serve as tangible evidence of system progress to the user. These can be done only if the top modules are in place. Finally, top-down coding improves programmer morale, and does so for the same reasons that it pleases users: tangible output.

Is there any justification for bottom-up coding? One justification may be that lower level modules are critical in some sense, perhaps performing tricky calculations. It may be important to get these working soon. But more often the reason is related to the use of people resources. A structure chart often looks like a pyramid, with many modules at the lower levels. A manager with many programmers to keep busy, or a tight schedule to meet, might decide to assign the lower level modules to several programmers early in the project. Compare top-down and bottom-up coding in Figures 12-1a and 12-1b.

In addition to the standard coding approach via a programming language, many aids are available to programmers in the form of software. Modern installations are embracing these software tools because they help programmers work faster and better.

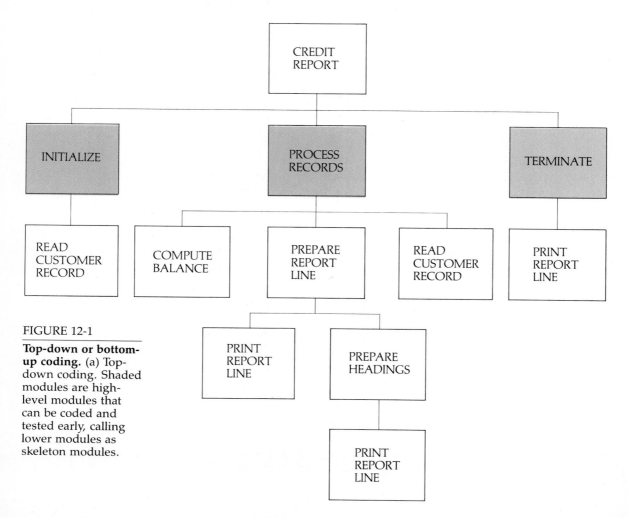

FIGURE 12-1

Top-down or bottom-up coding. (a) Top-down coding. Shaded modules are high-level modules that can be coded and tested early, calling lower modules as skeleton modules.

◇ SOFTWARE TOOLS

The subject of software tools has become extremely important in the computer industry. A **software tool** is any software product that can be used to facilitate the programming process and improve productivity. The key word in that definition is the last one: *productivity*. The industry pays good wages to programmers, so managers are always looking for ways to increase productivity. Shortening the programming process helps achieve that goal.

The programming process may seem at first to be simply coding—that is, actually writing the logic in a programming language. But coding is only part of the process, and the shortest part at that. Other key elements are program design, testing, and documentation. All of these activities are labor-intensive, and therefore subject to improved productivity. Hundreds of

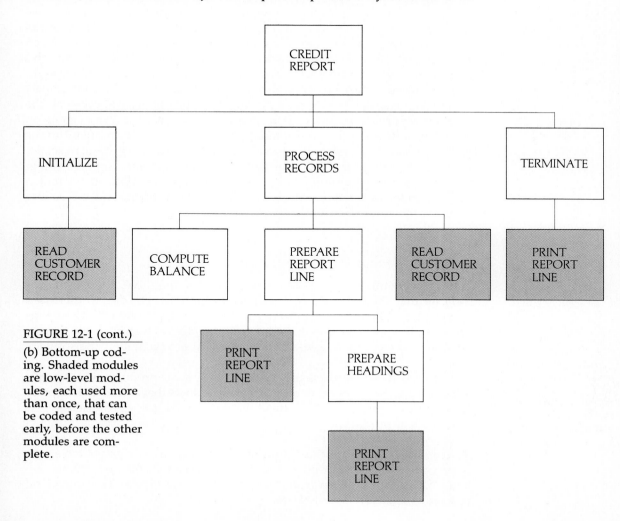

FIGURE 12-1 (cont.)

(b) Bottom-up coding. Shaded modules are low-level modules, each used more than once, that can be coded and tested early, before the other modules are complete.

software tools are available, many for large systems and others for micro-based systems. A number of software tools have been developed to aid in coding, testing, and documentation. These have been accepted by a broad segment of the industry. Fewer automated tools have been developed for system design and these are not widely accepted. Let us consider categories of software tools.

Software Tools for Design

Software tools for system design have fared less well than tools for other parts of the programming process. Their lower acceptance rate is probably due to the fact that design is a more creative function and is less amenable to being fit into a preplanned software package.

Even so, design tools that integrate several of the design functions are gaining acceptance. These typically include the drawing of structure charts and data flow diagrams, creation of a dictionary to cross-reference design components, facilities for designing screen definitions and reports, and provision of word processing narratives to describe programming specifications. These factors are interrelated. You can use them to see the impact of a design change. For example, changing a data name will cause all the affected reports to be listed. It is also worth noting that design tools on microcomputers are sometimes mouse-driven, so that you can point to the screen icon of the graphics you want drawn. The tools are general in scope, designed to help in the logical design of systems without tying the coding to a particular programming language or methodology. With such a tool, a project leader can better manage the development team, and the coding phase will be more efficient.

Software Tools for Development

Although the impact of software tools is not yet definitive, some estimates indicate that they provide cost savings as high as 30% due to increased productivity. Managers seem to agree, however, that productivity savings are directly related to how logic-intensive an application is. If the logic is very complicated, there is less improvement in productivity using software tools.

All tools described here are available as specific products with specific advertised names. There are often several products that accomplish approximately the same goal. Since they change rapidly, we shall discuss only categories of products.

Precompiler.   The input to a precompiler is programmer-generated short-hand code that is converted to a full-blown version in the appropriate language. This is a time-saver appreciated by programmers. The shorthand process is relatively easy to learn.

Code Generator.   Online code generators are among the latest in a long line of sophisticated tools. This type of program is often menu-driven and operates in question/answer format. A typical question is what the name of your driver module is. This may be followed by the code generator asking the names of the modules called from the driver module. And so on. A code generator must be supplemented with "user code," that is, the code not covered by the general processes of the generator. Typical user code includes editing input data or formatting output data. Code generators are appreciated because coding can be such a time-consuming and error-prone activity. The big savings, in the long run, however, comes at maintenance time, because a data processing shop that uses a code generator gets consistent code that is easier to maintain.

Report Generator.   A report generator is a special kind of code generator. The programmer writes a sort of shorthand version of code, with special report parameters. This shorthand program is processed by the report generator program, which produces a fully coded program to generate the desired reports. Report-writing capabilities are paramount in a batch system. Early system utility programs generated reports in a primitive way, but recent report writers are more powerful. Although some training is required, report writers are relatively easy to use.

Fourth-Generation Languages.   Most programmers have grown up with high-level languages like COBOL, tagged "third-generation" languages. Fourth-generation languages differ from third-generation languages in that they are nonprocedural. That is, a user of such a language need only convey what is needed, not how to accomplish the task. It is certainly faster to say what you want—for example, a report of all sales representatives in California who exceeded their quotas in 1985—than to write the detailed logic to produce the report. Fourth-generation languages have some syntax, and training is required to use them. They are not appropriate for every task, but they are quick and useful for applications that are not complex.

Split-Screen Editor.   There are many occasions when it would be convenient to work on two or more files at once. These files could contain program source code or data or text. An example is when you make a change that affects two related programs, and you want to make sure the changes are compatible. Rather than changing one and then printing it out for reference as you change the second, you could use a split-screen editor to view and change both files at once. Each file would show in a separate window on the screen; the files can be placed either side by side or one above the other, as you can see in Figure 12-2. A flip key can be used to go back and forth between the files.

Optimizing Compiler.   An optimizing compiler takes standard source code and generates highly efficient object code. This kind of compiler is not

a. Side-by-side split screen.

FIGURE 12-2

Split-screen editor.
(a) Side-by-side split
screen. (b) Above-
below split screen.

b. Above-below split screen.

usually used during the debug phase, because an optimizing compiler takes
significantly longer than a standard compiler. Instead, the program is checked
out in the normal way and then submitted to an optimizing compiler.

Cross Compiler.   Suppose you want to write a program that will run on a
microcomputer, but you want to write it in a language whose compiler is
too large for the small computer. You could write the program on a larger
computer and then use a cross compiler to generate object code that will
run on the microcomputer.

Software Tools for Testing

The program testing process historically has consisted of checking output and searching through program listings for errors. These exercises were augmented by language debugging tools such as traces of logic flow and displays of certain data values. Now the process has been enhanced with some very sophisticated automated testing tools. Collectively, these tools can significantly reduce testing time.

Online Debugger.   Among the most powerful software testing tools is the online debugging program. You can use this tool to plan breakpoints (stopping places) in your program. As the program runs online in checkout mode, you can monitor it from a terminal. It stops at each breakpoint, allowing you to examine the contents of storage locations and even to change the contents of those locations before it proceeds to the next breakpoint. Another important feature is the ability to step through the program line by line and watch what happens. This is the equivalent of having total control over the running of the program, asking for the next line when you are satisfied with what the last one did. This tool is exceptionally helpful in detecting logic errors.

Logic Analyzer.   This type of product will tell you about logic errors that you might miss if you depend only on running test data. For example, a logic analyzer will tell you if there are program statements that can never be reached. A logic analyzer can also tell you if you have statements inside a loop that are not affected by the looping process, that is, statements that should be removed from within the loop. These errors, and others like them, are unlikely to be detected in the early stages of debugging unless you use a logic analyzer.

Abnormal End Debugger.   When a program in checkout makes an unscheduled stop, the programmer must play detective and find out where and why. For years programmers have searched through program dumps that displayed the contents of memory at the moment the program ended abnormally. Newer software tools are more specific. Typical information supplied by abnormal end debuggers includes the name of the module where the program stopped, the last program line executed, and quite specific information about the problem. For example, instead of a cryptic numbered error message (OC4® is a message many programmers on IBM systems have seen), a programmer is given the contents of the fields that caused the problem, such as nonnumeric data on which the program was trying to perform calculations.

Command Language Tester.   The command language most noted for needing testing is IBM's Job Control Language (JCL)®. Since a missing comma in the JCL can cause the entire job to fail, it is expedient to pretest the JCL by itself. Many tools exist to do this. Some recent tools go further than checking parameters and punctuation. For example, they can check existing libraries to make sure you do not try to create a file that already exists.

Software Tools For Documentation

The historic complaint about documentation has been that it is always out of date. The most popular automated documentation tool, of course, is word processing, which we discussed in Chapter 4. Several automated documentation products address the problem of keeping the documentation synchronized with the actual version of the program. Among the most widely known and used documentation aids are the graphic products such as logic and system flowcharters. Some of these products can draw charts automatically from the source code. Also, documentation writers include program analyzers that can prepare skeleton document outlines from the source code.

Problems Using Software Tools

Despite the praise we can heap on these software tools, the road to their successful use is not smooth. There are three problems involved: acceptance, training, and familiarity. Acceptance and training are manageable problems. Managers need to apply change theory to get reluctant programmers to accept new tools. This takes time and effort but is by no means impossible. Training can be planned and executed with a reasonable effort. But the familiarity issue is a thorny one. The next situation illustrates the point.

Elizabeth Campbell is a computer professional for a major bank with international connections. Elizabeth works in the credit card division, which has several large automated systems and dozens of smaller systems. Although each professional computer employee has the title programmer/analyst, a highly skilled person like Elizabeth is used primarily as a senior analyst. In this role, Elizabeth divides her time between special assignments and project lead positions. Her special assignments include representing the company to vendors and other related industries, and evaluating hardware and software products.

But it is in her role as project lead that Elizabeth encounters some problems. Although she excels at organizing and monitoring projects, she has trouble keeping up with all the software tools the projects employ. The bank has been at the forefront in the search for productivity and has embraced a variety of software tools. Each of these products requires competence by the user, but it is hard to be knowledgeable about tools that one uses infrequently. Therein lies the problem for Elizabeth and other senior people: It is difficult to perform special assignments and project lead duties, *and* be familiar with all the software tools. But project lead decisions sometimes rest on such familiarity. In fact, during systems testing, Elizabeth sometimes switches to programmer mode and needs to be familiar with the tools in order to make program adjustments. The industry is grappling with this issue and coming up with a variety of answers, such as designating tool specialists to act as consultants.

Every programmer knows that coding is the easy part. Now we shall turn to testing, which is more difficult and more time-consuming. Testing must be planned well because this is the point at which you have the opportunity to undo the mistakes made in the coding.

◇ TESTING THE SYSTEM

Testing is a feverish activity but it is also very satisfying. The results eventually show that data and logic are behaving in predictable and correct ways. Testing usually proceeds from unit testing (testing a single module or program) to system testing (sending data successfully through the entire system) to volume testing (using large amounts of live data) (see Figure 12-3). Checking off item after item, then whole modules, programs, and finally the entire system is a tangible reward for all your efforts. But testing must be planned very carefully.

The Test Plan

The analyst will plan the testing strategy well in advance. The following are the key elements in the plan:

Assignments. This is the time to name names. Individual programmers are usually assigned responsibility by the project leader for testing their own modules and programs, but some organizations assign special testing personnel for this purpose. Some person or persons, usually the analyst, will be designated as responsible for system and volume testing.

User Participation. This is the time to bring key users into the process as participating team members. Although formal training is usually considered part of the implementation phase, it is often a great service to train some users early so they can join the testing activities. You might think that it would undermine user confidence to see the system in its initial state of

FIGURE 12-3

Testing the system. After individual modules and programs are tested during unit testing, they are tested together during system testing. High-volume data is used for the final testing phase, volume testing.

disarray, but the opposite is true. As users see the system taking shape and improving daily, their faith in the system strengthens. Also, users recognize their own usefulness, as they see the system change in response to their suggestions.

Test Conditions.   The list of conditions to be tested may be very long, even for a medium-size project. Time must be scheduled to decide what the test conditions are and to coordinate among the various participants. Here are some typical test conditions: Make sure that the salary rate is numeric; see that a transaction that updates the master file is reflected on a certain output report; check the transaction count.

Test Data.   Since data is needed for unit and system testing, someone has to create that data. Many different types of transactions will be needed, as well as test files and databases.

Testing Schedule.   The analyst needs to prepare a detailed schedule of what will be tested when. It would be unwieldy to list each test condition by date, so testing is usually scheduled by module or program.

Computer Time.   Many organizations have computer schedules that are close to the saturation point. Whether easy or difficult, arrangements must be made for computer testing time.

Testing Documentation.   The organization may use a formal testing documentation guide. If it does not, you can make your own. Typical column headings are as follows: condition tested, program name, data used (part number or social security number, for example), run date, expected results, actual results, and columns for sign-off initials. Note the King Books testing documentation in Figure 12-4.

As you can see, creating a test plan is a sizable task. Once the plan is in place and testing begins, then the analyst will monitor the ongoing testing documentation carefully.

Testing Documentation

Testing documentation is the analyst's best friend. Testing documentation is useful before, during, at the end, and long after the project is complete. Advanced documentation, as part of the test plan, ensures that testing is thorough and complete. As project testing progresses, the documentation provides a valid measurement of how much has been done and what is yet to be done.

At the end of testing, you have the completed testing documentation to prove that the testing was all it was supposed to be. Statements like, "Gee, didn't we test for that?"—known to strike fear in the hearts of analysts—

KING BOOKS

FIGURE 12-4

TESTING DOCUMENTATION						
PROGRAM ____EDIT____						
CONDITION TESTED	DATA	EXPECTED RESULTS	ACTUAL RESULTS	DATE TESTED	RUN NUMBER
PURCHASE ORDER AMOUNT NOT NUMERIC	P.O. 26701 AMOUNT 24 B.06	'PO AMT NOT NUMERIC' ON ERROR REPEAT	AS EXPECTED	10-31-88	7
INVALID PURCHASE ORDER DATE	P.O. 26742 DATE 13-12-88	'INVALID PO DATE' ON ERROR REPEAT	NO MESSAGE. PROGRAM ERROR. RUN AGAIN.	10-31-88	7
INVALID VENDOR NUMBER ON PURCHASE ORDER	P.O. 26811 VENDOR # 447-21113	'INVALID VENDOR NUMBER' ON ERROR REP	AS EXPEC		
DUPLICATE PURCHASE ORDER	P.O. 267B				

Testing documentation. This sample testing documentation form is filled in with results from testing an edit program that is part of the King Books system.

no longer apply. When questions arise in the months immediately following testing, the analyst has a source that can be checked. Everyone knows just exactly what has been tested and what the results were. Let us turn now to the first phase of testing: unit testing.

Unit Testing

Considering a module or a program as a unit in the system, **unit testing** is the initial testing of a module or program using test data. Test data, rather than a sample of real data, is often preferred because real data is too normal most of the time. Suppose that a particular error condition occurs infrequently. You cannot simply hope that it will come along in a set of real data. Instead, you must plan test data that will include that condition, so you can see for yourself—and document—that the program handles the condition correctly.

Some programmers approach the testing process as a challenge. They enjoy planning the most difficult data possible, finding satisfaction in seeing that the program can handle the data correctly. It is certainly practical to test a program thoroughly. A well-tested program is an asset to the system, the users, the company, and the programmer's reputation.

Sometimes unit testing can produce unexpected results. Lois Merry, an experienced programmer, was getting ready to unit test a general ledger program that was part of a large accounting system. Lois had been distracted by other tasks and had not yet prepared test data. Her colleague Lee Nguyan was ahead of schedule and offered to make up some data for Lois's program. Lois accepted the favor, expecting her program to breeze through the test data. She was stunned when her program could not begin to process the transactions, which had been deliberately prepared out of order. It was humbling to realize that the test data she would have prepared for herself would have passed through the program easily. Lois and Lee made a pact to prepare data for each other for the rest of the project. Not everyone will enter into test data pacts, but all programmers must somehow identify what needs to be tested in the program. Many organizations, it should be noted, do not allow programmers to make up their own test data.

Part of the fascination of programming is the number of opportunities it provides to introduce errors and the challenge it provides to avoid them. Humans, of course, cannot avoid errors altogether although it is commendable and cost-effective to try. After human effort to avoid errors, the next line of defense is the ability to locate any errors that were inadvertently introduced. Every program module must be tested and each condition within the module must be tested. The programmer will be looking for various types of errors in these areas:

- Input/output
- Data formats
- Record keys
- Logic control structures
- Calculations, including truncation and overflow
- Table construction and overflow
- Comparisons, especially mixed data modes
- Error processing

Recall the top-down and bottom-up methods of coding modules. We noted in that discussion that this kind of coding facilitates testing. Here is an example of testing using bottom-up coding, in which high-level modules in skeleton form call completed lower level modules. Suppose you are writing a program to process bank transactions. Since several different actions can occur (deposits, withdrawals, transfers, loans, and so on), each action can be represented by a different type of transaction. A high-level module can sift through the transactions and call a separate processing module for each transaction type. Thus, coding in bottom-up style, your skeleton calling module initially will do nothing but call the right module. Meanwhile, if two routines such as those that process deposit and auto loan transactions are coded, they can be checked out in parallel.

After programs have been checked out separately, they must be checked out together. This is the function of system testing.

System Testing

The first consideration in system testing is data flow. **System testing** tests the flow of data through the entire system. Information that is output from one program may be input data for some subsequent program. If those two programs were written by different people (or even if written by the same person!) there can be communication problems and misunderstandings. In other words, the data expected by the second program may not be quite what was produced by the first program. These situations can be alleviated by clear design specifications, but it is unlikely that they will be eliminated altogether. Hence, system testing.

Again, test data is prepared. This is an area in which users can be particularly helpful. They know what kind of data comes into the system and what kind of output is anticipated from that data. Although system testing must be extensive, a subset of system testing, called **acceptance testing**, is sometimes used. At this point, the system is sufficiently intact that users on various levels can actually watch it in action. That is, you can demonstrate to users and their managers that data is behaving in a certain and correct way in the system. As this occurs, users can initial the appropriate lines on the test documentation, indicating their "acceptance" of the testing.

Consider the King Books case for an example of acceptance testing (see Figure 12-5). Assume that one condition that must be tested is that the system show it can accept multiple separate invoices from the same vendor and produce a single check that combines the payments for all the invoices. The test documentation would show the invoice numbers of the transactions entered and briefly describe this test condition as "Combine invoices." Expected outputs would be listed on the documentation as (1) the invoices on the transaction report, (2) the check itself, and (3) the listing of the check on the check register report. Showing these correct results to the designated user should convince him or her to accept this particular test condition, and to indicate acceptance by initialing the test documentation.

KING BOOKS

ACCEPTANCE TESTING DOCUMENTATION				
CONDITION TESTED	DATA	TEST RESULTS	DATE TESTED	INITIALS
MULTIPLE INVOICE, SAME VENDOR	INVOICES 21-146-429 24-198-146 VENDOR IS BANTAM BOOKS	• BOTH INVOICES IN TRANSACTION REPORT • COMBINED CHECK FOR B.B. • CHECK SHOWS ON REGISTER REPORT	12-1-88	GN
QUERY VENDOR STATUS	VENDOR # 4621-9941 (SIMON AND SCHUSTER)	• S.S. ADDRESS SHOWED ON SCREEN • CORRECT STATUS	12-1-88	GN

FIGURE 12-5

Acceptance testing documentation. Note that this form is slightly different from the testing documentation form in Figure 12-4. In particular, there is a place for users to initial their acceptance.

Volume Testing

Testing is an inexact science. It is not possible to test for every possible condition or data combination because the number of possibilities is simply too large. And even if such testing were possible, it would not be economically feasible. These kinds of words are sometimes shocking to novice programmers, who tinker with a new program until they "know" it works. What they really know, of course, is that the output appears to be correct for the data they have chosen. Exhaustive test data is not used in most classroom situations. Another consideration, of course, is complexity. Programs produced in the workplace are apt to be several factors of difficulty beyond programs used as learning vehicles.

Although we acknowledge that it is not feasible or economical to perform 100% testing, we can certainly perform significant testing. **Volume testing** is the process of sending large amounts of real data through a system to test its capacity and its ability to perform well under these circumstances.

Volume testing is notably different from unit and system testing because normally it uses only real data. There are two reasons for this. One is that it is impractical to create high-volume data from scratch. But a more important reason is that it is now time to see what kinds of data combinations can be generated by large amounts of real data. Some of these combinations will undoubtedly be unexpected. Unexpected data combinations may cause problems that must be fixed. In fact, even at this late date, some program changes may be necessary.

Volume data runs also test the capacity of the system. There can be several surprises, not all pleasant. It is not uncommon to discover, for example, that tables whose size was thought to be adequate have overflowed. Similarly, record storage capacity may prove insufficient. Perhaps real data will deliver hashed disk addresses that have an unacceptable number of duplicates. It may be that an online system staffed with the maximum number of terminal users slows down to a crawl. That is, just as you are approaching the finish line, big problems may block your path. Not all systems have all these problems. Some have none at all. Advanced planning can help the analyst avoid most of them.

◇ THE DEVELOPMENT DOCUMENT

In previous chapters we discussed the contents of the preliminary investigation, analysis, preliminary design, and detailed design documents. These documents were somewhat similar in format and style. They tended to build one upon another. The development document does not fall into the same category as the others. The development document is actually a set of bound program listings accompanied by logic charts and narratives. Testing documentation also is considered part of the development document.

◇ ON TO THE FINAL PHASE

The implementation phase is an outgrowth of the development phase. In fact, some implementation tasks, such as training, may have already begun. The final, upcoming phase will bring the project to a climax. The effort you made in the development phase is about to pay off in the form of a new system in place for the users.

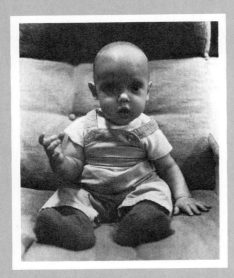

FROM THE REAL WORLD

Software Tools as Child's Play

The William Carter company is known to parents. They make the famous Carter baby clothes. In fact, Carter is the largest children's apparel manufacturer in the country, in business since 1865. Data processing is not new to Carter; the company acquired its first computer in the 1950s.

In 1980, Carter turned its attention to automating its manufacturing divisions. The major systems that needed to be "up and romping" were garment specifications, inventory control, production planning, material requirements, and production planning.

The new aspect of this project development was the use of a code generator. A COBOL code generator called System-80™ was selected. Data processing director David Bumpus used three experienced computer professionals but was able to supplement that staff with several people just out of college. Within two weeks they were writing online programs using System-80. They were able to use System-80 to generate about 70% of the code. From 1980 to 1984 the programming staff of eight generated 500 manufacturing applications programs.

The biggest advantage of using a code generator, they found, was the short time it took to develop applications. But there were other advantages as well. The code generator forced standardization of file structures, so that program maintenance time has been reduced.

—From an article in *Computerworld*

SUMMARY

- The analyst role in the development phase varies according to organization policy, management style, the size and nature of the project, and analyst preference.

- Activities associated with programming are design, coding, compiling, planning test data, testing, documentation, scheduling, design reviews, and coordination meetings.

- The programmer may receive detailed design specifications or may complete the unfinished design activities.

- Choice of coding language may be installation-dependent or may be based on the suitability of the language to the application.

- Top-down coding is based on the principle of coding the high-level modules first and leaving the lower level modules in skeleton form, to be filled in later. Bottom-up coding begins with some complete lower level modules, while the high-level driver modules are merely skeletons that call the lower modules.

- A software tool is any software product that can be used to facilitate the programming process and improve productivity.

- Software tools for design typically include the drawing of data flow diagrams, creation of a data dictionary, and provision word processing narratives to describe the programming specifications.

- Software tools for development include the following. A precompiler converts shorthand code to a full-blown version in the appropriate language. A code generator program turns programmer input parameters into usable code. A report generator is a special kind of code generator that allows the programmer to write a shorthand version of code, using special report parameters, that is processed by the report generator to produce usable code. Fourth-generation languages are nonprocedural languages that can be used to write programs quickly. A split-screen editor allows a user to view and update two or more files on separate windows of the screen. An optimizing compiler takes standard code and generates highly efficient object code. A cross compiler uses one computer to generate object code that will run on another, usually smaller, computer.

- Software tools for testing include the following. An online debugger allows the programmer to stop an online program at preplanned breakpoints, to monitor the progress of the program. A logic analyzer tells the programmer about program logic errors. An abnormal end debugger supplies information about the problem if the program makes an unscheduled stop. A command language tester pretests command languages before they operate on programs and data.

- Software tools for documentation include word processing and graphics software to generate supporting charts.

- The test plan includes considerations for assignments, user participation, test conditions, test data, the testing schedule, computer time, and testing documentation.

- Testing documentation is useful to plan the testing, to measure testing progress, and to document the results of testing.

- Unit testing is the initial testing of a program or module, using test data. Types of program errors to test for include input/output, data formats, record keys, logic control structures, calculations, table construction, comparisons, and error processing.

- System testing tests the flow of data through the entire system. A subset of system testing is acceptance testing, tests that can demonstrate to users and managers that data is behaving in a certain and correct way in the system.

- Volume testing is the process of sending large amounts of real data through a system to test its capacity and its ability to perform well under these more realistic circumstances.

- The development document consists of bound program listings, logic charts and narratives, and testing documentation.

KEY TERMS

Abnormal end debugger	Precompiler
Acceptance testing	Report generator
Bottom-up coding	Software tool
Code generator	Split-screen editor
Command language tester	System testing
Cross compiler	Test plan
Development document	Testing documentation
Fourth-generation languages	Top-down coding
Logic analyzer	Unit testing
Online debugger	Volume testing
Optimizing compiler	

REVIEW QUESTIONS

1. Describe the role you might expect an analyst to take in the development phase in the following situations.

   A large insurance company has a policy of hiring only programmer trainees. Since these people usually leave in a couple years and are replaced, the staff is usually consistently inexperienced. However, experienced analysts prepare detailed specifications for the programmers.

An analyst consultant has been hired to provide a microcomputer system, with a variety of packaged software, for a small vending company.

An in-house analyst has several small projects going at once for the quality control group of an aircraft company. Most of the programmers have worked in the quality control group for three years or more.

A bank has a strict policy that analysts do no coding.

2. Describe the programming-related activities. How might the order in which they are done vary?

3. Which do you think you might prefer—top-down or bottom-up coding? Both at once? Reasons?

4. What software tools have you heard about or read about in the computer literature? Do you know any brand names? What kinds of tools do you think you might like to try if given the opportunity? What advantages, in general, would software tools have for you personally?

5. Differentiate among unit, system, and volume testing. Why are all three necessary?

6. Describe testing documentation. Discuss its usefulness before, during, and after testing.

CASE 12-1
AN ELEGANT SOLUTION

Although primarily a business applications programmer with experience in COBOL and BASIC, Don Phillips had always wanted to write some programs in Pascal. Don knew that some people with computer science backgrounds consider Pascal to be the language of choice. Don's boss Jim McDermott was less enthusiastic, feeling that Pascal, with its limited input/ output capabilities, was not particularly suited to the applications in their event-ticketing business.

Undeterred, Don bought some Pascal books and starting teaching himself on his own time. He had often heard Pascal described as an "elegant" language, and his studies confirmed that it was. When Don was assigned a package of statistical programs to track ticket sales, he felt he had the perfect Pascal application. Although he worried a little about his manager's reaction, Don thought Jim's reluctance stemmed from the extra time it would take to code in an unfamiliar language. Don solved that problem by doing a lot of coding at home on his own time.

Still, the programs were late. Don had been given a lot of freedom in the past and no one had checked closely to monitor his progress. When

Jim did take a closer look, he was less than pleased and ordered the programs redone in COBOL.

1. What other reasons would Jim have against using a new language in the organization?
2. Do you think disciplinary action is appropriate in this case?

CASE 12-2
THE BIG TESTING COVER-UP

Dane Chaffin came to programming via the science library of a large chemical company. Self-taught, he had spent a lot of time reading anything he could get his hands on about computers, including manuals used by the computer professionals for his own company. He also had two microcomputers at home on which to experiment. Before long, he had convinced his boss at the library to install micros there and to use them to hook into national databases specializing in chemical information. For an amateur, his accomplishments were pretty impressive.

But Dane did not want to be an amateur. He wanted to get out of the library business and be a full-time computer professional. By being persistent and using his connections inside the company, he eventually managed to make a lateral transfer to the data processing organization as a programmer. There he went to work on a large system project where he was required to go through the usual testing procedures.

Things did not go well. The problems began in Dane's programs. They just did not check out the way they should. Many conditions listed in his test plan did not give the expected results and he had trouble figuring out why. He needed more time, but time was running out. He thought that some of the conditions were almost right, so he penciled them in as complete on the testing documentation. The "almost" and "not quite" syndrome continued, with accompanying notes in the test logs.

When it came time for system testing, Dane actually changed one of his programs to simulate acceptance of a transaction that his program could not process—he just passed it on to a subsequent program. From there the deceptions continued. Dane felt as if he were powerless in a downward spiral. At one point a colleague became suspicious but Dane convinced her that everything would be all right as soon as he had a little more time.

The day of reckoning came with volume testing. Everything, or so it seemed, went wrong, and most of it was eventually traced to Dane's programs. A great deal of time and money, not to mention credibility, had been lost.

1. How could Dane get away with this? Would tighter controls have alerted people to the problems earlier?

2. The story seems to imply that Dane's somewhat sketchy background may be partly to blame. Could his background have been a contributing factor? Could this sort of thing have happened to a person trained as a computer professional? Discuss.

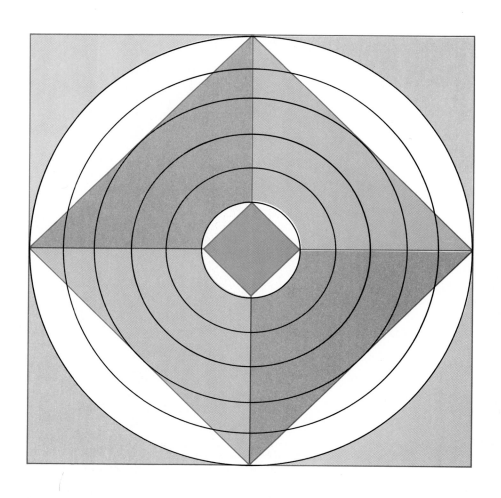

Chapter 13

Systems Implementation

The park department of the city of Seattle decided to automate the maintenance function of its 14 major parks and dozens of smaller parks. Working with park department managers, analyst Catherine Green determined that they needed to be able to track their maintenance needs on a weekly basis and disperse personnel accordingly. The requirements included planning extra staffing for a number of significant park events, and making changes on a last-minute basis. Processing was centralized through the park department headquarters. The system was more complicated than originally envisioned and needed a lot of flexibility built in.

Catherine suspected that this project would be difficult to implement and she was correct. Part of the problem was that the personnel were in different geographical locations every day, and even the supervisors using the new system moved around every day. Although Catherine planned carefully and trained everyone thoroughly in the use of the system, emergencies continued to arise during the implementation phase. For example, the second weekend the new system was in effect, the annual food-tasting fair, an event that usually attracted 10,000 people per day, was held at Green Lake. This year, over 13,000 people showed up each day, complicating parking, traffic, and cleanup problems.

Supervisor Marv Moore at Green Lake phoned the central office manager to have him locate on-call backup personnel through the new automated system. But the central office manager was on temporary duty and could not produce the correct system output. Marv finally used the old manual system, with which he was more familiar, and made several phone calls to locate workers. That work data was not entered into the automated system and it was months before the computerized reports were totally accurate. The users' trust in the new system was further diminished.

All sorts of unexpected problems develop during the implementation phase. Sometimes old psychological problems resurface, as novice users worry about actually coming to grips with an automated system. These fears can be manifested in the form of avoidance or even sabotage. Analysts have problems, too. The implementation period can be very intense. The tangible parts of prior phases have been documentation and presentations and programs, all promises of things to come. But now the plans and promises must be fulfilled in a system that works. At this point, you may feel that your reputation is on the line. It is time to produce, time to deliver the system. As in all phases of the systems life cycle, planning is a critical ingredient.

◇ IMPLEMENTATION PLANNING

You will want to prepare a written plan detailing how implementation will progress. As we proceed through the discussion of system implementation, you will see that the planning itself must begin well before the implementation phase. The plan will contain the general categories of activity and, within each of those, a detailed list of tasks that need to be performed. Beside each task write the name of the responsible person and the expected completion date. General categories of activity are as follows:

- Training
- Equipment installation
- File conversion
- Procedures conversion
- System conversion
- Evaluation

The analyst can use the expanded list to monitor progress as the implementation phase proceeds.

Before we enter a detailed discussion of the implementation phase, it is useful to note a few practical considerations. The first is that your credibility will be greatly enhanced if problems that arise are handled promptly. There will be times, of course, when you cannot show immediate results, but it is always possible to acknowledge concerns in a timely manner. That is, ideally you should take care of the problem right away; failing that, at least give the user written confirmation that the problem is being investigated. If complaints and problems are shunted aside for "more important" concerns, users will become discouraged quickly. Genuine interest and quick response are incentives that spur users to offer further advice and to participate fully. This kind of attention to user concerns often means a considerable time commitment from the analyst. You probably will be working long hours.

KING BOOKS

Implementation Plan Outline

The following general notes are the first draft of plans for implementation of the new system for King Books.

Training. King Books managers and office finance staff will be given formal training by analyst Dave Benoit. Classroom training at Universal Computer Services (UCS) will be for everyone. This will be followed by monitored practice for the finance staff on the terminals in the King Books offices. Dave will prepare a user guide, practice sets, and practice data. Dave will also introduce software tutorials prepared by the vendor.

Equipment installation. The batch reports and checks can be processed by the existing mainframe system at Universal. Since these reports will be printed on-site at King Books, a storeroom near the finance department will be converted to house the printer and supplies. Certain desks in the finance department will be converted to workstations for the online operation.

File conversion. This task is less strenuous than usual because the vendor files are fairly static and, once keyed, will have few changes. This keying activity will begin as system development begins. Files for purchase order, invoices, and vouchers will be generated from transactions that occur during system conversion.

Procedures conversion. Existing manual procedures for accounts payable are sketchy and inadequate. They will be replaced with procedures that are incorporated into the user guide.

System conversion. The system will be converted by phases. First, the users in the finance department will be able to query vendor data. Next, new purchase orders will be placed on the automated system and the invoices and vouchers that they cause will be automated also. The manual system will continue to be used for the documents not originally automated, until all paperwork for them is processed.

Evaluation. King Books chief accountant Phyllis Wasson will work with a representative from an independent consulting firm to evaluate the system.

One more point. Although you are responding to user needs as the system is implemented, this is no time to consider new requirements. As their knowledge of the system improves, as it will daily, users will come up with lots of new ideas. These must be deferred to the system maintenance function.

Since implementation cannot begin until all involved personnel are trained in the use of the new system, we begin our detailed discusssion with the subject of training.

◇ TRAINING: PSYCHOLOGY MEETS TECHNOLOGY

As any teacher knows, there is a certain psychological factor in teaching. You cannot just throw information at students and expect them to absorb it. This is especially true if the students are supposed to emerge from the class sessions with skills to accomplish something. If potential users of a system are to gain skills and confidence, then their training must give them practice in the skills that build the confidence. As we shall see, documentation for users is important but not sufficient by itself. They need lots of practice.

There are several facets to preparing training sessions. You need to consider who the trainers will be, what tools the trainers need to do their jobs, who needs to be trained, how the training will be presented, and how the training will be evaluated.

Preparation for Training

In most cases, the "training staff" will be you. And, in fact, you will probably prefer it that way, because you have an abiding interest in seeing that the users understand the new system. This type of in-house training is common and often works out well, since you understand the system. Also, the same communication skills that make you a good analyst probably make you a good teacher. And, lastly, many or all of the users know you already. If the group of trainees is large, you will need some help. You may be able to discover a natural ally, a trainee to whom the others seem to turn. You will probably find yourself dispensing more information to this person, knowing that he or she will be passing it along to the others when they return to their group environment.

If the number of trainees is very large, you may want to "train the trainers," that is, train a person from each user group or department. These trainers then instruct their own colleagues. You will need to pick these trainers carefully. Their communication skills are important, but it is even more important that they have the confidence of their peers.

A totally different approach is to use vendor training, especially if it relates to purchased software. Sometimes this training is free, that is, included in the price of the software. If there is a fee, the cost may be less than that of analyst time in-house. The overriding consideration, of course, is the quality of the training; useless low-cost training is no bargain.

Training Tools

The traditional training tool is user documentation. There was a time when such documentation was delivered to the user in the expectation that it was necessary and sufficient. Even after user documentation became more slick and easy to understand, trainers did not always understand that users also need to see demonstrations and try the system out themselves. Now that is changing, and so the classroom environment is changing, too.

If trainees are to practice using the new system, then the training staff needs training tools. Some written materials will be necessary, but more is needed. If training sessions involve several students at a time, then an early consideration is classroom facilities. You may have access to a training center equipped with terminals, modems, microcomputers, printers, and other equipment. It is more likely, though, that you will be using a conference room for some lectures and then move to other locations where terminals, or whatever students need for practice, are already in place. The trainers may also need software, either tutorials or the actual system software.

Trainers will need to prepare practice sets and, possibly, practice data. They may also need such standard classroom items as videocassette recorders, tape recorders, and overhead projectors.

The most important training ingredient is "hands on" time.

Trainees

One of your early tasks will be to identify who needs training. Managers in the user organization will be able to help you select qualified personnel. Trainees will be anyone who interacts with the system or who directly oversees people who do. Some of the people included are people who prepare input data, data entry personnel, people who retrieve information, people who need to understand system-generated reports, production control clerks, operators, and managers.

Another consideration is the trainees' schedules. Can they be spared for all-day sessions? Should classes be only half days? And so on.

Learning by Doing

If trainees could say what is in their hearts, they would probably say something like this: "I won't read a big thick manual, but I like to have someone show me how a system works."

You can do this for them. You can give them demonstrations. Then you can give them the user document. But instead of expecting them to wend their way through the document unassisted, give them exercises to do, using the document for reference. They need to learn their way around the documentation. However, they are no more likely to just read the document than they are to "just read" a dictionary.

What do they need to know? Here are some typical tasks you could demonstrate and then have them practice. Note that many of them need advance preparation of files and data.

- Enter proper identification and authorization.
- Enter a customer payment.
- Retrieve an account balance.
- Run a spreadsheet program.
- Make corrections on an error report.
- Enter data on a point-of-sale terminal.

In addition to learning these types of skills, users may need to be able to demonstrate adequate speed. This is certainly true if customers will be waiting for results. You need to plan in advance that trainees must be able to enter a certain number of transactions within a certain time frame.

Purchased microcomputer software and word processing software often come with tutorials, which are usually easy and pleasant to use. You can plan written answer sheets that match the tutorials. This gives both you and the trainee evidence that he or she has reached a certain level of expertise.

The trainee needs frequent closure—that is, frequent successful completion of measurable tasks. Each successful event builds assurance to go on to other tasks. Eventually the trainees will reach their goal of using the new system with confidence.

Training Evaluation

You may want to give some sort of achievement test to measure trainee performance. This may need to be considered confidential, to minimize the nerves of those being monitored. At the least, you will want to have the trainees complete a form, indicating their opinion of the training and its value to them.

As the analyst, you need to know in what areas they are weak and what else they need to know. You can use this information to improve future classes and as a basis for follow-up training.

◇ CONVERSION

Conversion to the new system involves several major activities that culminate in the conversion of the system itself. These activities may be pursued in the same time frame: equipment conversion, file conversion, and procedure conversion. When these are done, you are ready for system conversion.

Equipment Conversion

Equipment considerations vary from almost none to installing a mainframe computer and all its peripheral equipment. If you are implementing a small or medium-sized system on established equipment in a major data processing organization, then perhaps your equipment considerations will involve no more than negotiating scheduled run time and disk space. If you are purchasing a medium amount of equipment, such as terminals and modems, then you will be concerned primarily with delivery schedules and compatibility. A major equipment purchase, however, demands a large amount of attention to detail.

For a major equipment purchase, you will need site preparation advice from vendors and other equipment experts. You will need to know the exact dimensions and weight of the new equipment. There have been infamous stories of the new computer being too big to get through the door; one computer, in fact, had to be hoisted by ropes through a window. Electrical capacity and wiring hookups must be considered. You may need to consider flooring: for example, there will probably be a need for raised artificial flooring to hide cabling and make access easy for repairs to large computers and related peripheral equipment. Finally, most medium-to-large machines need air conditioning and humidity control.

Microcomputer systems are far less demanding, but they too need some site planning, at least in terms of space available, accessibility, and cleanliness.

Meanwhile, if you are waiting for equipment, where do you test the new system? There are several possible answers, none of them convenient.

You could rent time on a similar computer in another installation. You could buy time on the same configuration from a full-service software house or equipment leasing company. Or you could have your own computer in place before testing begins.

File Conversion

File conversion can be a considerable problem. File conversion is often costly, time-consuming, and tricky. The problem is particularly difficult if you are automating a large manual system for the first time. There are usually many file drawers of information that need to be keyed to machine-sensible form. Finding the personnel and budget to make that change is only part of the problem. The real challenge is the timing.

When are you going to do it? If you wait until just before the switch-over to the new system, then you may have to assign a major task to an array of temporary personnel. One of the reasons this may not work well is that these people are not really part of the team and are less motivated to key the data correctly or to follow up each keying problem until it is resolved. For example, if addresses are being keyed in, should appendages such as Junior or III or PhD be included in the name? It is critical that file information be correct, so even small decisions are important.

Suppose you decide, instead, to use company data entry personnel and schedule their time over a period of months before the actual system conversion date. You could plan for them to key so many records per week. But there is a major problem with this approach. These may be active files, files that are in daily use and updated accordingly. The still existing manual files can be updated, but the keyed data cannot be updated because the programs that will do so are not yet ready. In other words, keyed records will stay static until the new system is in effect. Thus you have files that are out-of-sync: one set updated and the other not. A solution to this problem is to have update transactions keyed and stockpiled during this process. Then, later, when the new system is able to process them, use the transactions to update the file.

A far easier process can be used if the system already is automated. Many large paper-processing systems, such as those for banks and insurance companies, were automated as batch systems many years ago. Systems projects today often involve converting an old automated system into something that will better meet user needs. This means that the files already are in machine-sensible form and can probably be put into the required new formats by writing programs to do so. Although this can be tricky, too, it is, nevertheless, a giant step away from keying manual records.

Files for microcomputer systems can come from a variety of sources. They may not be converted because they did not previously exist. Some microcomputer files are relatively small and may be keyed in by the user of the anticipated application. An example of a small file is a list of phone numbers that can be dialed automatically by engaging software to place the call through the microcomputer's modem. Other files are not so small but are keyed by the the users in small amounts as the data occurs. An example is a file of classified advertisements for a small newspaper. Some data files, such as those used for spreadsheet programs, may be built by the user keying from reports produced by another computer system. And, finally, some microcomputer files are transported without keying from existing mainframe files. These can be anything from the employee vacation list to corporate financial data.

Procedures Conversion

Part of the documentation process is converting old procedures to new procedures. These are the new user documents and operator documents whose details we discussed in Chapter 4. These should be ready before training begins.

System Conversion

An interesting phenomenon often occurs when it is finally time to convert to the new system. Users who have been pulling out all the stops to get the system ready suddenly develop a paralysis. They do not want to make the final move to leave the old system. A series of delays develops. There is always some reason why the actual conversion is just around the corner. There is always something that must be done before the actual system conversion can take place.

This sounds somewhat like the old fear-of-change syndrome from the analysis phase. And, in fact, it is the same problem. Many people can face change as long as it is sometime in the distant future, but not when it is imminent. This may be the time to present the acceptance criteria you prepared during preliminary design. Using the testing documentation, you can demonstrate that the system does indeed meet all the acceptance criteria.

System conversion may not be as dramatic as all that. You may move smoothly into the change, especially if you have chosen a conversion method that introduces changes gradually. There are four basic approaches to system conversion, ranging from instant to gradual to long-term. Note the comparison of system conversion approaches in Figure 13-1. They are described in the following sections.

Crash Conversion.   As its name implies, a crash conversion is one that happens quickly. By decree, one day the old system stops and the new one begins. This usually is not very practical. A crash conversion method probably should be made only if the old system is nonexistent or so poor as to be unusable.

Pilot Conversion.   The pilot conversion method uses a chosen group of users to be the first to convert to the system. A typical application is an

System conversion.
In crash conversion,
everyone uses the
new system immedi-
ately. In a pilot con-
version, if there are
three user groups,
one begins the new
system as the pilot
group, with the
others following (in
this case) one per
month. In a phased
conversion each
group begins with
part of the new sys-
tem and adds more
parts gradually. In a
parallel conversion,
all groups use both
systems (old and
new) for some time,
until they finally let
go of the old system.

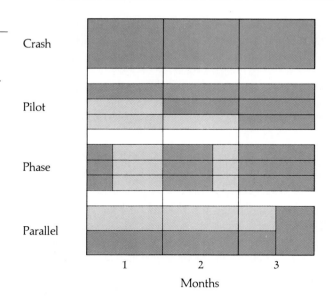

inventory system for a chain of sporting goods stores. The employees of
one store could be designated as the pilot group for the new system. There
are two key advantages here. The first is that the users thus selected can
be the ones that are the most supportive and the most enthusiastic. They
are the ones most likely to make the system succeed. The nervous can be
brought on board later, possibly encouraged by the pilot group that preceded them. A second advantage is that a limited group of people is easier
to manage, and so their needs can be met quickly. A group of satisfied users
is a powerful attraction to others.

Phase Conversion.   A conversion method that introduces only certain phases
of the system at a given time is appealing for several reasons. People can
get used to incremental changes gradually. If something goes wrong, there
is less chance that the consequences will be major. The entire process is
more manageable. Confusion is reduced. Consider a new microcomputer
system for a small agency that uses trained volunteers to staff a women's
shelter. The microcomputer will be used for word processing, for maintaining files of donors, and, together with a spreadsheet, for budget planning.
These separate operations are well suited to being implemented one at a
time. But many systems cannot be automated in piecemeal fashion. All or
nothing is required.

Parallel Conversion.   If a system processes data that is critical, then parallel conversion may be appropriate. The parallel conversion method processes
both the old and the new system at the same time and continues to do so
until users feel sufficient confidence in the new system to drop the old. An

appropriate application for parallel conversion is any system in which failure is not just inconvenient or expensive, but intolerable. Systems monitoring hospital patients or aircraft in flight fit that criterion, as do less dramatic systems such as bank cash machines. Of course, this dual operation is expensive. It is also difficult to manage. Since your goal is to demonstrate that the new system works, you will normally do this by comparing it to the old. But such a comparison is rarely a straightforward process since the systems may have different inputs and outputs, or at least different formats.

Sometimes it is appropriate to use a combination of conversion methods. Consider that phase conversion gives some of the system to all of the people, while pilot conversion gives all of the system to some of the people. But what if you gave some of the system to some of the people? This combines phase and pilot methods and may work well in some situations. Its main advantage is that the risk factor is lowered, since a chosen group of users receives selected parts of the new system, and all this can be carefully monitored. But, since nothing is ever guaranteed, let us consider some of the pitfalls of implementation.

◇ IMPLEMENTATION PROBLEMS: WHAT CAN GO WRONG?

Consider this theory: An experienced analyst who has planned the implementation phase with care should have the satisfaction of seeing the system come into being in a predictable, smooth fashion. This theory, as reasonable as it sounds, does not hold up in practice because there are too many things that can go wrong. It would be more plausible to say that an experienced analyst who has planned the implementation phase with care should be able to handle any problems that arise.

One of the key elements in handling implementation problems is the on-site availability of the analyst or some other knowledgeable person. The analyst can answer questions and coordinate new developments as they occur. Despite training, there will be many questions in the beginning. There will also be many unexpected developments. Here are a few samples of things that have gone wrong in systems implementation.

- A brokerage firm planned to send customers a printed notice of a new service being offered. The computer-generated letter was supposed to appear on dignified preprinted forms that did not arrive in time from the printer. A last-minute program change was needed to place the brokerage return address on the letter.
- Just as the sales tracking system for a publishing firm was to be implemented, two of the three trained clerical staff quit. The analyst

had to make time to give special training sessions for the replacements.

- The hardware for the new microcomputer system arrived in several boxes. The analyst consultant set about putting it together, only to discover that the cable for the microcomputer/printer interface was missing. Several phone calls yielded the information that the cable that was supposed to be included could not be found. Several more phone calls produced an alternative cable.
- A new neighborhood bank planned an unusual marketing gimmick —sending local people a "free sample of their product." They planned to mail computer-generated letters with a crisp dollar bill enclosed in each. At the last minute, top management balked at having a machine stuff the envelopes with the money, so the analyst and some users stayed up all night to stuff the envelopes by hand.

There are many other possible problems, some related to hardware and software, and many related to people. The latter are the most difficult to resolve. A manager in the user organization may feel there is no incentive to make his or her employees use the system correctly. Or clerks using point-of-sale cash registers may be too slow at first, causing customers to become irritable.

All these problems eventually will be solved. The system will be implemented. Then you can relax a little and feel some satisfaction. When the system has been operating more or less successfully for a period of time— weeks or months, depending on the size of the project—it is time for evaluation.

◇ SYSTEM EVALUATION

System evaluation is the process of measuring the existing system against expectations. The evaluation is usually made by someone other than an analyst associated with the project. A disinterested party is best, because this person has no preconceived notions or vested interests. In other words, an uninvolved person will have more objective perceptions of the system and how it is meeting its goals. The evaluator could be another analyst from the organization, or possibly an outside consultant. A team of two people— a technical person and one representing the user organization—would be even better.

This is the time to review the system requirements and the anticipated benefits. These should be compared with what is actually happening. Have the requirements been met? Are the benefits being realized? If not, why not? Tangible figures are needed to make a case one way or the other. Some examples:

- Is there a reduction in account errors? What percentage?

- Has absenteeism diminished? How much per average employee?
- Can the system operate with fewer employees? How many people have been transferred elsewhere? How many attrition slots deliberately have not been filled?
- Have sales increased? How much?
- Has product shipping time decreased? How many days or hours per average item?

The evaluation process may be lengthy if the system is a complicated one. Evaluators will need to gather statistics and do some interviewing and observation in order to assess system functioning and user satisfaction. When the evaluation report is made, the analyst and the user organization will decide what changes need to be made to tune-up the system to its full potential. Since the system has been designed, developed, and implementated, these changes fall under the category of maintenance, our final concern.

◇   SYSTEM MAINTENANCE

The point at which implementation is finished and maintenance begins is not always clear. At some point, the new system—available, being used, and processing data—must be declared implemented, even if it is not yet perfect. At some point, the providers of the system and the buyers of the system must agree that the system has, in fact, been delivered, and sign some agreement to that effect. The "sign off" or "buy off" or "delivery" papers may vary, but the intent is the same.

System maintenance is the continuing process of correcting, modifying, and improving the system. Some people define the systems life cycle to include maintenance of the system as one of the phases. Maintenance activities include aspects of other phases of the life cycle, such as defining new requirements, designing, coding, testing, and so forth.

System maintenance is a major activity in most data processing organizations. Over the years the number of systems that are completed and put into production continues to grow. This means that an increasing percentage of resources, especially programmers and other computer personnel, must be devoted to maintenance of existing systems. Note the change in the ratio of development time to maintenance time in Figure 13-2. Thus, the percentage of people available to develop new systems gradually decreases. Some installations devote as much as 80% of their resources to the maintenance function.

Maintenance of a system may become dormant eventually if there are no changes. Some routine systems continue producing the same information for years and need no attention until an unexpected program error causes a production problem. Other systems continue to bubble in a whirlwind of activity that never seems to subside.

FIGURE 13-2

Maintenance function. As new systems are completed and turned over for maintenance, an increasing percentage of computer resources must be devoted to this function.

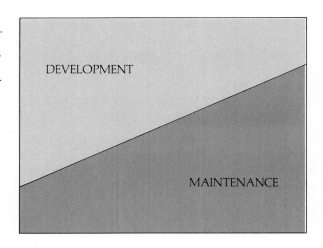

◇ THE VERY END

The system has been signed off. You are done. Or are you? You are finished with the original design and development of the system. But you may be asked to stay on the system for a period of time to implement the changes that users want and need. These are probably the same changes that were rejected somewhere in mid-development, when the users were told to wait until maintenance. Sometimes the list of anticipated changes grows so long that, even before development is complete, users and analysts alike are referring to maintenance as "the second stage" or some similar name.

But perhaps it is time to move on. One of your goals throughout the systems life cycle was to put in place a system that could be run without you. So now you can begin again. On another system.

FROM THE REAL WORLD

Training First, Micro Second

At United Technologies Corporation (UTC), they think they have their priorities straight. Company policy, from the top down, is that you go through a training program before you ever set eyes on your own new computer. This is in stark contrast to the traditional pattern of getting a microcomputer for an executive and then trying to get that executive to use it. At UTC, it's first things first.

United Technologies is a giant conglomerate whose subsidiaries include Otis Elevator, Pratt and Whitney Aircraft, and Sikorsky Helicopters. UTC has ordered 1,100 IBM Personal Computers for its top executives. Before executives get their micros, however, they must attend a 3-day training program. UTC wants people to be able to actually use the equipment once they get through the course and avoid having the computers become status symbols or dust collectors.

The executive training program involves extensive hands-on experience with the IBM PC. Most executives have never even touched a keyboard. The element of fear is removed right at the beginning. Before the trainees quite realize what they are doing, they are actually using the computer. By the third day it is all coming together in a simulation of an everyday business situation.

The course centers on integrated software, a combined spreadsheet, database, word processing, and graphics package. The minicourse begins with the basics and culminates with a project in which the trainees apply their newfound computer know-how. Each executive is "given" $20 million and an opportunity to play movie producer. They have to draw up a budget and bid for certain items, then go back and make adjustments. They use the database, word processing, spreadsheet, and graphics functions of the software. They really get involved and apply everything they have been learning to day-to-day kinds of work.

UTC has embraced a strictly hands-on approach. The pilot test of the new course included about an hour of computer literacy. The executives insisted that it be taken out. They were not interested in learning general information *about* computers. They wanted to turn the thing on and find out how it works.

—From an article in *Popular Computing*

SUMMARY

- Implementation planning must be started well before the implementation phase. The written plan includes information on training, equipment installation, file conversion, procedures conversion, system conversion, and evaluation.

- During implementation, problems that arise should be handled promptly. But this is not the appropriate time to consider new user requirements. New requirements must be postponed until system maintenance.

- The training staff will probably be the analyst alone, although training by vendor is another possibility. If the number of trainees is very large, the analyst may want to "train the trainers" so that they can instruct their own colleagues.

- Although trainees need documentation on how to use the system, they also need to practice using the system. Training tools needed may be terminals, modems, microcomputers, printers, visual aids, software, practice sets, and practice data.

- The trainees will be anyone who interacts with the system or who oversees people who do.

- To help trainees learn by doing, the trainer can give demonstrations and then assign exercises that require the trainees to use the documentation for reference.

- Training evaluation can be in the form of an achievement test or, at least, a survey form indicating the trainee's opinion of the value of the training.

- Equipment conversion can vary from almost none to installing a mainframe and all its peripheral equipment. For a major equipment purchase, some site preparation considerations are dimensions and weight of the equipment, electrical capacity, flooring, and air conditioning. Microcomputer site planning includes space availability, accessibility, and cleanliness.

- File conversion may be costly, time-consuming, and tricky. Automating a large manual system for the first time means many records must be keyed to machine-sensible form. This keying effort can be done at the last minute with temporary personnel or over a period of time, saving update transactions to be used when the new system is able to process them. If the files are already in machine-sensible form, they can be put into the required new formats by programs. Microcomputer files may be keyed or transported from existing mainframe files.

- Part of the documentation process is procedures conversion—that is, converting old procedures to ones that match the new system.

- System conversion approaches are called crash, pilot, phase, and parallel.

- Crash conversion happens quickly, as the old system is simply switched to the new system. This is not usually practical.

- Pilot conversion lets a chosen group of users be the first to use the system. These first enthusiastic users can encourage those who follow.

- Phase conversion is a method that introduces only part of the system at a time, letting the users get used to incremental changes.

- Parallel conversion means that both the old and new systems operate at the same time and continue to do so until users feel confident enough to drop the old system.

- Although an experienced analyst plans with care for the implementation phase, there usually will be things that go wrong.

- System evaluation is the process of measuring the existing system against expectations. The evaluation is usually made by someone other than the analyst on the project.

- System maintenance is the continuing process of correcting, modifying, and improving the system. Maintenance is a major activity in most data processing installations.

KEY TERMS

Crash conversion	Procedures conversion
Equipment conversion	System conversion
File conversion	System evaluation
Implementation planning	System maintenance
Parallel conversion	Training evaluation
Phase conversion	Training tools
Pilot conversion	

REVIEW QUESTIONS

1. The chapter alludes to the fear-of-change syndrome once again. Why, at this point in the project, would users be afraid of change?

2. Describe, in a general way, possible implementation planning for the following systems. For each system, consider training, equipment, files, procedures, system conversion, and evaluation. Since the details are very sketchy, make any needed assumptions.

   A system for an amateur pilots association will keep track of files of planes that are available for rent from members. The file will be kept by office staff on a newly purchased microcomputer.

A group of dentists in the same downtown medical building have contracted with a service bureau to purchase and share an in-house minicomputer, with terminals and printers in each office. The system will provide standard packages for accounts payable and accounts receivable, and permit the dentists themselves to use a spreadsheet package for budget planning.

A food brokerage employing 42 people has used a service bureau for the payroll function, delivering time cards to the bureau and receiving paychecks back. Last year the brokerage purchased a small minicomputer and gradually moved most of its financial activities onto the computer. Now the company wants to move the payroll function also, using a purchased software package on their own machine.

An aircraft engine company has used its in-house computing staff to automate engineering drawing records. The clerical personnel, who are used to working with manual records associated with each part and assembly, will now input data to the mainframe system via terminals and use computer-generated part and assembly reports.

A credit union, working with a consultant, is purchasing a total set of credit union software packages, an in-house minicomputer and associated peripheral equipment, and terminals for each agent.

3. A group of experienced computer users that is switching to a new system will be quite different from a group that is about to use a computer system for the first time. Discuss the differences in your approach to training these two groups.

4. Make up an appropriate application for each of the four system conversions mentioned in the chapter: crash, pilot, phase, and parallel.

5. In the systems described in question 2, think of things that might go wrong at implementation time.

CASE 13-1
I WON'T READ IT—SHOW ME

This particular case took place during the implementation of a laboratory records processing system. Several dozen users had to be trained on 18 online inquiry and update functions. The most important element of the training was a very thorough users manual produced at a cost of 150 analyst hours, plus word processing, copying, collating, and binding costs. In the manual, each of the 18 functions was described in detail in several pages of outline-style text.

The users were gathered together in small groups. Each group was given a cursory demonstration on a terminal and then referred to the written manual for more detailed information. Questions were answered at these sessions and for several weeks thereafter. Telephone calls to the analyst were frequent.

This may seem like a successful training procedure, but here is what actually happened. During the initial intense training period, only 10% of the trainees actually read the manual. Instead, they listened carefully to demonstrations, and many took notes. These notes were referenced continuously, and the users discussed the new functions among themselves. Within each group were one or two who understood quickly and became local heroes by helping everyone else in their department.

Most of the learning was "show and tell," not "write and read." Although all questions asked during the initial period were answered in the manual, the most frequently consulted reference was a piece of paper with the name and telephone number of the analyst/instructor. Every terminal had one of these taped on it. Half a year later, no one else had read the user manual. Update pages to the manual had not been filed. In fact, several users could not find their copies. New users were trained by experienced users, again using show and tell. The most referenced paper during this period was a list of transaction codes. The list had been jotted down by a user during one of the training sessions and had been photocopied and passed along to others.

Three years later the users' manual still remained unread. Training was still by word of mouth. Cheat sheets abounded. One-page summaries of the hard-to-remember codes appeared on each screen. When the cheat sheets became too dog-eared and coffee-stained, users stuck them into glassine folders.

1. This story seems to imply that manuals are of no value. What went wrong here?
2. Is there a way to combine training in the use of the manual with demonstrations and hands-on training? Discuss.

CASE 13-2
AMATEUR HOUR AT CONVERSION TIME

Krider Moorage Management has contracts to handle 17 separate boat moorage facilities in Long Island Sound. Since moorage space is always at a premium, prices are high and the moorage business is lucrative. At a minimum, moorage management involves keeping the facility in working order (electricity, hoses, parking, security gate, and so on), and tracking the paper flow of payments, title transfers, waiting lists, invoices, and taxes. Owner Alan Krider was considering taking on other moorages, ones that catered to owners who wanted to live aboard. Before he tackled those extra problems, though he felt he needed to streamline his paper processing operations.

Each moorage, from the smallest at 60 slips to the largest at 450 slips, has always operated autonomously through a local on-site manager. All paperwork has been handled locally, with summary reports going to Alan on a scheduled basis. Alan wanted to retain some local autonomy but, at the same time, give himself more management visibility and overall control. After talking with four consulting firms, Alan chose McHale Software Systems (MSS). Alan and analyst Boyd Fletcher decided on a network of microcomputers, one per moorage facility and another for Alan's headquarters office. Information could be transferred among the microcomputers and Alan could query status files in any of the moorage facilities from his own computer.

The phases proceeded fairly smoothly, with Boyd working through a local vendor to determine and assemble the hardware and software packages. Boyd contracted out most of the conversion activities, arranging to send the moorage managers to a class sponsored by the vendor, and hiring a temporary service company to key the data files. The only troublesome point was that a few of the managers were grudging participants in the automation process.

Boyd decided that, since there was no old system to speak of, the new system could be brought up all at once, and he set the date for June 1, the beginning of the company's fiscal year. Meanwhile, Boyd left for Europe for an extended vacation on May 15, leaving the new system in the hands of his assistant.

June 1 was a disaster. Nothing went right and nothing could be fixed. The assistant did not assist. On June 15, Alan called in a consulting firm. The whole operation eventually was salvaged but it was early September before it was running to everyone's satisfaction.

1. Alan was heard to remark that Boyd had not "stayed close to the product." He was not referring to his vacation. In what ways could Boyd have monitored the conversion more closely?
2. What changes in conversion procedure would have increased the probability of success the first time around?

PART 5

SYSTEM CONTROL

Enlightened project management, from start to finish, is the heart of project control. The text presents a thorough discussion of project management, including the topic of estimating, which even a new analyst must master. Once the system is in place, it must be safe from accidental or intentional harm. The security chapter gives a complete presentation of all security-related issues.

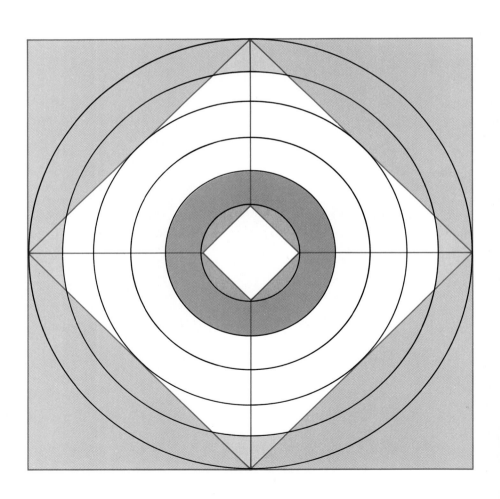

Chapter 14

Project Management

===

When Susan Nelle was hired as a programmer trainee at Metro Transit, she was immediately caught up in a myriad of technical details. Although knowledgeable in two programming languages and related topics, she felt she had to run fast just to keep up. Gradually she picked up the tools she needed to program effectively.

She worked on maintenance programs for a while, then did some one-shot developmental tasks. She was accountable to her supervisor for both of these activities, planning her own schedules and submitting oral progress reports that her supervisor charted. This effort seemed but a footnote to her technical activity.

When she joined her first full-fledged systems project as a programmer, however, all that changed. The planning, the schedules, the budgets, and the reports seemed to take on an overwhelming importance. She began to wonder if her project manager's only concern was time and money. Although she later was able to put these activities in perspective, the attention to management activities did not diminish. Susan found out that to be a part of the business world meant taking some responsibility for business functions. The rationale and techniques for project management will become clear in this chapter.

◇ THE RISE OF PROJECT MANAGEMENT

Now that you understand the systems life cycle, you can begin to appreciate what it would take to manage an entire project. With today's rapidly expanding computer resources and ever greater demands for new systems, the subject of project management has come into critical focus. More and more industries have accepted the concept of project management as the tool to make efficient use of corporate resources.

For our purposes, we can say that **project management** refers to the process that covers system planning, priority setting, effective design, and awareness of the project status. The difference between being a first-line supervisor managing a group and a project manager managing a project is akin to the difference between owning and renting. A project team is a group of people temporarily chosen for their special skills and availability. Their charter is to complete a given systems project, after which they may disperse and be assigned to other project teams. Meanwhile, the project manager may move on to manage another project or may put his or her technical skills to use in a non-managerial role. Once a project manager, not always a project manager. In this chapter we shall examine the nature of management, the special needs of project managers and their teams, and the techniques of project management. We begin with a brief overview of the management function.

◇ THE TASKS OF MANAGEMENT

The tasks of management are well established and have been discussed endlessly in everything from professional journals to pop psychology magazines. Whether managing an assembly line or a bank or a systems project, the challenge is the same: Use available resources to get the job done on time, within budget, and to user satisfaction. Let us begin with a discussion of how managers do their jobs.

Classic Management Functions

Any beginning business book will tell you that the basic functions of management are planning, organizing, staffing, directing, and controlling. All managers perform these functions as part of their jobs. The level of commitment to these functions, however, varies with the level of the manager (see Figure 14-1). Whether managing General Motors or a small Chevrolet agency, top-level managers must be primarily concerned with the long-range view, with **planning**. What do consumers want or need to buy? What factors may change their wants and needs in five years? In ten years?

Middle-level managers must be able to take a somewhat different view because their main concern is **organizing**. The middle manager will prepare

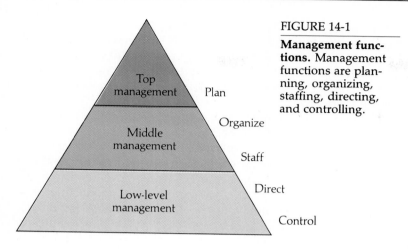

FIGURE 14-1

Management functions. Management functions are planning, organizing, staffing, directing, and controlling.

to carry out the visions of the top-level managers, assembling the material and personnel resources to do the job. Note that these tasks include the **staffing** function. To a production manager, this may mean putting assembly lines into place, staffing them with skilled labor, and ordering materials and subassemblies from other manufacturers.

Low-level managers, usually known as first-line supervisors, are primarily concerned with **directing** and **controlling**. Personnel must be directed to perform the planned activities. These managers must also monitor progress closely. In the process, the supervisor—an assembly line supervisor in this case—will be involved in a number of issues: making sure that workers have the parts needed when they need them, checking employee attendance, maintaining quality control, handling complaints, keeping a close watch on the schedule, tracking costs, and much more.

Can these skills be transferred laterally to the systems arena? Yes and no. Some people say that a manager is a manager and the subject of that management is not germane. Hence, we have the top-level manager from a soft drink company shift to running a company that manufactures personal computers. Shifts at the very top levels can be successful because long-range planning skills, their principal activity, are more universal than the skills needed at the lowest detail level. Shifts into technical areas at lower levels may not work well.

The Project Manager

You can plan on problems if a systems project manager is drawn from nontechnical ranks. The project manager position is comparable to that of a first-line supervisor, with some notable variations. The first is that the people and the subject matter—the project—change at regular intervals. The second is that there is a distinct need for technical skills. A project manager who has not personally programmed or specified or designed or tested a system has no firsthand knowledge of what his or her people are doing. The project manager is thus unable to respond to delays or user concerns and, further, is at the mercy of an incompetent staff or external forces. In short, a person who does not understand the computer environment is rendered ineffective.

The **project manager**, sometimes called the **project lead**, is a special breed of manager, filling both technical and managerial functions. Project managers often come from the ranks of senior systems analysts, who have extensive analysis and design experience. They are strong in traditional analyst skills. These same skills may lend themselves well to the usual management functions of planning, organizing, staffing, directing, and controlling. You can also expect an experienced analyst to have good communications skills. These skills will go a long way toward helping the project manager handle the personnel tasks associated with a project.

Desirable characteristics for a project manager vary slightly from those of a systems analyst. The first characteristic on the list is leadership. A successful analyst will have some leadership qualities, but the project manager should be a superior leader, guiding team members in a way that inspires their confidence. Next, the project manager will have business acumen, that is, an ability to view the project from the perspective of overall company goals. A third important quality is a talent for "keeping a hand on the product," that is, maintaining an awareness of the project status at all times. This awareness is represented primarily in the form of schedules, budgets, and user needs.

A project manager may be assigned full time to a medium or large project, or may have several small projects at the same time. In either case, a key responsibility is staffing and organizing technical personnel.

◇ MANAGING TECHNICAL PERSONNEL

"Give me ten good analysts and programmers and I will build the best computer system in the company," said Liz Marden. Liz is probably right, but there are a few other factors to consider. Liz and her ten good people will probably cost the company half a million dollars a year in salaries, space, vacations, and employee benefits. With ten people of designated exalted station, who does the fetch and carry work, the kind that is plentiful on every project? Any of the chosen ten will be overqualified for these chores. If Liz succeeds, what does she do with the ten people (who love challenges, impossible deadlines, and so on) after the project is finished—put them on maintenance? Clearly, the composition of teams needs careful thought. This is one of the main issues related to managing technical personnel.

Let us examine this dual challenge for the project manager, first finding suitable team members and then putting them together in a workable way.

Finding the Right People

As a project manager, one of your functions is to staff the project. In fact, staffing involves filling slots, hiring, firing, and training. You may have a pool of analysts available to you, from which to choose team members for an upcoming project. As sensible as that sounds, however, it is almost never true. You often have to take whoever is available, that is, whoever is just finishing up work on another project. This limits your choices and gives you a person who is still locked in to the old project and may have a hard time breaking away.

If you have the option to hire from within or to truly hire new people from the outside, the interview is all-important. Although the resume may guide your choice of interviewees, and references may back up that information, the real information will be obtained in the interview. Give only passing attention to the usual interview considerations such as whether the interviewee looks you in the eye or understands the mission of your company. Emphasize technical questions—specific, closed, technical questions.

When hiring new people, the interview process should include technical questions.

Ask questions directly related to the work the interviewee will do. Many managers who have been burned by hiring the interviewee with the strong handshake keep repeating their error. Go for the technical questions and you will not regret it. Good team members, of course, are not guaranteed by technical expertise. Other factors must be considered; technical expertise is emphasized because so many employers seem to ignore it.

Having found good people in one way or another, you must now organize them in some appropriate manner.

Planning the Systems Team

The composition of the systems team varies over the life of the project. In the beginning phases of the life cycle the team will be user-heavy, particularly if several organizations are involved. As the project takes on a more technical flavor, some users will drop off the team and more technical people will be added. By the time the project has entered the development phase, the team will consist mostly of analysts and programmers.

Assembling the team may appear to be a straightforward matter, but there are several philosophies of teamwork that deserve consideration. Remember that a systems project team is not like the standard group of people working for a supervisor in a business organization. The latter group, although having the same manager, may have few functions in common. A team, however, as indicated by its name, is a group of people working together toward a common goal. A good manager may be able to elicit from the members a high level of dedication to the team and the project. Team members work for success on *this* project—each project provides a new opportunity to be a hero. This enthusiasm could be called by its common name: team spirit.

The project manager must know how to place people in teams. There are several possible ways of doing this. Picking the right method will depend on the roster of people available, the nature and size of the project, company policy, and the personal leadership style of the project manager.

Chief Programmer Team. The title "chief programmer" was first used formally by Harlan Mills, who promoted the idea of using just one person's ideas for a project. That one person is called the chief programmer. Since that person is likely to be a senior analyst, the "chief programmer" name may seem a little dated, but the concept is still in use. If you imagine a rather small project with five or six team members, having just one set of ideas may seem a natural state of affairs. But there are projects large enough to have dozens of team members, and this situation sets the stage for a potpourri of ideas that do not mesh. By using just one person's system concepts, and other people in supporting roles, the integrity of the system logic is maintained. Typical supporting roles are assistant (or alter ego) to the chief programmer, language specialist, administrator, documentor, tester, and so forth.

As you can see from this description, a chief programmer team is most suitable for a very large project. In fact, some large projects are broken into a series of subteams, each having a chief programmer. The chief programmers must then meet on a regular basis to coordinate ideas.

Specialist Teams.   Although a subset of the team remains constant throughout the life of the project, specialists act as a supporting cast. Specialists may be assigned to one team at a time or, if the projects are in need of only minimal services, more than one team at a time. Specialists can fill needs in areas such as data communications, network architecture, access methods, and database management.

Specialist teams are often a necessity for projects that involve a variety of technologies. An example is a system involving distributed databases. The average analyst cannot be expected to work closely with the user and also possess detailed expertise in all the hardware and software aspects of such a system.

Peer Group Teams.   Teams designated as peer groups—that is, groups of equals—are considered more democratic but may be more difficult to manage. True egalitarianism may dilute responsibility and weaken the reporting structure. Thus, even if the team structure is internally a peer group, you need a single point of responsibility for reporting purposes. This person is the project manager. This type of team may stay together from project to project, with members taking turns as project manager. Peer groups are successful for small or medium-sized projects, when many or all of the

Meetings between members of a systems team do not always take place in an office.

team members are experienced and when management has a commitment to an egalitarian philosophy.

Even as the team is being assembled, the project manager is handling the daily activities of the project, answering questions, acting in a liaison role between users and programmers, and monitoring schedules and budgets. Managing a software project is a complicated task.

◇ MANAGING SOFTWARE PROJECTS: AN OVERVIEW

As a project manager, one of your most important tasks will be preparing estimates of resources needed to complete the project. Estimating is a difficult job and one that continues from phase to phase throughout the project. A poorly estimated job is doomed to failure in both schedule and budget. However, the best estimates in the world will not hold up if the project is not carefully monitored, so controlling work in progress becomes equally important. Finally, the user must be satisfied with the product. This satisfaction will not just happen automatically, nor can you just hope for the best in the end. Every aspect of the project must be carefully coordinated with users, to ensure their ongoing satisfaction.

Let us now examine these critical project management elements, first estimating, then controlling, and, lastly, meeting the needs of the user.

◇ ESTIMATING: THE HARD PART

Judy Carrithers, an analyst with three years experience, recently asked her colleague Rob Speaker how to estimate how long a project will take. Rob was a little surprised by the question, because he knew Judy had been making estimates for some time. Judy was equally surprised by Rob's answer. He said that you never really get any better at estimating because you are forced to pull numbers out of the air before you have a thorough understanding of the project. Both Judy and Rob need help. The help they need is more detailed than either of them imagine.

Estimates Based on an Educated Guess

New analysts often make estimates with a certain élan based on self-confidence. They invariably underestimate. Thus burned, they then make a direct shift to the side of caution. Even so, their estimates are likely to be wide of the mark. We have noted that many organizations have an unattractive record in the area of project management, with schedule slippage and cost overruns predominant. Some organizations think this is normal. "If you don't understand delays you just don't understand data processing," said one stalwart who seems to find failure inevitable. The key reason for a string of disappointments is not the nature of data processing, however, but poor estimating techniques.

Perhaps we should say, instead, *no* estimating techniques. Top-of-the-head judgments loosely based on experience, coupled with hunches and wishful thinking, hardly constitute a "technique." As uncharitable as this description sounds, many organizations prepare their estimates in roughly this manner. One common approach is to base estimates on undocumented experience, especially if the looming project "looks like" or "feels like" a previous project with which the estimator has had some experience. Some analysts automatically pad their schedules with a safety factor. On the other hand, some managers suspect analysts of padding and arbitrarily shave a certain percentage from the original estimate. Another management strategy is to agree on a schedule with the analyst and then pad it (without telling the analyst) before turning it in at higher levels. This gives the manager a little cushion if the project is late. These cat and mouse games are distressing, to say the least, and one wishes there were a better way. There is.

Estimating Techniques That Work

Estimating can be accurate. Some estimating techniques, however, are more accurate than others. **Historical estimating** extrapolates data from historical records. In other words, the organization keeps careful records on projects, accumulating statistics related to project activities and associated time intervals. These records provide a reasonable basis from which to plan estimates for future projects.

Although more formal than playing hunches, there is a danger in historical estimating. The weakness of this method is that no two projects are alike and that what applies to one may not apply to another. A rather large array of variables, from the experience of the personnel to the complexity of the project, may render the historical records useless. Still, in organizations with stable personnel and a degree of similarity in its projects, the historical method has proven satisfactory.

Another approach is **standard formula estimating**. Such formulas, some supported by industry giants like IBM, propose weights for certain aspects of the project. Are the technical personnel experienced or are they green? Are the users sophisticated or are they naive? In both cases, the "untried" factor weights the estimating formula more heavily. From a software point of view, you could be more specific. For example, you could rate a split-screen output more heavily than a single-screen output. And so on. If you are going to use an existing formula of some kind, you may find it both attractive and repellent. Its attraction lies in its formality, which makes it less susceptible to local fudge factors. You may be repelled, however, for almost the same reason: It may be too inflexible to bend to the special needs of your organization and your project.

A superior estimating technique is known as **checklist estimating**. Its hallmark is that it is comprehensive but still flexible. You will see that it resembles the standard formula method just discussed, but it has more advantages and fewer limitations. Note these true statements:

- The basic problem with project estimates is that they are too low.
- The reason for low estimates is that they are usually made at too high a level, a single number for a large task.
- Estimates done at the detail level will always have more hours than estimates done at the summary level.

From these statements, you can conclude that detailed estimating is needed.

For each project task, a checklist is prepared. The checklist should contain every measurable detail associated with the task. For example, rather than estimating the time to code and test an edit program, estimates will be prepared for checklist subtasks such as studying specifications, conferencing with analyst, planning the program design, studying manuals, coding, keying the program, desk-checking, planning test data, keying test data, and on and on, the more detail the better.

When checklists for each task are complete, a typical hour estimate is assigned to each item. Estimating teams produce better estimates than one person working alone. (An individual may be reluctant to admit how long something really takes.) If possible, include a user on the estimating team—let the user see first-hand why a project takes so long. Once the initial estimates have been made, they need to be adjusted according to factors affecting this particular project. These factors can relate to a variety of issues: a learning curve for new software, an inexperienced team member, machine turnaround time, and other environmental concerns.

Since there may be hundreds of items on the checklist, many companies have prepared standard checklists that project managers can modify to suit themselves. Although clerically tedious, the checklist estimating method has been found to be the most effective.

The quality of the estimating process will influence the credibility of the project team and the project manager when the work is complete. Few tasks are more important. But even a well-estimated project can fall victim to a variety of problems that impact the schedule, including requirements changes and key employees quitting. As all managers come to learn, a project must be monitored closely if the plan is to become reality.

CONTROLLING WORK IN PROGRESS

Organizations that produce on-time projects have certain characteristics in common: careful estimating, close monitoring, and the flexibility to deal with the problems that inevitably arise. Estimating has been discussed in some detail. The third characteristic—flexibility—is related to the leadership qualities of the project manager and the resources provided by the organization. The project management activity we consider in this section is monitoring. There are several approaches to project monitoring. Some managers find one more useful than another, and some managers use them all.

The common element in the monitoring techniques is finding a way to measure progress. Verbal reports such as "It's coming along" or "The coding is 90% done" are meaningless and unacceptable. A more specific and tangible device is needed. Measuring devices that relate directly to planned estimates are the Gantt chart, the milestone chart, and the PERT chart. Another commonly used device is the periodic progress report. These devices can be used in different ways. A common use for the project manager is to get individual reports from team members, relating to their specific responsibilities. Then the manager combines these into a report to be submitted to higher management. Let us examine some reporting techniques.

Gantt Chart.   A Gantt chart is both a planning and a reporting device, as we noted earlier in Chapter 7. In its simplest form it is just a horizontal bar chart, laying out activities on a time line. It is common practice to have a master Gantt chart for the entire project and individual Gantt charts for subtasks assigned to individuals. Some organizations like to measure progress by running a parallel line along each activity bar, sometimes in a different color.

Milestone Chart.   Some managers consider a milestone chart to be the key to controlling a project. A milestone chart differs from a Gantt chart in that it marks specific events with a due date, as opposed to designating a period of time for activities. Rather than noting the end of a phase, the milestone chart notes certain activities whose completion is tangible and obvious. Milestones should be staggered throughout the systems life cycle. A sampler of milestones culled from various phases of the project might be: questionnaires tallied; requirements document approved; report formats accepted by users; new terminals installed; user training complete; and files converted. Note that each milestone mentioned is specific and measurable.

For purposes of illustration, consider a project to publish a textbook. Major milestones are listed in Figure 14-2. The chart represents the time at the end of period 14. The open inverted triangle indicates scheduled completion of an activity and the shaded inverted triangle indicates completion. As you can see, copy editing was late and so was type setting. The last three activities have not yet begun.

As a monitoring device, a milestone chart can be marked in a variety of ways. Marking completed milestones in black while highlighting late milestones in red is one way to draw attention to problem situations. Some organizations make such charts large and keep them on office walls where the wayward items can be seen by all. This exerts more pressure than any words could.

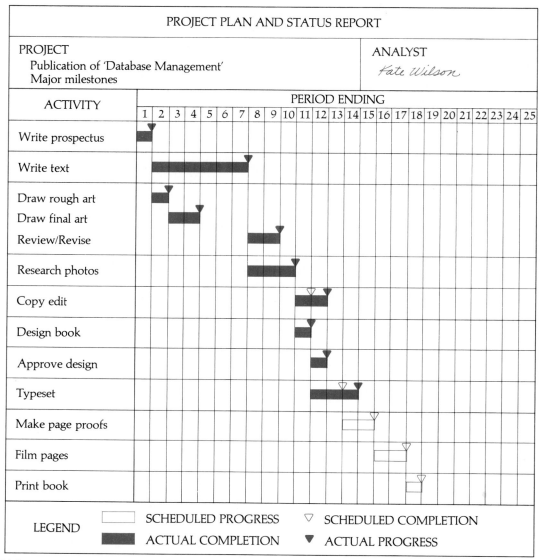

FIGURE 14-2

Milestone chart.

PERT Chart.   PERT is an acronym for program evaluation review technique. An interchangeable name is CPM, for critical path method. A PERT chart is a network of events connected by activities. An **activity** is the use of people and other resources over a period of time, usually measured in terms of weeks. Activities are represented on the PERT chart by arrows. An **event** marks the beginning or end of an activity, and is represented on the chart by a circle. An event often corresponds to a milestone. See Figure 14-3.

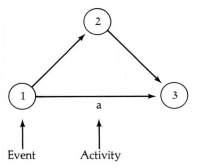

FIGURE 14-3

PERT chart elements.
An activity is repre-
sented by an arrow
and an event is repre-
sented by a circle.

The PERT chart is constructed by drawing the relationships between events. The main idea in the construction process is to note which activities are contingent on the completion of prior activities. Now to the heart of the matter: The manager drawing the chart looks for the longest path, the greatest number of weeks, from beginning to end. The longest path is called the **critical path.** It is called critical because any schedule slippage in the activities on this path will cause slippage in the entire project. Delays in activities on other paths may not impact the project at all.

A PERT chart is developed and used to monitor the progress of various system events *in relationship to each other*. Note the book publishing activities in Figure 14-4. These are the same activities we looked at in the milestone chart, but activity dependencies can be seen on the PERT chart. For example, activity 10–11, film pages, must be complete before activity 11–12, print book, can begin.

Other relationships are a little trickier to depict. For example, activities 2–3, draw rough art, and 2–4, write text, must both be complete before activities 4–6, review/revise, or 4–5, research photos, can begin. In order to show this dependency, a dummy activity, 3–4, is introduced. Nothing is really happening in activity 3–4; it is there only to note the requirement to complete 2–3 before beginning 4–5 or 4–6.

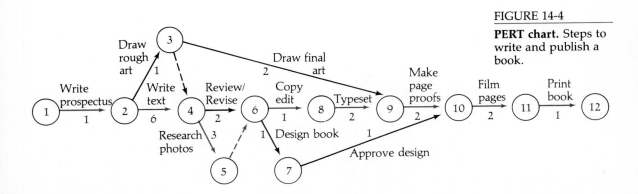

FIGURE 14-4

PERT chart. Steps to
write and publish a
book.

Note the critical path in Figure 14-4: 1–2–4–5–6–8–9–10–11–12. The critical path is normally highlighted in some way.

A PERT chart is most useful when there are so many activities that it is difficult to keep track of the interaction among them. PERT was developed, in fact, for the United States Navy Polaris submarine program, a rather large undertaking with thousands of subtasks. A PERT exercise would not be appropriate for small or, possibly, medium-sized projects.

Progress Report.   Also referred to as a status report, a progress report is a brief document that accounts for the project during the period of time since the last report. If a progress report is submitted weekly, as is common, then the reporting period is the previous week. Generally, the report will cover anything of significance that has happened during the last reporting period. Its purpose is threefold: It is a communication vehicle, an historical document, and a management control device.

A progress report is concise and to the point, as opposed to a meandering stroll through the thicket of project details. Topics customarily covered in the report are:

- *Accomplishments*. The list should be as specific as possible, but not include the details of *how* each item was accomplished.
- *Problems*. Mention all problems encountered, especially the unexpected, and their resolutions.
- *Follow-up items*. Report on items raised from the last report, by either you or your manager, and their progress or resolution.
- *Schedule progress*. This brief narrative will probably be accompanied by one of the reporting charts described earlier, marked appropriately.
- *Budget report*. Describe the budget used on various aspects of the project during the reporting period. If the progress report is submitted frequently, however, it will probably not be the vehicle used to report budget expenditures.
- *Plans*. Note any deviation expected in the next reporting period. An example would be task revision due to expected vacations.

Some organizations like to use a printed form for progress reports. Note the sample in Figure 14-5. Many organizations have automated some or all of their reporting techniques, which is a very appropriate use of the computer.

◇ PROJECT MANAGEMENT SOFTWARE

Project management is a major category of software. Software vendors, eyes gleaming, speak of project management software as "the next spreadsheet," meaning they expect demand for it to be comparable to the demand for the popular financial analysis tool. Project management software has become

PROGRESS REPORT	
Project Name	Report Date
Project Number	Reporting period: From _____ To _____
Project Manager	

Original Due Date	Estimated Due Date	Comment
Budget To Date	Cost To Date	Comment

Completed Activities

Problems

Action Items from Last Period

Action Items for Next Period

General Comments

FIGURE 14-5

Progress report. This progress report format is representative of those in common use.

a fundamental business tool. Indeed, there are already over 300 such packages, from the inexpensive program that displays only the length of each project task, to programs that offer complex project planning and scheduling tools.

Project management software can be used to manage any type of project, from construction to manufacturing to moving to new facilities. But our discussion here centers on using such software for systems analysis and design projects.

What Project Management Software Can Do

A **project management software** package is one that allows a manager to quickly and easily enter information on a project, and then see when various portions of the job will be completed and which resources are being used on different tasks. These outputs can be seen graphically or in tables. And, like spreadsheets, project management software can be used as a what-if tool to project the impact of delays or revisions, or to try out plans on-screen before implementing them. For example, a manager can determine how much overtime will be necessary to bring a project in on time if one stage slips by a week.

Once a project is underway, the project manager may discover that a task cannot be completed on time or that certain resources cost more than originally expected. For example, a project manager may have fewer in-house programmers available than anticipated and decide to contract out some of the tasks, an event that increases their cost by 20%. Or perhaps certain machine resources are available for testing only after a certain date. How will these changes affect the rest of the project? Can the project still be completed on time and within budget? Project management software allows you to make these and other, perhaps compensating, changes and then view the resulting project status.

These screens are samples from *Microsoft Project*, a popular project management software package.

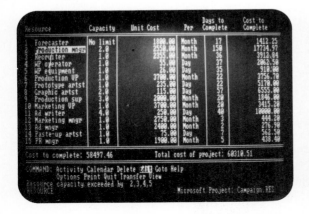

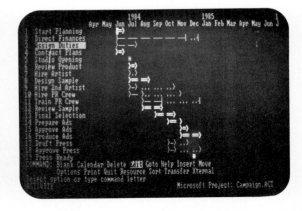

How Project Management Software Is Used

Project management software is not new. Various programs have been available on mainframes for some time. But this power is now available via the personal computer in a form that, unlike the obtuse mainframe software used by professional project planners, is easy to learn and use.

Part of the success of a project depends on making optimal use of time and resources. Project management software can help managers oversee complex projects effectively in three ways. The first is by dividing the overall project into manageable tasks. The second is by assigning resources (people, machines, materials) to each task. Many packages let you define resources in a table before assigning them to tasks. The third is by reassigning resources to tasks as a situation changes.

The basic concept of project management software is relating tasks to resources. Each project is broken down into tasks. In a project to select an accounting software package, for example, some of those tasks might be: interview users, write requirements, locate vendors, prepare request for proposals, review proposals, evaluate vendors, select vendor, test system, and train users.

Descriptions of tasks and resources are entered, using the project management software. The program can create graphs that represent project status and the flow of work. Gantt-type bar charts on a time line are the most common. But PERT charts are also used because they are an excellent device for visually depicting the dependent relationships. If a task cannot be accomplished, for example, until several earlier tasks are finished, each of those tasks is linked by a line to the subsequent task.

BOX 14-1	**Putting Project Management Software to Work**

Step 1: Identify project resources.

Step 2: Establish a list of the tasks necessary to accomplish the project.

Step 3: Identify important midproject milestones.

Step 4: Assign the necessary resources to each task, identifying what resources a task needs and for how long.

Step 5: Note the prerequisite relationships for each task.

Step 6: Enter the project data, review it for accuracy, modify accordingly.

Step 7: Print project time and relationship charts.

Step 8: Review, update, and reprint as necessary.

Time and Cost Projections

The basic job of project management is monitoring time. You can, for example, display graphs to show completed, partially completed, and incomplete tasks. Or you can key different approaches to using the time resource. Typical what-if time scenarios that can be played out are: preplanning more help during critical stages of the project, comparing the costs of speeding up one phase of the project versus adding overtime later in the project, and doing jobs in-house as opposed to contracting them out.

Managing the project calendar is a critical time item. Since time is a fundamental element of any project, project management software uses calendars to monitor time devoted to the tasks that make up the project. A project may be scheduled to start on a certain day or it may be triggered by other tasks. You can define a projectwide calendar to account for normal workdays, vacations, and holidays. On some software you can even change the length of the workday.

But costs are an important consideration also. Many project management packages permit files to be transferred to spreadsheets for a more detailed look at cost analysis. Also, you can display or print project costs, usually via graphics, which are especially useful. Typical cost figures are resource costs, cumulative costs over the life of the project, and a comparison of budgeted and actual expenses.

A successful project is usually the result of good planning based on good information. With project management software, you can increase the quality, quantity, and timeliness of the information you use.

◇ MEETING THE NEEDS OF THE USER

Users often are not sure what they want or need. As we have noted before, the systems analyst is the catalyst for helping the users define what their needs are. It is the responsibility of the project manager to see that user needs are met—that is, that the systems team actually produces what the user expects. Some mechanism must be put in place to determine whether or not this is taking place. The project manager wants to be sure that needs are being met and that quality standards are being upheld.

These goals can be accomplished in various ways as the project progresses. Two variations on this theme are user reviews and structured walkthroughs. A user review is a formal process associated with phases of the systems life cycle. A structured walkthrough is a formal process in which program design or code is subject to peer review. Both help to identify project problems and keep the project on target. Let us examine these two processes in more detail.

User Reviews

In Chapter 4 we noted that the end of each phase of the systems life cycle is marked with a document of some kind. Each of these deliverable documents can be examined to show how the project is unfolding. A **user review** is a formal process in which work on the system is subject to user scrutiny. A practical way to set up user reviews is to schedule them to follow the appearance of the deliverable documents. That is, the documents can serve as a basis for discussion and action at the user reviews.

Some organizations make user reviews part of the normal presentation process. They work like this. As the end of a life cycle phase nears, the associated document is also nearing completion. Using the analysis phase as an example, the analysis and requirements document should be ready for production. As the analyst, you plan distribution of the document and schedule the presentation/review. All interested parties should get a copy of the document at least a week before the meeting. Participants come to the meeting prepared to discuss the document.

The first order of business at the meeting is the analyst presentation, which is usually much more general than the document. Attendees who have reviewed the document first should have no difficulty following the presentation, even if they have not been involved in prior project activities. After the presentation, the document may be discussed in detail. A quality review process, then, becomes the ideal vehicle for active user involvement in all the phases of the project. For the users who do not need to be intimately involved in project details, this review exposure requires a minimum expenditure of time.

There is a catch, however. Once this review process is set up, authority must be given to the user reviewers to block or delay the project's progress if—as evidenced by the deliverables—the project is not satisfying the agreed-upon requirements. Using a system of memos is just not an adequate attention-getter. Review memos tend to be ignored. The project team must redo the deficient results. This action, of course, changes the schedule and the delivery date.

But bad news is not the expected result. The project manager and various team members normally have a figurative finger on the pulse of the project. In particular, they have antennae to pick up negative vibrations early on. It would be surprising for a project manager to get caught with unsatisfied users at a review. Now we move on to a more detailed type of review, one involving the approval of knowledgeable peers—that is, the structured walkthrough.

Structured Walkthroughs

The **structured walkthrough**, a formal review process by a peer group, has become a staple of good-management literature. It has not yet become a standard in industry, but it is safe to say that a large percentage of large installations endorses structured walkthroughs.

Structured walkthroughs are used primarily to review program design and program code. The analyst who prepared the specifications will be present for a design review. Other reviewers usually are programmer colleagues in the same organization. One reviewer is appointed the review leader. Neither users nor management are invited, because their presence might be inhibiting. The free exchange of ideas should not be squelched by the prospect of judgments or performance evaluations. Besides, every reviewer knows that his or her turn will come soon, so it truly is a forum of equals.

Each reviewer is given a copy of the design or code in advance of the meeting, so he or she will have ample time to look it over. In the review, comments are made about anything that is not clear, will not work, or does not meet organization standards.

One of the key goals is to make programs more readable in order to reduce maintenance costs. This laudable objective is usually achieved, since reviewers have to be able to read and understand the program in order to comment on it. Most organizations use a formal walkthrough report to indicate overall satisfaction (minor changes only) or a need for major changes and another review. (See Figure 14-6.)

Every major study of the structured walkthrough process has shown that the time is well spent and the process cost-effective. The payoffs come in the form of reduced testing time, higher quality programs, programmer learning experience, and easier program maintenance.

The Enforceability Issue

A delicate point to consider is the enforceability issue. Here the project manager must tread a very fine line. Analysts and programmers are professional people and expect to be treated as such. If the review processes with users or peers are too rigorous, they may feel that they are being treated unprofessionally. However, without some enforceability structure in place, some key project activities will not happen or will be unacceptable.

In planning a compromise between these two positions, the project manager must decide what standards will be the basis for measurement. Begin with enforcing the *standards* of the organization. Two examples might be structured program techniques and consistent documentation standards. The next issue is *quality*: to make sure that the design is true to user requirements and that the creation of a technically feasible system is ensured. A third item is to ensure that project planning and reporting provides *accurate information* for management. These typically represent what the project manager chooses to enforce. Managers make it happen by invoking measurement standards at the same time they treat project team members with professional respect.

DESIGN REVIEW REPORT	
Project Name	Review Date
Project Number	Project Manager
Presenter	
Design/code description	
Findings/recommendations	

Resolution

☐ Accept   ☐ Reschedule review after major changes as above

☐ Accept with minor changes as above ☐ Reschedule review after redevelopment

Review participants

_____ _____

_____ _____

_____ _____

Attachments_____ _____

Leader signature

FIGURE 14-6

Design review report. This report is representative of those used for structured walkthroughs.

Although not easy, it is possible to manage projects so that the most important goals are met: on-time schedules, reasonable expenditures, and a product that satisfies users. But what if the project does not turn out that way? Read on.

◇ PROJECTS IN TROUBLE

Despite the abundance of good advice and computer assistance in managing projects, not every project moves along successfully. Most problem projects can be recovered. A few, however, deteriorate to project failures. Let us first look at the bright side.

Recovering Out-of-Control Projects

Monitoring a project is no guarantee that it will progress satisfactorily. But a central idea of project management is that problems can be identified early. Early identification often leads to early solutions. The best cure is prevention of serious problems.

But suppose that the project is headed for trouble. There are a variety of sources for the problems, from poor estimating to the departure of key project people, but "trouble" usually boils down to the fact that the project is behind schedule and overbudget. If users are unhappy as well, the problems are even more serious.

Approaches vary, but here are some ideas to consider. Check the scope of the project. Has it been expanded, perhaps bit by bit? If so, trim it down to the original size and let the additions be part of maintenance. Sometimes, though, a project turns out to be much more complicated than originally thought. This may be traced to insufficient analysis efforts. Another common malady is changing user requirements. In these two cases, it is often not possible to trim the project, but perhaps it can be divided into more manageable phases. That is, identify the basic requirements to be met in the initial system and delay others for later delivery.

If the project is out of control because it was not monitored adequately, then close controls should begin now. In fact, higher management will usually demand frequent reports on troubled projects.

Have you been slowed down by new technology? Perhaps it would be worthwhile to get expert help fast. But the most commonly used approach, adding more-of-the-same personnel to the project, usually meets with limited success. By the time the alarm is sounded and more people are brought in, the original people are deeply involved in the project. To use new personnel effectively, the experienced people will have to stop what they are doing and spend a significant amount of time training them. This, of course, slows the project down further. Many organizations respond, instead, by placing the experienced project people on overtime hours. This can be helpful for a limited period of time, but the returns diminish if overtime is seen as the normal work mode.

Another common approach is to give a project privileged status, so that resources, including computer time, are more readily available. Again, this works on a limited basis, but not when everyone is privileged.

Lastly, the problem of unhappy users needs to be faced directly. They need to be given attention, particularly in the form of listening. Regular meetings should produce plans for problem resolution. All this should be documented and reviewed with the user organization.

Project Failures

Although some projects are unsuccessful, we hear little about the blunders. Some projects, however, are such spectacular failures that they make headlines in the national computer press. Author Bob Glass has made a name for himself by chronicling the unhappy but true stories of projects that have failed. With names changed to protect the innocent (and the guilty), and a heavy dose of humor, Mr. Glass presents case after case in which computer experts with the most serious intentions make fools of themselves. A person can take a lesson from these. In fact, a person can become a little nervous thinking about what error is around the corner, just waiting to turn the project into a cause célèbre in the media.

The stories Mr. Glass tells have several common threads. Although the companies and details vary, a recurrent theme is that the project is poorly estimated and poorly monitored. Managers seem reluctant to question those involved and act instead on blind faith. Many projects he describes are simply too ambitious for the resources of the user and the project team. Occasionally the story will be of a fast-talking analyst or consultant who hoodwinks users and their own management. (One startling case that comes to mind is a highly trumpeted and long-awaited phony "nuclear computer system." Many believed; many had to scramble to recover their reputations.)

In all the examples, steady management and consistent management techniques are clearly lacking. Today's emphasis on project management does not preclude failures, but it does provide an environment for success.

◇ CAN YOU REALLY DO ALL THIS?

A chapter of this magnitude may be overwhelming to the beginning analyst. Also, it is somewhat unappealing. Most people would rather do the work that requires thinking and creativity, than feel nagged to be accountable. To you we say: You will get used to it and it will not be a large burden. Others see the management tasks described here as a stepping stone to career advancement. To you we say: You are probably right.

FROM THE REAL WORLD

PC Flies High with Airborne Freight

If you watch television, you have probably seen the advertisements for the Airborne Freight Corporation. A lean man holding a package and wearing an Airborne uniform jogs past other uniformed men representing the company's competitors. It is an image of a company constantly in motion.

That ad is also an apt metaphor for the entire priority delivery industry. Following an Airborne package during a typical 24-hour period gives the impression that the entire company, from managers to truck drivers to sorters, is always in motion. The quick delivery industry is intensely competitive, and Airborne is in second place.

But when you're not number one, you have to try harder, so Airborne has offered a sensational deal to its customers. All high-volume customers are offered an IBM Personal Computer and custom software to automate their mailrooms and allow them to keep track of packages' routes.

Airborne's menu-driven program is easy to use. The main menu offers such options as preparing an airbill for a package or express mail. After choosing to create a label, for example, the operator keys in the destination and the weight of the package. Other needed information, such as the sending company's address and billing codes, is already on file and available to the computer. In less than 12 seconds, the printer makes three copies, including the label that actually goes on the package. The system performs many other chores, including tracking packages and producing summary reports of the day's activities.

Using a communications device, customers can hook into Airborne's main computer to trace a package. Once the operator keys in the package number, the computer can supply the latest tracking point, such as central shipping or destination city.

Airborne continues to install a new PC about every two weeks. Not a single customer has asked to have the PC taken away. The customers are happy.

—From an article in *PC Magazine*

SUMMARY

- Project management refers to the process that covers system planning, priority setting, effective design, and awareness of the project status.

- Classic management functions are planning, organizing, staffing, directing, and controlling.

- The project manager, sometimes called the project lead, fills both technical and managerial functions. In addition to analyst skills, a good project manager will be a leader, will possess business acumen, and will maintain an awareness of the project status at all times.

- Staffing a project usually involves choosing from available in-house analysts. When interviewing potential new employees, emphasis should be placed on technical questions related to the work to be performed.

- The systems team may be organized in one of several ways. A chief programmer team uses the system concepts of one person, with others in supporting roles. Specialist teams are composed of a core of constant people plus people who are specialists in some particular area such as data communications. A peer group team is a group of equals, usually people with experience.

- Estimates based on an educated guess are usually underestimated. Historical estimating extrapolates data from historical records and uses that data as the basis for estimating future projects. Standard formula estimating proposes weights for certain aspects of the project, to compensate, for example, for a learning curve for new personnel. Checklist estimating uses a detailed breakdown of each task, assigning estimates at the lowest possible level; this method usually produces longer but more accurate estimates.

- Several devices can be used to control work in progress: Gantt charts, milestone charts, PERT charts, and progress reports.

- A Gantt chart is a bar chart, laying activities out on a time line; it is used for planning and reporting on a project.

- A milestone chart presents tangible activities with a certain due date; as a monitoring device the chart can be marked to draw attention to problem situations.

- A PERT (program evaluation review technique) chart is a network of events connected by activities. An activity is the use of resources over a period of time. An event marks the beginning or end of an activity. The longest path on a PERT chart is called the critical path.

- A progress report, also called a status report, is a brief document that accounts for the project during the period of time since the last report. Topics customarily covered on the progress report are accomplishments, problems, follow-up items, schedule progress, budget report, and plans for the next reporting period.

- A project management software package is one that allows a manager to quickly and easily enter information on a project, and then see when various portions of the project will be completed and which resources are being used on different tasks.

- A user review is a formal process in which work on the system is subject to user scutiny. User reviews often coincide with the completion of project phases.

- A structured walkthrough is a formal review of program design and code by a peer group.

- To enforce standards, the project manager should maintain the standards of the organization, promote quality, and provide accurate information for management.

- The best cure for projects in trouble is early detection. Other forms of assistance are trimming the project to the original scope, delaying some aspects of the project to later phases, getting expert help if the technology is new, using overtime, and granting privileged status for services.

KEY TERMS

Activity
Checklist estimating
Chief programmer team
Controlling function
Critical path
Directing function
Event
Gantt chart
Historical estimating
Milestone chart
Organizing function
Peer group team
PERT (program evaluation
  review technique chart)

Planning function
Progress report
Project lead
Project management
Project management software
Project manager
Specialist team
Staffing function
Standard formula estimating
Structured walkthrough
User review

REVIEW QUESTIONS

1. If you were a project manager, what would be your strengths? Weaknesses?

2. Check out project managment software brand names at your local computer store or in the library. Compare the different brands on these

features: number of disk drives required, tutorial availability, maximum tasks per project, resource list, Gantt chart, PERT chart, critical path highlighted, holiday calendar, and vendor support hot line.

3. What kinds of things would you expect to be stressed in a structured walkthrough in a company that advocates formal procedures and structured programming?

CASE 14-1
NON-MANAGEMENT AT UNIVERSITY HOSPITAL

Jeff Hildstrom was a button-down type of guy who had spent most of his working life in the most conservative life insurance company in the Northeast. Beginning as a systems analyst, he had progressed to project manager and then on up through the ranks of higher management. Part of his success, he felt, was knowledge of and respect for good management techniques. Employees who worked for Jeff knew where they stood and knew that they had better meet schedules.

Jeff decided it was time to move on and look for new challenges. He applied for several jobs but was captivated by a position at the local university hospital. The hospital directors interviewed him several times in depth and finally offered him the job of Director of Data Processing. The challenge he was looking for was about to begin.

Here are some of the things Jeff discovered in his first week on the job. A project that originally had been scheduled for 11 work years had already taken 8 work years and was less than half done, and no regular management reporting system was in effect. A system to report patient medications had 153 different reporting modules and worked to no one's satisfaction. The oldest and largest system, for patient billing, showed 150% personnel turnover in the past year. First- and second-line supervisors regularly abused flex hour policy, coming in late, leaving early, and playing volleyball before eating on their extended lunch "hour." Many programmers and analysts took similar liberties with time commitments. At the first presentation Jeff attended, with two managers higher than he in attendance, the presenting analyst showed up wearing faded jeans supported by red suspenders, and a denim shirt.

1. Perhaps no case study you have read is less believable than this one, but, except for name and place changes, it is true. If you were Jeff, what would your initial actions be?
2. What kind of reactions to the changes would you expect from the employees? How long would you expect it to take before the changes are in place and the organization is running smoothly?

CASE 14-2
HANDS-OFF MANAGEMENT

At Bole Chemical, Inc., Ann Allison had been a project manager once before and had been monitored closely by higher management. She saw herself relaying policies and orders from above, letting the team members know what management expected. After that successful experience, Ann was assigned to manage another team, and this time she was given much more free rein.

Ann felt that the better part of management was letting people know that she had confidence in them, and expected their best efforts. She was uncomfortable with close monitoring action, which seemed like nagging. Her relationship with the team members was one of supportive mentor. She accepted their biweekly progress reports with little question.

Ten weeks later, toward the end of the analysis phase, she heard rumblings from the users. Ann knew that the project had already slipped by three weeks, but was surprised to discover unhappy users, too. The user review at the end of the analysis confirmed her fears that the project was in trouble.

1. Why do you suppose Ann took a different management approach once she was on her own?
2. Should Ann tell her boss about these developments?
3. How can Ann get the project back on track?

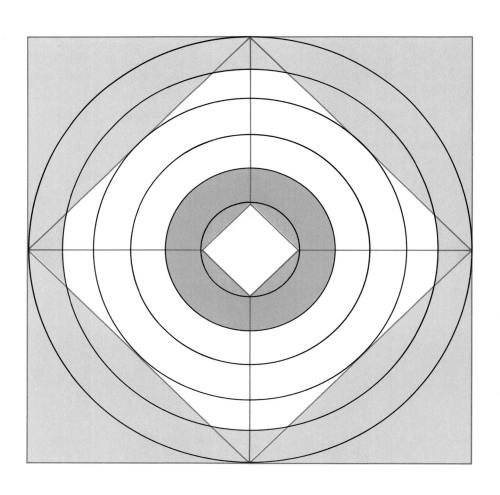

Chapter 15

System Security

Forget about systems for just a moment and consider your own personal security. You probably have locks on your doors and are careful about your whereabouts at night. You keep your car locked, especially if it has a tape deck in it. You probably have medical insurance.

But perhaps you also have some of the careless habits shared by so many of us. Maybe you have hidden a key in some obvious place such as under the doormat. Perhaps you leave the keys in the car when you briefly step into a convenience store. You may not bother with seat belts for short trips.

Few things in life are guaranteed, but you may really groan if you become a victim of your own carelessness. That old it-can't-happen-to-me feeling is still with us, even though we may reject it intellectually. You can see where this is leading: You cannot let carelessness make your computer system a victim either. As we shall see, the price is too steep.

BOX 15-1	**The Rules of the Road**

Everyone knows that the rules of the road have to be taken seriously. So do the rules for using a computer.

Two of those rules are basic: Everyone who uses a computer has a responsibility for the security of the information in that machine. No one who uses a computer has the right to violate anyone else's security.

—From an IBM advertisement

◇ SECURITY: A CRITICAL ELEMENT IN SYSTEMS DESIGN

Security is a system of safeguards designed to protect a computer system and data from deliberate or accidental damage, or access by unauthorized persons. Security was mentioned in the systems life cycle chapters, especially the design chapters. The references were brief, but not because the subject was unimportant. The subject of security is so important, in fact, and so comprehensive, that it deserves a chapter of its own. But security cannot be draped over a new system as an afterthought. It must be integrated into the design and development of the system. In other words, just because sheer volume of information forces the subject into a single place in a book does not mean that your actions should reflect this artificial organization.

Let us now consider what could go wrong and what to do about it.

◇ THREATS TO SYSTEM SECURITY: WHAT COULD GO WRONG

Think for a moment about what could go wrong in a computer system. Your experience may make you think of program bugs first. If you are hardware-oriented, your first thought may be of a disk crash. Or perhaps you are more aware of the incessant headlines about hackers invading systems. These things—and most hardware and software problems—are tangible and within a controllable range. The possibilities for things that can go wrong, not to mention the possibilities for total disaster, go well beyond this small list, however. Here are a few things to think about.

Hardware and Software Problems. The list is long: program bugs, operating system failures, incorrect data, deleted data, disk crash, power loss, communications mismatch, and so on.

Natural Disasters. Even though natural disasters are not preventable, you still need to be aware of their possibility or, in some cases, probability. The list includes hurricanes, windstorms, tornados, floods, volcanos, earthquakes, fire, and lightning. (As computer installation workers in the northwest vacuumed the fine white Mount St. Helen's ash from their equipment, they must have mused that they never thought of volcano insurance.)

Natural Forces. Less devastating, and more predictable, are such forces as ice, snow, mud, and drought, any of which could adversely affect the physical environment of your computer system.

Social Forces. The system could be impacted by a variety of factors. You may be aware in advance that some of these are especially related to your system. These include strikes, bomb threats, riots, vandalism, terrorist activities, accidents, and libel suits.

People Problems. Dishonest or disgruntled employees can wreak havoc in a number of ways. High on the list is fraud, followed by theft, espionage, sabotage, mischief, or just plain carelessness. Other sources of people problems are customers, the competition, prank-driven youngsters, and dishonest members of the public at large.

The list may seem overwhelming but it is possible to plan wisely and well to protect your system.

◇ PHYSICAL SECURITY

Let us begin with the physical environment of your computer system. It can be critical. A commodity exchange lost $500,000 in a single afternoon due to computer downtime caused by a faulty cooling system. To avoid problems, computer room environmental control systems should have cooling and moisture-removal capacities that meet or exceed the requirements of the hardware manufacturers.

In addition, the computing facility needs to meet minimum requirements to protect it from calamities and intrusions. As you can see from the physical environment security checklist, care needs to be given, for example, to prevention of fire and flood damage.

Some organizations have emergency power to switch to if something happens to the main power supply. Another consideration is an **uninterrupted power supply (UPS)**, a battery unit that prevents computers and related equipment from crashing in the event of a main power line failure. UPS, as you may suspect, is more expensive than emergency power for use after the crash.

☑ BOX   Security Checklist: Physical Environment and Hardware

1. A computer facility should be on high ground and on at least the second floor to avoid flood damage. Also avoid nearby pipes that might spring a leak.
2. Sensors should detect smoke or fire. The threat of fire damage should be minimized by a sprinkler or (better) halon system.
3. No one should smoke or eat in the computer room, and users should be cautioned not to set their coffee cup on top of the office microcomputer.
4. Critical files should be in fireproof vaults.
5. Copies of important programs, files, documentation, and so on should be stored off-site.
6. Your computer equipment should not be visible from the street. There are computer junkies who like to hang out near the machines. They may or may not be harmless.
7. Large facilities should install TV cameras as monitors.
8. The computer room facility should have emergency power.
9. A backup facility should be available in case of extended downtime.

◇ SOFTWARE SECURITY

Software security is a relatively new idea. The computer industry has always paid more attention to data security, since data is, in one form or another, always before the public. As we shall see in the next section, extreme vigilance is needed to keep data secure. Software security, however, was recognized as a problem when the advent of microcomputers thrust software into the hands of the masses.

Despite increased industry vigilance, microcomputer software continues to be copied as casually as people make copies of music tapes. Both are illegal. If your system produces original software to be used on microcomputers, you need to get it copyrighted and lock-protected. Lock protection refers to the encoding of the software to prevent—or at least discourage—copying. Marketed software for any size computer should receive the same treatment.

☑ **BOX**   **Security Checklist: Software**

1. There should be a formal procedure for changes to programs in production.
2. A control copy of each production program should be maintained and a comparison should be run irregularly against the copy in use.
3. The production control clerk should scan the output job control summary to make sure that no unauthorized program was run.
4. Programs should provide counts of activities—number of transactions read, and so on.
5. Software to be marketed should be copyrighted and protected with software locks.

◇ DATA SECURITY

Quickly now, which computer system resource is most difficult to recover or replace: communication lines, data, equipment, people, physical space, or supplies? This is not a trick question. Communication lines, equipment, physical space, and supplies are all tangible, insurable, and replaceable. People are more difficult, but good documentation and imported experts could make the problem bearable. That leaves data, which can be extraordinarily difficult to recover if no backup has been made. Can you envision losing three years of design data for a new space vehicle? Or how about your accounts receivable files—imagine the advertisement in the newspaper: All Persons Owing Money to Elliott Marine Supply—Please Write and Tell Us How Much.

☑ **BOX**   **Security Checklist: Data**

1. Data forms should be prenumbered and accounted for.
2. More than one programmer should have access to a given file to prevent an unquestioned monopoly.
3. Reduce opportunities for casually picking up classified information by encouraging a neat work area.
4. Sensitive obsolete documents should be shredded.

Data is the most critical element of a computer system. Data also may be the most valuable asset of the company. Protecting it from loss, destruction, distortion, and prying eyes is a top priority.

Data Protection Techniques

In addition to the data protection techniques listed in the security checklist for data, note the special items here.

Backup Files.   You need backup files in case anything unfortunate happens to the originals. Copying master files and databases regularly is standard practice in any responsible computer installation. The copies, usually on tape, should be removed from the premises and stored in a safe place. Microcomputer users need to make copies of their diskettes. If files are on hard disk, those files should be copied onto diskettes.

Security Packages.   Security software packages are available for purchase, either independently or through the hardware vendor. These packages act as part of the operating system and limit access to program and data files. They are usually quite sophisticated and require active participation and maintenance by the computing organization.

Secured Waste.   An employee holding an important-looking document approached the office shredding machine, a dignified woodgrain box, and asked the women at a nearby desk for instructions. She showed him the input slot and the buttons and returned to her work. She looked up a few minutes later to see him still standing there, looking puzzled. "How many copies does it make?" he inquired. Despite this regrettable story, shredders that tear papers into unrecognizable small strips are becoming commonplace in offices that have special security considerations. Another secured waste device is a simple locked waste bin.

Audit Trails.   Modern auditors are familiar with computers and what they can do. In some organizations, auditors play some role in system design, so that their future auditing assignments will be easier. Audit trails begin with input data. The system must be able to trace any given transaction through the system, from data entry, through processing, to files and output reports. Note that if the data is captured at the source, a transaction journal, a file of all online transactions, must be kept. Also, if output is questioned, there must be a way to trace output back to the original input data.

Communications: Keeping Intruders Out

Any computer system that can be reached through the public phone system is vulnerable. This is complicated further by popular local area networks

(LANs), which make corporate financial information, customer files, and other sensitive data potentially available to every employee workstation.

However, the computing community is alert to these dangers and has taken steps to reduce electronic trespassing significantly. In addition to the security access checklist in the box, give special consideration to the following.

Passwords. The first line of defense against system interlopers is the careful use of passwords. If users, including programmers, are allowed to assign their own passwords, should they be cautioned to choose a password that does not identify them in any way. Names of spouses, children (remember Joshua in *War Games*?), or pets are discouraged, as are words like PEACE or PASSWORD. The latter, by the way, is no novelty. The advantage of user-assigned passwords is that they are known only to the user. System-assigned passwords are coming into favor, however, because they avoid cuteness and also force real password changes at regular intervals.

As noted in the box on access security, there are a variety of ways the system can act on invalid passwords. In addition to these, another fairly

---

☑ BOX Security Checklist: Access

1. All computer facility doors should be locked.
2. Locks, or combinations, should be changed frequently.
3. Keys should be recovered from terminated employees and locks changed.
4. Intruder alarms should be standard in the computer area.
5. Access to the computer room should be extremely limited and on a need-to-be-there basis.
6. All terminals and microcomputers should have lock devices.
7. All system users should be assigned passwords of at least four nonvowel letters. Passwords should be changed at least every month.
8. Callers from remote phones should not be allowed to log on directly. Instead the computer should hang up and call back the phone number listed for the authorized user of the password.
9. Users should be limited to three wrong password tries before the computer hangs up.
10. A terminal unattended for more than ten minutes should cause the computer to hang up.
11. Data access rights—read only, write, delete—are assigned and related to user access codes.
12. Use cryptography to prevent wiretapping.
13. Have a security specialist make unannounced visits to check the premises for hidden microphones and transmitters.

simple technique is to record the time of each unsuccessful log-on attempt. This method was helpful in tracking down the infamous Milwaukee 414s, a group of teenage hackers who penetrated files in banks, hospitals, and even Los Alamos National Laboratory. As an extreme measure, a computer system could be programmed to trace a telephone call after the caller makes a series of unsuccessful attempts.

Cryptography. Thwarting data stealers who know how to tap transmission lines involves scrambling the data so it is unrecognizable and, presumably, unusable. Cryptography is the process of encoding data to be sent over the lines, and then decoding it at the receiving end. Cryptography routines use complicated mathematical routines that take time and thus add to the cost of transmitted data.

In 1977 the National Bureau of Standards adopted the **Data Encryption Standard (DES)**, a set of mathematical principles for encoding data using a 64-bit key. DES produces gobbledygook so resistant to tampering that it is considered virtually undecipherable to all but the most sophisticated computers. Note the example in Figure 15-1.

The United States Treasury Department has issued a directive that requires all government agencies, as of June 1988, to use DES to send or receive electronic funds transfers. An **electronic fund transfer (EFT)** sends money via a computer network from one account to another. So, if the government is involved, all monies sent electronically must be encoded using the DES encryption standard. This, of course, also affects any private companies that deal in such transactions with the government. For example, social security recipients can arrange to have their checks deposited directly in

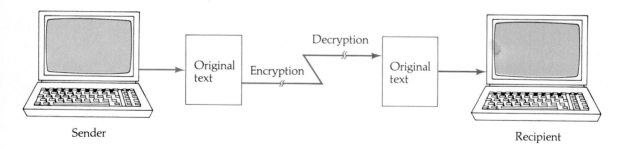

Sender   Recipient

FIGURE 15-1

Encryption. The original text in the memo is encrypted into unreadable text before it travels over communication lines. Then it is decrypted back to the original text before being given to the recipient.

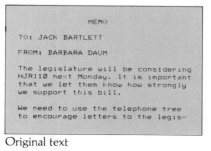

Original text

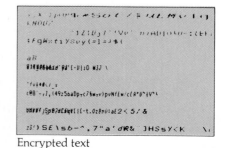

Encrypted text

their bank accounts. Instead of getting a check in the mail which then must be deposited in the bank, the monies are sent electronically to the recipient's bank.

Data Traveling Across Borders.   It may seem exotic but many companies, large and small, find markets for their products and services in other countries. This often means that data travels the same routes, allowing users to take orders and exchange messages with customers and other vendors. If you find yourself establishing a system that will span map lines, it would be wise to check the communications regulations of the other countries and to investigate any special security problems they may have.

◇ SYSTEMS CONTROL: IDENTIFICATION AND ACCESS

The computer industry was in its infancy when installation managers began to realize that it was a good idea to keep unauthorized people out of the computer room. Programmers, who were at the least pests and at the most potential embezzlers, topped the list. Soon, however, keeping people physically separated from the hardware was not enough. We all know that people can access computers from remote locations.

Various means, from the simple to the sublime, have been devised to give access to authorized people without compromising the system. They fall into four broad categories: what you have, what you know, what you do, and what you are (see Figure 15-2).

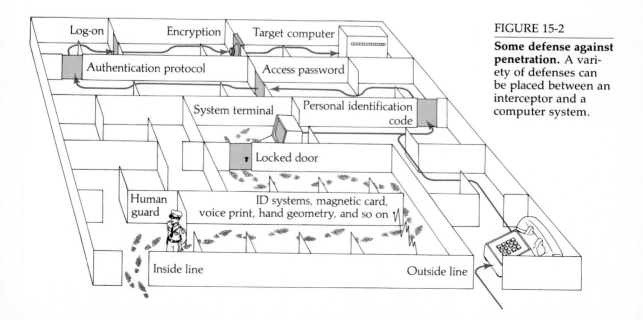

FIGURE 15-2

Some defense against penetration. A variety of defenses can be placed between an interceptor and a computer system.

Authorized access to an automated teller machine is a combination of what you have—your plastic card—and what you know—your identification number.

What You Have.   You may have a key or a badge or a plastic card to give you physical access to the computer room or a locked-up terminal. A card with a magnetized strip can give you access to your bank account via a remote cash machine.

What You Know.   Standard what-you-know items are a system password or an identification number for your bank cash machine. Cipher locks on doors require that you know the correct combination of numbers.

What You Do.   Your signature is difficult but not impossible to emulate. Signatures lend themselves to human interaction better than machine interaction.

What You Are.   Now it gets interesting. *Biometrics* is the science of measuring individual body characteristics. Fingerprinting is old news, but voice recognition is relatively new. Both of these techniques are now in use as part of procedures that allow a machine to recognize a properly authorized human. Newer still is the notion circulating in scientific communities that

Entry to the computer area is controlled by a cipher lock, shown at left.

BOX 15-2	Don't Leave Your Body at Home

You may forget your identification card or keys, but the most important security feature you have is what you are, and it goes with you at all times. Donn Parker, acknowledged security guru for the computer industry, sees identification moving away from what a user has or knows to what the user *is*.

Some tamperproof elements are the unique blood vessel pattern in the retina of the eye, the length of the finger, or the lines that crisscross the palm. It could, in fact, even be the pattern of brain waves.

☑ BOX   Security Checklist: Personnel

1. Employees should be screened before hiring.
2. Job functions should be separate.
3. Personal cameras should be prohibited in the computer area.
4. Programmers, analysts, and users should be prohibited from entering the computer room without authorization.
5. A disk-stored user profile should be defined for each programmer and user, indicating to which programs and files that person may have legitimate access.
6. Duties should be rotated so that no system is totally dependent on one person.

☑ BOX   Security Checklist: Documentation

1. Documentation standards should be established and enforced, without indicating in detail how to do it but, instead, stressing what needs to be done.
2. Complete documentation should be maintained for all existing systems.
3. Backup documentation should be maintained off-site.
4. Access to documentation should be controlled.
5. Documentation should not leave the company premises.
6. Photocopy machines should be carefully controlled.

---

we could measure other body parts. In France, for example, they are studying the shape of the ear. Some scientists are studying foot attributes, but one wonders about the efficacy of removing tennis shoe and sweat sock at, say, the cash machine.

A system will sometimes use a combination of these techniques. A pattern favored by top secret military computer installations, for example, uses a set of double doors with a little hallway in between. You get past the first door by slipping your personal identification badge in the slot. Once inside the hallway, a speaker tells you to repeat four randomly chosen words from the 16 words prerecorded in your voice. A match on all four opens the second door into the installation.

◇ DISASTER RECOVERY PLAN

This is a good time to mention that many organizations have full-time employees whose job description includes the word *security*. If this is the case where you work, you will know it, because computer security officials make their presence felt. In fact, their continued presence is sometimes a source of griping among programmers who find security safeguards an inconvenience.

The security issues decribed so far, although extensive and time-consuming, can theoretically be considered as the system is being designed and developed. Major security provisions such as disaster recovery and risk analysis, however, will probably be done by resident or free-lance security experts. These tasks are mentioned here, not as part of the individual systems project, but to round out the complete security picture.

A **disaster recovery plan** is a method of restoring data processing operations if those operations have been suspended due to major damage or destruction. There are various approaches. Some organizations might revert temporarily to manual services, although this has become quite an unattractive alternative. Others arrange to buy time at a service bureau. If a single act, such as a fire, destroys your computing facility, it is possible that a mutual aid pact will help you get back on your feet. In such a plan, two or more companies agree to lend each other computing power if one of them has a problem. This is of little help, however, if there is a regional disaster and many companies need assistance. A deterrent to this type of agreement is that most organizations use their computing resources at the maximum level and can not absorb extra processing quickly.

And we do mean quickly. No time can be wasted in most organizations. It has been estimated that a bank that loses its computer processing capabilities will be out of business—literally—in three days. Banks and other organizations with survival dependence on computers sometimes form a **consortium**, a joint venture to support a complete computer facility. Such a facility is completely available and routinely tested but used only in the event of a disaster. A **hot site** is a fully equipped computer center, with hardware, environmental controls, security, and communications facilities. A **cold site** is an environmentally suitable empty shell in which a company can install its own computer system.

The use of such a facility, or any type of recovery at all, depends on advanced planning, specifically the disaster recovery plan. The underlying premise of such a plan is that everything except the actual hardware has been stored in a safe place somewhere else. The actual storage location should be several miles away, so it will be unaffected by local physical forces such as a hurricane. Typical stored items will include program and data files, program listings, program and operating systems documentation, hardware inventory list, output forms, and a copy of the disaster plan manual.

The disaster recovery plan will have these provisions:

- *Priorities*. It will list which programs must be up and running first. A bank, for example, will give greater weight to account inquiries than to employee vacation planning.
- *Personnel requirements*. It will establish procedures to notify employees of changes in locations and procedures.
- *Equipment requirements*. It will list equipment needed and where it can be obtained.
- *Facilities*. Since most organizations cannot afford consortiums, the plan will locate an alternative computing facility.
- *Capture and distribution*. It will decide how input and output data will be handled in a different environment.

Some computer installations go beyond making plans and packing a few items to a nearby warehouse. They actually practice for a disaster much as a school practices for a fire by having fire drills. At some unexpected moment a notice is given that disaster has struck and the locals must run the critical systems at the designated alternate facility.

◇ SPECIAL SECURITY CONSIDERATIONS FOR MICROCOMPUTERS

Mark Bernstein came into his office one morning, snapped on the lights, and gasped these words: "It's gone!" His computer was gone. Right out of his office. There is an active market for stolen microcomputers and their internal components. As this unfortunate tale indicates, microcomputer security issues can be pretty basic. Mark's computer would not have been such an easy target if it had been in a microcomputer lockup cabinet or bolted to the desk. Both these options are available and effectively prevent physical removal.

In addition to theft, microcomputer users need to be concerned about the computer's environment. Microcomputers in business are not coddled the way bigger computers are. They are designed, in fact, to withstand the

| BOX 15-3 | **Dress Code for Computers** |

Your micro can be smartly attired in its own secure environment. A glossy cylindrical box has been designed by Artisan House to conceal personal computer components and accessories and keep them under lock and key.

FIGURE 15-3

Diskette security problems. Diskettes are subject to many security risks.

wear and tear of the office environment, including temperatures set for the comfort of people. However, most manufacturers discourage eating and smoking near computers and recommend some specific cleaning techniques, such as vacuuming the keyboard and cleaning the disk drive heads with a mild solution. The enforcement of these rules is directly related to the awareness level of the users.

Most microcomputer data is stored on diskettes, which are vulnerable to sunlight, heaters, cigarettes, wastebasket fires, scratching, magnets, theft, and dirty fingers (see Figure 15-3). Hard disk used with microcomputers is

A surge protector, as you can see here, looks very much like an exotic extension cord plug. If you plug all your computer devices into this surge protector, a sudden power surge could not harm your equipment.

subject to special problems, too. Consider these events, all of which have happened:

> An employee from the mailroom decided to have some "fun" while making his last rounds after the office was empty. He picked two or three floppy disks from the disk holder on each desk and distributed them randomly to the disk holders on other desks.

BOX 15-4   Copyproof Software

Hackers usually greet reports of new "uncrackable" software locks with an amused smile or even guffaw. They routinely send their code-breaking formulas to electronic bulletin boards, where user groups swap information. Under copyright law it is, of course, illegal to copy a software program except for archival purposes. Software manufacturers are stepping up the war on piracy. In addition to more sophisticated techniques, they are prosecuting perpetrators vigorously and lobbying for an increase in the penalty from $50,000 to $250,000.

This is a contest of grit and resolve. Since the manufacturers have the most to lose, it seems reasonable to predict that, in the end, their determination will outdo all others.

A pharmacy clerk wanted to format a floppy disk but neglected to mention which disk drive, so formatted the hard disk instead, destroying all the hard disk files.

A storm caused a power surge that destroyed data on hard disk.

Several precautions can be taken to protect disk data. One is to purchase a **surge protector**, which will prevent electrical problems from affecting data files. The computer is plugged into the surge protector, which is plugged into the outlet. Hard disk files should be backed up on diskettes. Diskettes should be under lock and key.

The awareness level of microcomputer security needs is gradually rising. In the following section, however, you will see that security measures, and the money to implement them, are directly related to the amount of the expected loss. Since the dollar value of microcomputer losses is often relatively low, microcomputer security may be less than vigorous.

◇ RISK ANALYSIS

There is no way to avoid all risk. The eruption of just one volcano is all we need to remind us of human frailty in the face of extraordinary phenomena. Security experts, however, like to zero in on the risk factor. **Risk factor** is the likelihood of having a particular problem. It is a numbers guessing game.

Why bother with guessing games, however educated they may be? There are two reasons: (1) to focus attention on inherent risks, and (2) to get funding to prevent or recover from bad events. People often are reluctant to spend money to prepare for things they hope will never happen. When the whole situation is presented in dollar figures, however, it becomes more concrete and more palatable.

Risk analysis assigns a dollar value to a potential loss and a percentage figure to represent the probability of that loss; then it multiplies these numbers to get the expected loss:

Expected loss = Dollar value × Probability

For example, the dollar value of hardware replacement might be $600,000 and the probability of such a loss (according to a given installation expert's opinion) only 15%.

Expected loss = $600,000 × 0.15 = $90,000

The same organization might also worry about unauthorized disclosure of highly sensitive data and have serious concerns about intrusion. If they think resulting lawsuits might net $2,000,000 and the probability of disclosure is now about 20%, the expected loss figure is $400,000. Comparing the two figures, $90,000 and $400,000, it seems clear that security efforts should be directed primarily to the disclosure area, and funded accordingly.

This brief example illustrates the principles behind risk analysis: focusing attention and receiving appropriate funding. A true risk analysis would, of course, be much more comprehensive, including all aspects of the computer system.

◇ THE LONG ARM OF THE LAW

Post-Watergate fervor produced a spate of laws designed to protect citizen privacy and open up the government's computer files on its citizens. The **Freedom of Information Act** allows citizens to have access to data gathered about them by federal agencies. The **Federal Privacy Act** stipulates that there can be no secret personal computer files. But these are essentially privacy matters—there is no federal security legislation.

Trying to Get Security Laws

Computer crimes in the twentieth century are being tried under nineteenth century laws. Some version or other of the Federal Computer Systems Protection Act has been passed around in the halls of Congress since 1977, but

---

BOX 15-5   **Snooping at the IRS**

In case you think Big Brother is alive and well at the Internal Revenue Service, be assured that no information could be accessed by random trespassers and precious little by other agencies. The reasons are an interesting amalgam of tight money and congressional apprehension.

A totally integrated electronic system sending documents such as W-2 forms directly to the IRS was rejected by a Congress nervous about potentially misused power. The flood of documents, therefore, must be keyed in by hand. The IRS does not have enough people to do the work, so not all of the cross-checking data is available.

A deliberate anachronism is a prohibition on data transmission. The IRS must package and ship computer tapes to and from the computer center. The advantage of this fragmented system is that it presents an obstacle to snoopers. No one can just sit down at a computer terminal and access the agency's files to get information about a taxpayer. The data must be requisitioned by the local tax center and physically retrieved by officials.

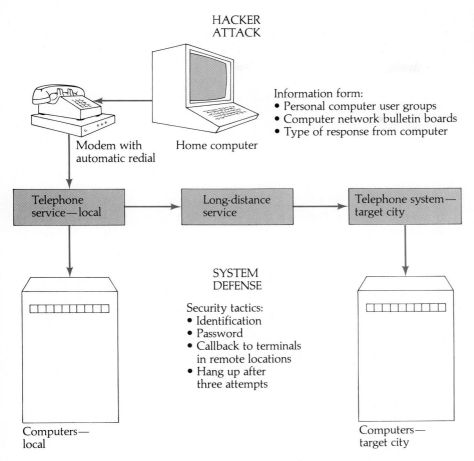

HACKER ATTACK

Information form:
- Personal computer user groups
- Computer network bulletin boards
- Type of response from computer

Modem with automatic redial   Home computer

Telephone service—local → Long-distance service → Telephone system—target city

SYSTEM DEFENSE

Security tactics:
- Identification
- Password
- Callback to terminals in remote locations
- Hang up after three attempts

Computers—local

Computers—target city

FIGURE 15-4

Hacker attack. There are several system defenses against hackers.

it has yet to pass. Meanwhile, the states have tried to take up the slack with a hodgepodge of laws that often are ineffective.

One of the key problems is that computers do not know state lines, and data hops across borders as easily as telephone lines do. This problem, together with a desperate need for consistency, is the basis of the need for a set of laws at the national level. Meanwhile, teenagers and others terrorize the data lines with impunity.

Cracking Down on Hackers

Hackers who are caught have a distinct advantage, and they know it. They are usually too young to receive the full force of the law. Most are given suspended sentences or a light wrist slap. Worse, the patchwork of state laws usually specify the need to prove malicious intent, if not outright fraud or embezzling. None of these cover the usual hacker sins of browsing uninvited through files, or nonprofit mischief. (See Figure 15-4.)

BOX 15-6	**Choosing Just the Right Password**

Quickly, now, choose a password. Make it one that is unlikely to be guessed by an intruder. Were you truly clever? Did you choose your dog's name? Your wife's initials? Perhaps the name of a rock group? If this list is getting uncomfortably close to your actual choice, then you are not so clever after all. In fact, you are a sitting duck.

Bill Landreth, himself an experienced teenage hacker, tells all in his absorbing book *Out of the Inner Circle*. Bill and his friends laid some simple traps that netted information about workers who had access to computer accounts. One of the hackers simply stood outside the company building after work and innocently passed out forms that he wanted completed "for a high school project." On the form were questions about childrens' names, the name of the family dog (!), hobbies, and so forth. The young hackers reaped a rich harvest of password-related information and went on a sightseeing tour of the company's computers.

Landreth also describes the "database hack," in which a hacker patiently uses a database of about 400 words known to be common passwords. A typical database will contain such entries as: Love, Sex, Secret, Demo, Games, Test Account, Intro, Password, Alpha, Hello, Kill, Beta, Dollar, Dead, System, Computer, Work, Yes, No, Please, OK, Okay, God (popular with system operators), Superuser (also popular with system operators), Aid, Help, the name of the company formatted in various ways, the letters A through Z, 200 to 400 first names, 10 or 20 names of pop musical groups, 5 to 10 names of automobiles, the digits 0 through 9.

A good password? How about HJUVFL? Or LDNGRF? Not to be confused with your dog's name.

—From Out of the Inner Circle

Slowly, things are changing. Laws are being written to allow vigorous prosecution of trespassers. In addition, schools are accepting responsibility for educating young computer users in the ways of the law. Officials seem determined that computer children will not grow up to be computer criminals.

◇ ARBITRATION

So far we have talked about the law in terms of prosecuting people who violate computer security. But there is another important issue relating to

| BOX 15-7 | **A Glossary for Computer Crooks** |

Although the emphasis in this chapter is on prevention rather than crime, the well-versed analyst will want to be familiar with computer criminal terms. Many of these words or phrases have made their way from the language of hackers into the general vocabulary.

Data diddling. Changing data before or as it enters the system.

Data leakage. Removing copies of data from the system without a trace.

Logic bomb. Sabotage method in a program that will trigger damage based on certain conditions and that is usually set for a later date—perhaps after the perpetrator has left the company.

Piggybacking. Using another person's identification code or using that person's files before he or she has logged off.

Salami technique. Using a large financial system to squirrel away small "slices" of money that may never be missed.

Scavenging. Physically searching trash cans for printouts and carbons containing not-for-distribution information.

Trap door. Leaving an illicit program within a completed program, allowing unauthorized—and unknown—entry.

Trojan horse. Placing covert illegal instructions in the middle of a legitimate program.

Zapping. Bypassing all security systems with a vendor-provided (and illicitly acquired) software package.

computers and the law. The issue is the resolution of differences among computer professionals and their clients. This also is a security matter.

A Need For Arbitration

Consider this typical, and true, story. Vivian Beardsley, a woman of above average intelligence, was summoned to serve on a jury to hear a case related to the installation of a computer system. Hart Manufacturing had contracted with Weaver Computer Consulting to provide a distributed processing system among Hart's central office and remote manufacturing plants. Weaver subcontracted part of the system to MRX, a communications company. The resultant system was late, with substantial cost overruns. Worse, the system did not meet Hart's needs adequately. Hart sued Weaver. Weaver sued MRX and countersued Hart.

Enter Vivian Beardsley, a juror in the first trial. The attorneys in the case took five days early in the trial to give the jurors background in the technical aspects of the case. After receiving basic information about computer input, output, processing, and storage, Vivian and the other jurors heard detailed explanations of baud rates, asynchronous transmission, network configurations, and the like. The point of this story is not the merits of the case or even who won but what Vivian said when it was all over. When asked how she could juggle all that technical testimony, she confessed, "None of us ever really understood any of it. We made our decisions based on the credibility of the witnesses."

Vivian and her fellow jurors are not to blame for this state of affairs. Our judicial system is not structured to permit only technical experts in such cases. But there is an alternative, and its use is growing: arbitration.

Arbitration is a legally binding procedure used to settle disputes between two parties out of court. A key feature of arbitration is that the arbitrators who will judge the merits of the case are technical experts. But arbitration is also an appealing alternative to the court system because it is usually considered quick, inexpensive, and private justice. And, because an arbitration proceeding is confidential, it is viewed as an especially attractive mechanism for resolving disputes arising in the computer industry, where trade and business secrets are important. So, in addition to its other features, arbitration can be seen as a security measure.

How Arbitration Works

The first step in arbitration is for the parties to agree to arbitrate, before or after the fact. Some companies include the standard arbitration clause as part of all contracts. The parties must notify a branch of the American Arbitration Association. The case is assigned to an association staff member, who provides each party with the same list of proposed arbitrators, local people technically qualified to resolve the dispute. The parties are given seven days to study the list, cross off any names, and number the remaining

BOX 15-8	**Standard Arbitration Clause**

Any controversy or claim arising out of or relating to this contract, or the breach thereof, shall be settled by arbitration in accordance with the Commercial Arbitration Rules of the American Arbitration Association, and judgment upon the award rendered by the Arbitrator(s) may be entered in any court having jurisdiction thereof.

names in order of preference. Using the marked lists, the staff member appoints an arbitrator or, in substantial cases, a panel of three arbitrators.

The staff member handles other details as well, such as scheduling an agreeable date and arranging for stenographic services. The parties line up lawyers and witnesses, and the process begins. The decision by the arbitrator, including monetary awards, is final, and fully supported by law.

Even with the best of intentions, people often perform less than they promise. Arbitration provides an orderly way for business people to settle controversies in private.

◇ ETHICS FOR TECHNOCRATS

Doctors take an oath, lawyers and accountants are morally bound by the professional–client relationship. What about computer professionals? Shall we raise our right hands and swear to protect data and not copy software illegally? In a word, yes.

A code of ethics is coming, perhaps not in the way just described, but coming nevertheless. Professional ethics for the computer industry is the subject of articles in technical periodicals, a key topic at conferences, and a lament in the halls of lawmakers. Any theme that gets this much attention usually results in action.

Most experts talk of self-regulation via the professional computer organizations. Some organizations, such as the Data Processing Management Association (DPMA) and the Association for Computing Machinery (ACM), already have a code of ethics. Handling this "among ourselves" is considered preferable to regulation imposed by a federal agency.

Some companies require their employees in computer-related jobs to sign a contract covering nondisclosure of company secrets and records, as well as how and when the employee can use hardware and software.

Just what a computer professional ethically can or should do is complicated by the issue of microcomputers, which seem to permit greater independence and freedom from rules. The microcomputer in systems is the topic of our next chapter.

BOX 15-9	**Code of Ethics: Data Processing Management Association**

In recognition of my obligation to my employer, I shall: Make every effort to ensure that I have the most current knowledge and that the proper expertise is available when needed; avoid conflicts of interest and ensure that my employer is aware of any potential conflicts; protect the privacy and confidentially of all information entrusted to me.

FROM THE REAL WORLD

Fire at the Westin Hotel

Although a smoky electrical fire cost $500,000 in damages and millions of dollars in lost revenues, the data processing at the Boston Westin Hotel barely missed a step.

Less than 24 hours after the midday fire drove guests and employees to the street, the data crucial to operating the "fully automated" luxury hotel—safely stored on diskettes—was being processed to meet payroll and accounting deadlines. The fire, among other things, knocked out the hotel's main and emergency power supplies. When the power went off, everything crashed, but the backup power supply was brought up immediately and the operation was back in business.

The computer even assisted in evacuating the hotel, providing fire fighters with printouts showing which guests were in which rooms. Most of the computer systems did not need to be in use, since no guests and few employees needed service.

A return to full operation required only the running of daily audits for the various departments for the days they were closed and the manual entering of payroll information for employees who worked during the shutdown.

—From an article in *Computerworld*

SUMMARY

- Security is a system of safeguards designed to protect a computer system and data from deliberate or accidental damage, or access by unauthorized persons.

- Threats to system security include hardware and software problems, natural disasters, natural forces, social forces, and people problems.

- Physical security may include suitable environmental controls, protection from calamities and intrusions, and an uninterrupted power supply (UPS).

- Software security refers to preventing the illicit copying of software, particularly microcomputer software, which is especially vulnerable.

- Data security refers to protecting data, which can be very difficult to replace if no backup files have been made. Data must be protected from loss, destruction, distortion, and prying eyes. Data protection techniques include backup files, security software packages for data file security, secured waste, and audit trails.

- Communications security measures include passwords and cryptography. Cryptography is the process of encoding data to be sent over lines and then decoding it at the receiving end.

- Data that travels across borders may be subject to different sets of regulations.

- Identification and access to systems can be controlled by what you have (key, badge, card), what you know (password, lock combination), what you do (signature), and what you are (fingerprint, voice). Biometrics is the science of measuring individual body characteristics.

- A disaster recovery plan is a method for restoring data processing operations if those operations have been suspended due to major damage or destruction. Organizations take a variety of approaches, such as reverting to manual operations, buying time from a service bureau, developing mutual aid pacts with other organizations, or joining consortiums.

- A consortium is a joint venture to support a complete computer facility. A hot site is a fully equipped computer center, with hardware, environmental controls, security, and communications facilities. A cold site is an environmentally suitable empty shell in which a company can install its own computer.

- A disaster recovery plan includes a list of priorities, personnel requirements, equipment requirements, facilities, designations, and methods for capturing and distributing data.

- Special security considerations for microcomputers include protection from theft, environmental controls, diskette protection, and protection from power surges. A surge protector is a device that will prevent electrical problems from affecting data files.

- Risk factor is the likelihood of having a particular problem. Risk analysis assigns a dollar value to a potential loss and a percentage figure to the probability of that loss, then multiplies those numbers to get the expected loss.

- The Freedom of Information Act allows citizens to have access to data gathered about them by federal agencies. The Federal Privacy Act stipulates that there can be no secret computer personnel files. There is no federal computer security legislation, but several states have passed computer security laws.

- Arbitration is a legally binding procedure used to settle disputes between two parties out of court. A key feature of arbitration is that the arbitrators who will judge the merits of the case are technical experts.

KEY TERMS

Arbitration	Natural disasters
Audit trail	Natural forces
Backup file	Passwords
Cold site	Physical security
Communications security	Risk analysis
Consortium	Risk factor
Cryptography	Secured waste
Data security	Security
Disaster recovery plan	Security software package
Electronic fund transfer (EFT)	Software security
Federal Privacy Act	Surge protector
Freedom of Information Act	Uninterrupted power supply
Hot site	(UPS)

REVIEW QUESTIONS

1. Define security.

2. What kinds of threats to security would you expect to find in these situations:

    A small shop where everyone does everything

    A shop where the lead computer operator has refused to take a vacation for two years

    An organization that has a dozen microcomputers, whose related disk files are stuffed in desk drawers or lying on desktops

    A system that takes backup files on an irregular basis

A hardware system fitted into a basement area near the water and heating systems

A shop that changes the cipher locks on the computer room once a year

3. Give examples of computer systems that could probably justify an uninterrupted power supply.

4. Discuss software security in terms of what you have seen or heard about software piracy. If you were the manufacturer or distributor of software, what steps would you want to take to protect your software? Would you consider legal action for known violations?

5. Why is data the most difficult resource to recover?

6. What computer system identification and access methods have you seen in use? Did those methods seem adequate for the system? Could you suggest alternate methods?

7. Describe, in general terms, a possible disaster recovery plan for a banking system.

8. Why is risk analysis a valuable exercise?

9. List the key advantages of arbitration over litigation.

CASE 15-1
THE EASY PASSWORD

Although some of these cases are exaggerated to demonstrate bad approaches, many reflect exactly what is going on in industry.

Lamont Pierce, who worked in accounts receivable, chose passwords that would be easy to remember. He did not do so lightly. This was his reasoning:

- Long nonwords like XRNPQF are impossible to remember.
- If he forgot his password he would have to submit to an embarrassing routine that included telling his supervisor and going through a security officer.
- If he wrote it down there was a good chance that someone else might find it.
- There was a requirement that passwords be changed every month.

Lamont solved the problem to his satisfaction by choosing two easy-to-remember words that reflected his hobby, SAIL and BOAT. He alternated these passwords monthly and continued to do so for 17 months.

Disaster struck on the eighteenth month. His files were invaded and changed in subtle ways that did not become fully clear until the next customer cycle produced complaints. The perpetrator turned out to be a disgruntled programmer who wanted to wreak a little havoc before he disappeared. He had nothing particular against Lamont—Lamont was just the handiest sitting duck, a person with guessable passwords.

Lamont, of course, regretted his naivete, but that was of little use after the fact.

1. What responsibility does Lamont's management bear for this fiasco?
2. Prepare a list of arguments for user-generated nonmeaningful passwords.
3. Prepare a list of arguments for system-generated passwords.

CASE 15-2
AS THE DISK TURNS

At the Bay City Crisis Center, 140 trained volunteers answer emergency phone calls on a 24-hour basis. These volunteers are supervised by a staff of professionals and supported by a computer time-sharing system that provides information on 1,200 local agencies and services. The paid Center staff does not include a computer professional. Martha Carrier, the director of the Center, has learned something about computers from her brother-in-law, who sells microcomputers.

Martha decided that the time-sharing services should be supplemented with microcomputers belonging to the Center. The needs assessment and purchase were done carefully. Hard disk was part of the plan because of the high volume of training materials. In a few months, the supervisory and clerical staffs were proficient in word processing and some had learned more. All of them were accumulating files on the hard disk. Although backup files on floppy disk had been suggested in the early training sessions, few people took the trouble to make them.

One Saturday a thunder and lightning storm passed through Bay City. The net result for the Crisis Center was extra emergency calls, a small leak in their roof, and—worse—ruined disk files. A power surge had destroyed most of the hard disk files. Some of the file data could not be recovered. Other data was rekeyed with a time loss of 87 hours.

1. Should the Crisis Center have used only diskettes for their files?
2. What advance security actions could have prevented the loss of data?

PART 6

SYSTEM TRENDS TODAY

Earlier sections of the book presented the systems life cycle as a set of traditional procedures tempered by modern methodologies. But systems work today also involves a set of related issues, which are discussed in this final section. These topics, in fact, are interrelated: microcomputers in systems, the information center, and office automation systems. The single thread that runs through these subjects is that users are learning to use computers directly. In this endeavor, they often need the assistance and direction of professional computer personnel.

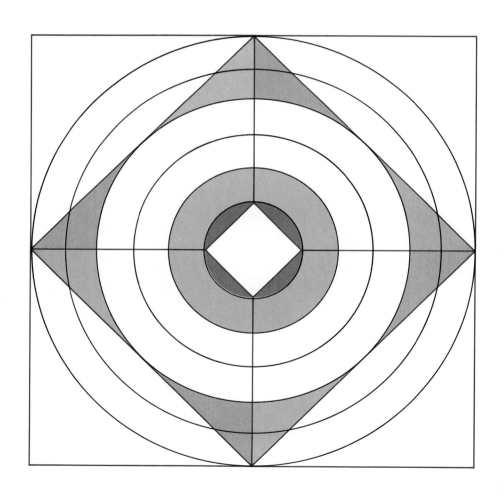

Chapter 16

Microcomputers in Systems

Bob Gantz was the analyst given the microcomputer investigation assignment at Kendell, Inc., a large advertising firm. Not just his boss, but his boss's boss and two company vice-presidents wanted some information on the proliferation of microcomputers, also known as personal computers, in user organizations throughout the company.

Bob already knew that the microcomputers were not officially under the jurisdiction of the data processing department. Encouraged by reasonable hardware prices, accountants, managers, and even copywriters had purchased their own machines. They had been convinced that they could use the computers without local expert advice and proved to do just that. They learned to use software packages to input and manipulate data, produce reports, and even draw graphs.

Trouble bubbled just below the surface, however. Bob traveled from group to group, interviewing computer users. He found two prevailing attitudes. One was a fierce desire to maintain independence from the data processing organization, the very reason they made their original purchases. They felt they had been mired in data processing department backlogs and rules too long and had no wish to return. The second attitude—in conflict with the first—was their need to get files from data processing to use with their own computers. This last stepped over the line into problems of data integrity, accuracy, and security.

Higher management at Kendell, partly because of Bob's report, decided to formulate microcomputer policy. In this chapter we shall examine why this is a difficult task.

◇ THE TREND IN COMPUTERS:
VERY BIG AND VERY SMALL

Although prefixes like mini, micro, and super have become common computer descriptors, you would struggle to form definitions that would be accurate for long. The market is too fluid (see Figure 16-1). However, a pattern has been clear for some time: There is expanded use of computers at the low and high ends, and the usages are related.

Microcomputer use is growing rapidly at every level in every kind of business. Users are demanding more and more from them, and the ultimate demand is that microcomputers milk the power inherent in machines at the other end of the spectrum, the very large machines. They, in turn, are in demand for the usual batch and online systems and, also, as support for microcomputers. The total amount of computing power available is doubling about every two years. Let us consider how the microcomputer got started in business.

◇ BREAKING NEW GROUND

Gone are the days when the only computer power around was in the hands of the professionals, and anyone wanting to partake had to appear as a supplicant at the gate. Many users have taken matters into their own hands.

Users Take the Initiative

Much has been written and said about the backlog of requests for computing services from data processing departments. Users took new and unprecedented steps when they gathered their own funds to purchase micros. The most compelling reason for the spread of personal computers was their low cost, an amount that can be squeezed out of normal equipment budgets. In fact, accounting departments have found microcomputer purchases hidden in various capital budgets. In one case, a microcomputer was disguised as a forklift. Furthermore, users have made purchases without the permission of any professional computer person. This does not mean, as is often hinted, that users are trying to use their micros to replace the data processing establishment. They want to supplement it. The facts could not be ignored. Microcomputers soon got respect.

Users as Microcomputer Pioneers

Is there a micro in your future? Is there a micro in your user's future? Odds are that the answer is yes to both questions. It is also probable that your user beat you to it. Many, many traditional programmers—and some analysts—know little about micros.

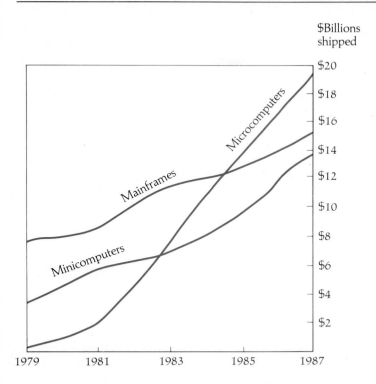

$Billions shipped

$20
$18
$16
$14
$12
$10
$8
$6
$4
$2

1979   1981   1983   1985   1987

Microcomputers
Mainframes
Minicomputers

FIGURE 16-1

The changing land-scape of the comput-er base. International Data Corporation provides figures that show dollars spent on microcomputers outreaching dollars for all other types of computer purchases.

At the outset, microcomputer users happily experiment with word processing, spreadsheets, databases, and business graphics. In time they come to realize that more is needed and cast their eyes on the corporate

---

BOX 16-1   Levels of Use

Users gradually change their view of the microcomputer, as well as the way they interact with it. Each change—that is, each movement to a new level of awareness—is an improvement.

Level 1: Machine. Seen as a mere machine, the primary concern is getting it to work.

Level 2: Tool. Using the computer as a tool, the concern is to make it work in ways that enhance work capabilities.

Level 3: Freedom. The computer can be a facilitator, a vehicle to free people to work outside the system, to do what-ifs on their own, and to get a competitive advantage.

mainframe computer. They do not want to approach it in the old way, however, through the data processing department. Instead, they want to approach the mainframe "directly," through their own microcomputers. Read on.

◇ THE MICRO–MAINFRAME CONNECTION

The media have trumpeted the micro-to-mainframe connection as an unsurpassed relationship, a wedding of cheap access to expensive power. But the move from micro to mainframe is not a short step.

Micro Problem, Mainframe Solution

From the start, microcomputers have been used by individuals as stand-alone devices, providing some computing power for users. But in terms of overall company information management, this approach has several shortcomings:

- Many micro operations are duplicated on other micros, creating unreliable duplication of data, in various states of modification.
- The spread of micros is difficult to monitor and control.
- Microcomputers are starved for fresh data. In most large companies, almost all the important business data is stored on mainframe computer systems.

Now consider what would happen if the data stored centrally with the mainframe computer was accessible to each microcomputer. The problems just listed would disappear fast because users would get fresh data, and control could be returned to the keepers of the data, the data processing department. But problems would remain.

Needing the Data

More and more tasks are being given over to the micros, but without a corresponding increase in data. That is the problem in a nutshell: machines— yes, data—no. Users need to make decisions using mainframe data that is often difficult or impossible to access from a microcomputer. Sending data from the mainframe to the micro is number one on the micro user wish list. To get data from a mainframe, your microcomputer has to pretend to be a terminal, because that is what the mainframe was designed to handle. Sending data from a mainframe to a microcomputer is called **downloading**. The reverse process, sending data from a microcomputer to a mainframe, is called **uploading** (see Figure 16-2). The micro and the mainframe are as far

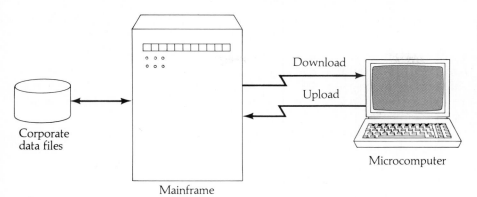

FIGURE 16-2

Downloading and uploading. Data sent from a mainframe to a microcomputer is said to be downloaded; the reverse is uploading.

apart in compatibility as they are in size. To bridge this gap, some rather complex conversion software is needed.

Political and Security Problems

Data processing managers are understandably nervous about letting micros interfere with vital corporate data. Some data processing managers also want to regain control of corporate computing power. This dilemma is being solved in favor of the micro users, simply because their demands are so great. As noted in Chapter 15, security is a problem whenever files are downloaded to microcomputers. One approach to solving this is to limit access to certain data. Another is to have only temporary updates of corporate databases through micros, letting them become true updates only after a review process.

Another possible approach is to download data to users on micros but not return that data to the corporate database. However, this, unfortunately, would create still another problem. Data on the micro could be processed and manipulated and put out on reports. Sooner or later, that report is going to be compared with a report made from the original corporate data—and they may not match. A manager might even present his or her own modified data as the official corporate version. These are the very problems the industry has tried to eliminate by switching from file processing to database processing.

Another direction the industry may take is to download data to certain micros at certain times. The data is sort of "checked out" to that micro user. No one else can update that particular data for the time being. Then the data is allowed to be uploaded, but only via certain strict editing procedures. In other words, "We'll take the data back, but only if you haven't hurt it." This is complicated, but right now it is the most likely solution.

In summary, the micro–mainframe link is hampered by compatibility and security problems. Nevertheless, there is no doubt that it is here to stay.

◇ NETWORKING: PLUGGED IN TO THE WORLD

While waiting for the micro–mainframe duo to reach perfection, another microcomputer application is tremendously successful: networking. The idea of networking, connecting several computers, is not new. What is new is local area networks connecting several microcomputers that would normally stand alone.

A personal computer, by definition, is a stand-alone, personal device. Even so, personal computers are not exactly islands, since floppy disks are easily shuttled among machines. But, as an organization grows, so does its need to efficiently store, retrieve, update, and correlate the data it collects and uses. As time goes by, users need to share their data with others in the office in a better way, and that way has become the local area network.

Local Area Networks

A **local area network (LAN)** is a medium for sharing hardware and/or information. A typical LAN consists of a software package and any of several types of hardware. In simple terms, LANs hook microcomputers together through communications media so that each micro can share the resources of the others. Each microcomputer in the network is called a **node**. Normally, all nodes will be in close proximity, probably in the same building. LANs can simplify communications among people as well as computers.

Here are some typical tasks for which LANs are especially well suited:

- A micro can read data from a hard disk belonging to another micro as if it were its own.
- A micro may print one of its files on the printer of another micro. (Since few people need a letter quality printer all the time, this more expensive printer could be hooked to only a few computers.)
- A micro may send a message to one or several micros. The message will show up when the user of the receiving micro checks for messages. The name of this activity: electronic mail.

As any or all of these activities are going on, the micro whose resource is being accessed by another can continue doing its own work.

Network Topologies

The physical geometric layout of a local area network is called a **topology**. Local area networks come in three basic topologies: star, ring, and bus networks. A **star network** has a central computer that is responsible for managing the LAN. It is to this central computer—sometimes called a **server**—that the shared disks and printers are attached. All messages are routed through the server. A **ring network** links all nodes together in a circular manner without benefit of a server. If one computer goes down, the network is down. Disks and printers are scattered throughout the system. A **bus network** assigns a portion of network management to each computer but preserves the system if one component fails. The majority of LANs are bus structured. Note the different topologies in Figure 16-3.

The Nonstandardization Factor

Suppose that, instead of 22 identical microcomputers, your office already is stocked with an assortment of 8 Apples®, 12 IBMs®, and 2 KayPros®. You want them to talk to each other and do all the things just mentioned. Now you have a problem. The various computers have different data formats and different operating systems. No manufacturer offers a networking product that will link all kinds of microcomputers. One solution is custom-made, expensive connections. Another is industrywide standards, which are coming but are not yet a reality.

A comprehensive network discussion could easily be a chapter or even a book by itself. This brief overview (and Case 16-1) is intended to demonstate when LANs might be a system solution.

◇ SOFTWARE FOR MICROCOMPUTER SYSTEMS

The microcomputer software industry is growing at a phenomenal rate. Software for personal use initiated the industry but software for business will keep it thriving. There are many potential software users and many untapped business software applications.

Micros in the Executive Suite

People have been making decisions forever, and they have been making them with the aid of computers for several years. But the role of computers in the decision-making process really took off when microcomputers entered the picture. Here was a tool that combined all the elements of decision-making: (1) data, (2) ways to manipulate the data, and (3) methods of presenting the data that could easily be used and understood by managers. For the first time, managers had the opportunity to try different what-if combinations before making a choice. Computers help managers manage.

Getting Smart with Software

The key software elements needed by most micro users are spreadsheets, graphics, database management, and word processing. The decision-making just mentioned is normally tied to spreadsheets, a financial analysis device that can display various combinations of figures and change the bottom line if one or more of them is altered in what-if situations. Graphics

Star network

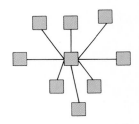

Ring network

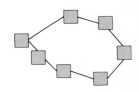

Bus network

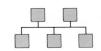

FIGURE 16-3

Local area network topologies. Typical network configurations are star, ring, and bus networks.

give managers a real sense of being in the driver's seat, as they present pictorially results that indicate present conditions and trends.

Database management software lets managers cross-check related information, accessing data in various ways. A customer file, for example, might be accessed using name or account number. Word processing is used to prepare and revise documents.

These separate software packages, sometimes called the "first generation" of personal computer software, have been geared toward clerical efficiency, especially the word processing and spreadsheets packages. They have made work easier and faster (see Figure 16-4).

The second round of personal computer software is leveraging brainpower. New spreadsheet packages, for example, can "backsolve" problems. That is, the software can figure out the middle figures by working backwards from desired totals. Power also comes from integrated software packages.

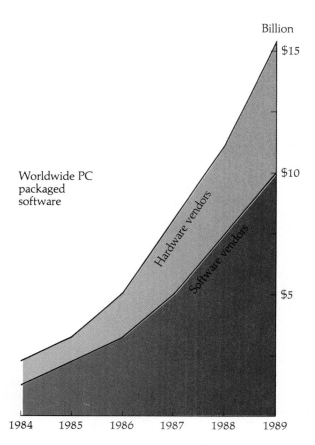

FIGURE 16-4

Packaged software for personal computers. Dozens of programs for personal computers have sold in the hundreds of thousands and a few have sold over a million copies.

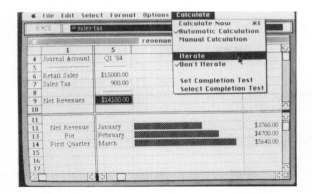

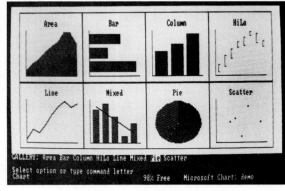

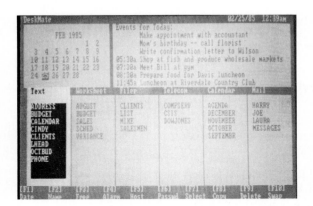

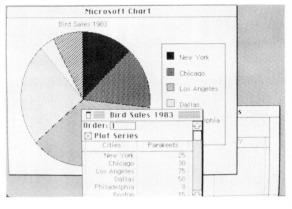

Personal computer software. Top left shows a screen from Microsoft's Multiplan, an integrated software package. Top right illustrates a variety of graphics using Microsoft Chart. Bottom left is a screen for DeskMate, a software package in the "desktop organizer" category. The bottom right screen is also from Microsoft Chart, showing its windowing capability.

Integrated software combines individual applications packages while at the same time coordinating the data among them. Common applications in an integrated package are word processing, spreadsheets, database access, graphics, and data communications. For example, one integrated package could be used to retrieve data from the database, present it in spreadsheet format, and then graph the results. Integrated packages are designed to provide communication and data exchange among different applications packages. Although the all-in-one packages have gained a significant market share, conventional programs dedicated to one purpose generally continue to dominate. The inherent weakness in integrated programs is that they are, like a Swiss army knife, fine to have when you need a little of everything but not particularly powerful at any given task.

Expert Systems in Business

The age of truly intelligent thinking machines is not yet here, but it is worth taking just a moment to note the progress that has been made in this area and to speculate on the impact of such machines on future systems. An **expert system** is a software package, used with an extensive set of organized data, that presents the computer as an expert on some topic. The user is the knowledge seeker, usually in a natural language format. This system

---

BOX 16-2 | **Expert Systems for the Experts**

Some expert systems have found wide acceptance in their respective expert communities. Examples:

Geology. When fed local data, an expert system named PROSPECTOR evaluates geological sites for mineral deposits.

Medicine. A system named PUFF offers expert diagnosis of pulmonary diseases. ONCOCIN selects therapies for cancer patients.

Computers. The Digital Equipment Corporation uses expert system **R1** to configure new VAX minicomputer installations.

Chemistry. A system named DENDRAL clarifies the chemical structure of complex compounds.

Genetics. An expert system named MOLGEN helps geneticists plan experiments involving structural analysis and synthesis of DNA.

Mathematics. High-level algebra is performed in a flash by MACSYMA.

goes beyond the usual user-friendly cliché—we are talking about software that is especially easy to use.

Expert systems have been available on large machines since the 1970s, but now they are moving to microcomputers. The cost of an expert system can usually be justified in situations where there are few experts but great demand. The main limitation of an expert system is that it requires a substantial amount of internal memory. In addition, a large amount of data, in terms of rules, facts, and source code must also be stored, dictating the use of hard disk. Though it will be a while before expert systems are routinely sold in stores and used in offices, the movement has begun.

Meanwhile, many analysts are discovering yet another application—using micros to revamp old systems. This process has been dubbed micro migration.

◈ MICRO MIGRATION

No, the micros are not migrating anywhere, but software is. The phrase **micro migration** was coined to describe taking old—really old—systems and converting them to run on microcomputers. To appreciate how wonderful this might be for some companies and some users, consider the pain involved in running the old systems. They are batch systems, written 15 or even 20 years ago, and are suffering from obsolescence. They have been patched and hand-massaged and despaired over, but hardly ever documented. Such a system is invariably assigned to the newest and most hapless person in the group, who never does find out exactly how it works before passing it on.

Given the right ingredients, an old system may make a fine candidate for being revamped and handed over to users for control on their own micros. Typical ingredients are as follows:

- *Old age.* The system must be so old that any reasonable change would be a distinct relief.
- *The right users.* The users of the system considered for migration are dying for a fresh approach and better control. Also, they may be computer-savvy, since they already have their own micros.
- *The right size.* Many decaying systems consist of programs that are very large indeed and could not be confined within the memory limitations of some micros. But some programs are smaller and some micros have 512K memory or more. (In fact, some "super micros" are moving into the megabyte range.)
- *The right data.* Are there a reasonable number of files? Are file interfaces with other systems manageable? Can record blocks be made an acceptable length?

Micro migration can be a very satisfactory experience for all involved if done carefully.

◇ TELECOMMUTING: HOME-TO-OFFICE LINK

Telecommuting is not for everyone. Nor is it for every business, especially those whose managers are very nervous about security. But, still, there is a niche for workers who want (need) to work at home. All you need is a microcomputer, a modem, and a telephone. This gives you the flexibility to work on your own computer or be hooked up to the company computer. This is **telecommuting**—"commuting" by telephone.

Telecommuters number only about a quarter of a million today, but that figure is expected to soar to 13 million by 1990. Contrary to the ballyhooed office-emptying of the 1980s, however, most telecommuters will work part time at the office and part time at home. Typical ratios will probably be two days at the office and three at home.

◇ MANAGING MICROCOMPUTER SYSTEMS

Micros have become so powerful and so numerous that they are sometimes seen as fragmenting entire corporate data processing systems. Managers face immense challenges in microcomputer acquisition and in integration of micros with existing systems.

Companywide Microcomputer Policy

Some companies have a very simple policy on microcomputers: There will not be any microcomputers. For those who do have micros, band-aid policy is sometimes in effect. Take the case at a major appliance manufacturing company. An engineer had a very solid case for using a micro in his job. He had experience on micros, too. But he ran into a brick wall in the person of Ed Oldham, the data processing manager, who lobbied vigorously against such a purchase. Meanwhile, Ed's boss decided that some executives needed micros and had them ordered without anyone lifting a finger. Ed provided no training or support for those people, however, so the machines are collecting executive dust.

The unfortunate case just described is not one of policy but of misapplied clout. In many companies, the days of an enthusiastic purchase out of your own budget have been replaced with political roadblocks.

Other companies have formed specific policy and a committee to administer it. It may seem that the freedom and élan of micros has given way to the old bureaucracy, but anything that grows eventually has to be regulated to avoid chaos. Policy applies to software purchases as well, because supporting a wide variety of packages is expensive. A typical policy might work this way. After the person making a purchase request gets approval from his or her own department, a description of the desired use and the support needed is submitted to a services group that has powers of decision.

BOX 16-3	Micros Invade Government

If you work with the United States government, you probably understand that you are working with a user organization that knows it has
clout. The General Services Administration (GSA) has encouraged
decentralization of computing power. A GSA-approved vendor list
includes 45 brands and 36 suppliers from which agencies can choose
hardware and software. Hardware and software are not enough, however. The federal government places primary emphasis on some sort
of continued central support, not just some initial training. Another
key requirement for getting on the list is offering the government a
below-list-price good deal.

Every agency from the National Science Foundation to the Air
Force is getting in on the act. To make it easier for government employees to get acquainted with, and order, micros, a series of retail stores
exist right in government office buildings. The first store was over-
whelmed with sales. The potential market is expected to be one-half
million microcomputers by 1990.

Both hardware and software may be somewhat standardized, so that
there is an "approved" list of micros and packages. Under such a policy it
would be easier to win approval if you can fit into the established standards
(say, a certain computer and specific in-use packages). These restrictions
have not been created out of perverseness but out of necessity. Let us say,
for example, that most users are working with Lotus spreadsheets but an
isolated person has an unknown product. Getting support for that person—
which will be needed sooner or later, especially if there are designs on the
mainframe—is difficult.

The Loss of Innocence

Users may be frustrated by the rising bureaucracy surrounding micro hardware and software purchases, but at least they understand what is going
on. Their vision is less clear on the placement of micros within systems.
The fact is that most micro users do not understand systems development.
They do understand that requests to the data processing organization mean
lots of time and money. And they may think that they are now free from
constraints that will complicate their lives.

However, they may not realize that as their list of applications grows,
so do the problems of integrating the new systems with existing systems.
In fact, eventually they will hit the same kind of integration problems that

slow down the data processing department. They will come to appreciate that it pays to spend time on the design stage and to document the system well.

Microcomputers and the Systems Analyst

We have not made a direct link in this chapter to the systems life cycle. The principles of planning and designing a new system are the same, however, whether the hardware consists of large or small computers. The subjects covered in this chapter are really background material to help you deal with microcomputers in a system environment.

As an analyst consultant, you could easily encounter a situation where a single microcomputer and applications software would be an appropriate solution. If you are working with large organizations, chances are excellent that microcomputers will form some part of the systems you design. You then face all the related problems of compatibility, mainframe access, networking, and security. Although you probably will not be an expert in all these matters, this chapter has raised the issues and, with them, your awareness level.

◇ ONE MICRO, ONE USER

Are microcomputers to be bought in mass purchases and used only by the big companies? What about the little guy? Does anyone use just one microcomputer anymore? Of course, the answer is yes, but that fact is sometimes obscured by the trade press.

The splash in the trade press correctly reflects the fact that microcomputers have made a powerful impact on corporate business. Headlines trumpet the news that Company X has just purchased 500 microcomputers that will access the mainframe, be networked together, and use every conceivable kind of software. Since many companies are thinking about taking such steps, the press supplies a steady stream of articles on micro–mainframe harmony, network options, and the like. A key reason for all these articles, of course, is that the technology is changing rapidly and the computing community needs information.

But it is worth reflecting on the single user, because this user is still a significant factor in the microcomputer picture. Although less glamorous, the uses for microcomputers in small businesses are many and varied.

- A small company that manufactures boat winches uses a microcomputer to keep track of customer accounts.
- An apple grower uses a microcomputer graphics package to produce maps of his orchards, drawing in valves, sprinklers, and irrigation

---

BOX 16-4 | **Electronic Perks**

The opportunity to buy a personal computer at a discounted price is a new perk for some employees. Several companies subscribe to the theory that a computer at home will make a better employee on the job. Managers who feel that personal computers will redefine everyone's roles in the coming years also think that anything they can do to help people learn is a good idea. Discounts range from 20% to 40%. The most common arrangement is to have a computer retailer offer better prices to customers who identify themselves as an employee of the company. The company picks up the difference.

---

channels. Copies of the maps, which can be changed easily to reflect new conditions, are distributed to orchard workers.

- A minister prepares sermons using word processing on a personal computer.
- A real estate broker uses a menu-driven software package to calculate regular and balloon mortgages, amortization schedules, and cash flows.
- A psychologist encourages new patients to use a question-answer software package on a personal computer in his office to divulge background information. Many patients actually prefer confiding to a machine.
- A bicycle repair shop uses a microcomputer to keep a calendar of sporting events of interest to its customers.
- A student takes correspondence classes using software on her own microcomputer at home.

You can probably think of other uses to add to this list. The personal computer will continue to be a stand-alone, personal device for many small enterprises and for people working at home.

◇ MICROCOMPUTERS COME FULL CYCLE

Programmers are getting into the microcomputer act. In fact, you could say that they started it. Professional hobbyists were the guiding force behind the original trend to small computers. For a time, they considered the small computer only a toy to be tinkered with on their own time. The "real" computers were at the office.

It was users marching to a different drummer who forced computer professionals to view microcomputers as a serious business tool. Now some computer professionals are scrambling to catch up with their users.

Microcomputers have made users a force to be reckoned with in a new way. Data processing organizations are responding to that force in several ways. One of those ways is the information center, the subject of our next chapter.

FROM THE REAL WORLD

Computers Move into Real Estate

Let's see. You're looking for a house in the Wedgewood district, with three bedrooms, in a range from $90,000 to $130,000. At the local real estate office, that information is easily keyed into the microcomputer, which is linked online to the Multiple Listing Service. In moments the screen is filled with listings of homes that match the specifications. Each line contains abbreviated information for one home: listing number, address, price, number of bedrooms, number of baths, and checkoffs for garage, pool, family room, and wheelchair access. Now you can pull the file of any hot prospect.

The method just described sure beats the old ways of relying on the agent's top-of-the-head knowledge or driving around for hours hoping to spot a likely house. But it does not stop there. Microcomputers are well established in every phase of real estate activities, from sales to amortization to office management. Suppose you now decide to make an offer on one of the houses and reach terms with the buyer, who agrees to carry the contract. These terms—loan amount, interest rate, down payment, number of years, and so on—can be keyed in and you will immediately receive output giving you, for each payment, the date, payment amount, amount toward interest, amount toward principal, and balance.

From a managerial aspect, the microcomputers can handle word processing, budgeting, financial projections, and bookkeeping. One challenge is keeping track of each agent's business by categories, such as number of current listings, number of listings sold, and commissions. Another is to monitor each piece of property for sale, keeping track of how often it has been shown, and its advertising and client costs.

Customers like the services. Realtors like the convenience. Managers like the cost savings.

—From an article in *PC Magazine*

SUMMARY

- The trend in computers is toward large computers, which are in demand for the usual batch and online systems, but also for support for the small machines, the microcomputers.

- Users took the initiative in bringing microcomputers into the business environment, but they soon wanted access to the corporate data associated with mainframe computers.

- Sending data from a mainframe to a microcomputer is called downloading; the reverse process is called uploading.

- In addition to the technical problems of the micro–mainframe connection, data processing managers are nervous about what might happen to data that is downloaded from mainframe to micro and even more nervous about taking the data back.

- A local area network (LAN) is a medium for sharing hardware and/or information. Each microcomputer in the network is called a node. Normally, all nodes will be in the same building.

- The physical geometric layout of a local area network is called a topology. LANs come in three basic topologies: star, ring, and bus.

- A star network has a central computer, sometimes called a server, that is responsible for managing the LAN. A ring network links the nodes in a circular way without benefit of a server. A bus network assigns a portion of network management to each computer but preserves the system if one component fails. Most LANs are bus structured.

- The role of computers in decision-making expanded with the advent of microcomputers and their business software. Key software applications used by users are word processing, spreadsheets, graphics, and database management. Integrated software combines individual applications packages while coordinating the data among them.

- An expert system is a software package, used with an extensive set of organized data, that presents the computer as an expert on some topic.

- Micro migration refers to the conversion of old systems to run on microcomputers.

- Telecommuting refers to working at home but being connected to the office via your microcomputer.

- Most companies have formed microcomputer policy to regulate the purchase and use of microcomputers and related software.

- Although there is an emphasis on large corporate micro systems, there are still a significant number of small businesses that use microcomputers on a stand-alone basis.

KEY TERMS

Bus network	Ring network
Downloading	Server
Expert system	Star network
Integrated software	Telecommuting
Local area network (LAN)	Topology
Micro migration	Uploading
Node	

REVIEW QUESTIONS

1. What problems are related to the micro–mainframe connection?

2. Name the common local area network topologies and differentiate among them.

3. How is integrated software different from a set of separate applications on the same disk?

4. Suppose you were in charge of formulating microcomputer policy for a company of 200 office workers. So far, there are three different types of microcomputers in use. There is some confusion in the office about what to do next.

5. Name five types of small businesses that could benefit from a microcomputer but probably would not need more than one.

CASE 16-1:
JUST TRUCKING ALONG WITH THE NETWORK

It is 8:00 A.M. Tuesday morning when service manager Fred Whiteman arrives in the office. Fred works for Harrington Equipment, a firm that sells and services trucking fleets. At Harrington, micros are numerous and they are all hooked together using a local area network.

Fred instructs his microcomputer to print out an inventory report. A customer calls to report that his truck is broken and must be fixed immediately. Fred can do it, but he might not return in time for a staff meeting with his boss, general manager Jason West. He calls Jason's office but gets no answer. So Fred uses his micro to send a message to Jason. When Jason's secretary comes in, the message will be waiting.

Before he goes out on the service call, Fred wants a copy of the broken truck's service record. It is in his own micro disk storage, but his printer is busy. No problem—he instructs his micro to print the record on another printer, attached to the micro in the conference room. Fred gets to the conference room just as it finishes printing, tears off the report, and heads out the door.

Meanwhile, salesperson Lesley Lewis answers a telephone inquiry about a generator. Even though the inventory records are kept by the service department, Lesley can access them through her micro in the sales department. She soon is able to report to the customer on the line that the part is in stock and makes her first sale of the day.

Diane Killorin, Jason West's secretary, turns on her micro for the day and finds two messages already waiting. One is from Fred, about the meeting. The other is for her, from the bookkeeper, saying that she has keyed in some old sales data to be used for financial projections. Although the bookkeeper keyed them in on her own micro, the files are available to Diane on her micro.

When Jason arrives in his office, Diane tells him about Fred's absence. Jason suggests that she check the calendars of the other attendees to see when the meeting could be rescheduled. The calendars are scattered among the Harrington computers, but they are easy to review because a scheduling program brings them all together. The particular ones Diane looks at from her computer are those belonging to in-house repair services, telemarketing, and advertising. The people in question are all available at 11:00 A.M. Wednesday. She checks that data with Jason, who tells her to draft a memo announcing the change and, from her micro, she sends everyone a copy.

The day has hardly begun, but Harrington employees have already made extensive use of their LAN.

1. How many microcomputers were mentioned in the LAN?
2. Make a drawing of the equipment mentioned. Show the network.

CASE 16-2:
JUST A MATTER OF HOOKUPS

At a major home products corporation, Northwest regional sales manager Pat Byers bought a microcomputer so she could do her own spreadsheets and word processing. The computer allowed her to bypass the data processing organization and generate reports in half the time, doing number crunching and text processing right from her desk.

She had all the power she needed to analyze market response to a new product or establish sales goals by region, right in the spreadsheet program. But she realized that she needed sales history figures in the corporate database to make a comparison between her region's sales and companywide performance. This meant the traditional wait for the data processing department to produce the printout, reported by region. And, once she had her hands on the report, she had to key the numbers into the format of her spreadsheet so they could be integrated into the final report. The duplication of effort just didn't make sense.

Pat read somewhere that it was possible to access the mainframe data directly, using her computer as a terminal. All she needed was a communications board and some software. She brought this topic up with her manager, who put her in touch with a systems analyst, Andy Clark. Andy explained that the data processing organization was looking into the possibilities, but there were still a lot of problems. He talked vaguely about networking and terminal emulation problems and security.

But Pat was impatient. Now that she had learned that her computer could actually be talking to the mainframe, she was anxious to see the program up and running. It's just a matter of hooking a few things up—isn't it?

1. How would you have explained the delay—or unavailability—of the micro–mainframe link to Pat?
2. How will the micro–mainframe link improve Pat's efficiency?

CASE 16-3:
NO MICROS MAY ENTER HERE

Booth-Wells, a Midwest glass manufacturer, has announced a formal policy to its employees: All access to the computer and its stored data will remain directly under control of the data processing department. What's more, no employee is permitted to generate his or her own company data. Finally, no department funds may be dispersed for the purchase of any computer hardware or software.

Booth-Wells, which tends to be a conservative company, took these measures in the interest of data integrity and security. Alarmed at the proliferation of micros in other companies, Booth saw this direction as trendy and the result of pressure from nonexperts. They decided it was appropriate to let those trained in the use of technology manage and control it.

The management at Booth did not ignore direct user computer use altogether, however. Sophisticated users were allowed to write programs in a fourth-generation language.

There were a few problems with this edict. One was that seven micros were already on the premises, purchased before the policy was announced. Their users continued to employ them, but more discreetly.

1. Add to the list of problems. You can probably think of half a dozen.
2. What will be the probable effect on the company and its employees if this policy continues for the next five years?

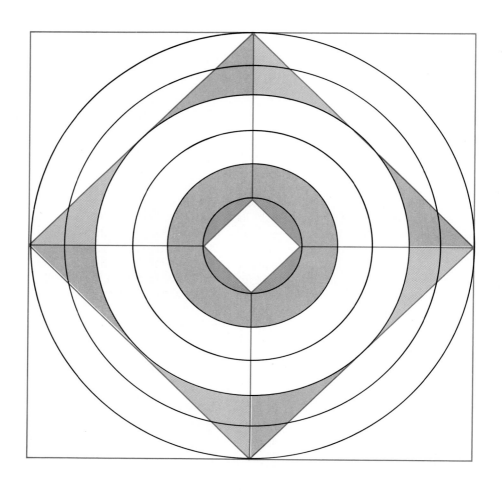

Chapter 17

The Information Center

Since she prided herself on relishing "new adventures," Kit McCormick was not intimidated by the idea of buying a computer for herself in the office. She was the first to do so. She had high hopes for the promised increase in productivity and an edge on the competition. Most of all, she hoped she could do some things for herself instead of always depending on the data processing department.

Flush with such ambition and her own department's blessing, she purchased a microcomputer and some business software. She took two classes provided by the vendor and then settled down to experiment. In a few months she was able to produce reports, prepare documents using word processing, and use spreadsheets to show her boss a variety of financial models

for a new product being considered.

Kit was not alone for long. In the next two years, microcomputer acquisition snowballed. All sorts of microcomputers and software packages appeared throughout the company. Keeping up with the literature, Kit and her colleagues began to approach the data processing department with requests for mainframe access, networking, software support, and training.

Chapter 16, on microcomputers, has already hinted that users are in need of such support. Not the traditional support through data processing department channels, but direct support in their own workplace on their own computers. One way of providing such support is through the information center.

◇   THE INFORMATION CENTER: "WALK-IN" ADVICE

An information center comes in many shapes and sizes and is known by many names. Industrywide, the generic term **information center** is used to describe direct user support. Users, of course, have been the reason-for-being of computers since the beginning and have received support as such. Support through an information center usually differs from traditional support in three ways:

1. Users are given support for direct use of their own equipment or user equipment at the information center.
2. Information center staff are devoted exclusively to giving user service.
3. User help is immediate.

So, there it is—just what the user has always wanted—a place to get instant help. For the company and the data processing department, the big payoffs are more user independence and some relief from user pressure. Let us look back, for a moment, to where those pressures came from.

Unmet Needs: The Visible and Invisible Backlog

The word **backlog** has an unhappy ring. Its literal meaning is an accumulation of work waiting to be done. Backlog is commonly used to refer to the scheduled data processing department work waiting for people resources. For users it means endless waiting and the feeling that nothing ever seems to get done. For the data processing department, where backlogs are often measured in years, it means not giving satisfactory service and always being too late with too little. Both of these perceptions are exaggerated, but they do not fall too far from the mark.

Over a period of years, as more and more computer applications were developed, more computer professionals were designated as their keepers. That is, the percentage of programmers maintaining old systems, as opposed to developing new systems, gradually increased. During that same time period, system users became more knowledgeable and made increasing demands for new systems. The burden became too much. Endless articles illuminated the way out of backlogs, but the problems remained pretty much the same. It was said, in fact, that there was not only a backlog, but also an **invisible backlog**, a set of user needs that users were not even bothering to mention because they knew how long the list already was!

Enter the microcomputer. For a relatively low cost, microcomputers provided fast response, availability, portability, and some excellent software. However, micros were poor at moving or printing large amounts of data, so total independence from the large computers and the data processing department was not the issue. After an initial period of timid reflection, users went after microcomputers with a vengeance. Here was the answer;

BOX 17-1	**Info for All at EPA**

The Environmental Protectional Agency (EPA) has established a first-class information center, with a variety of hardware, software, and services. The EPA, a federal agency, offers a full range of services at its headquarters operations. The information center is the single point for accessing and storing information.

The Center, as it is called, provides classroom and individual training, consultation, problem resolution, a library, and follow-up assistance to EPA personnel. The Center also helps to establish and administer special-interest groups such as those who are all using the same brand of microcomputer or the same software packages.

For those who do not have their own hardware, The Center provides microcomputers, plotters, word processors, and printers. It also stocks all the common software packages for spreadsheets, graphics, database management, and word processing. Users at EPA are well supported.

here was the way to get things done. And, most of all, here was a way to have some control and not be totally dependent on the data processing department. Meanwhile, the data processing department had some new problems of its own.

The Changing Role of the Data Processing Department

Even as you read this, there are some data processing managers digging in their toes, trying to keep the lid on the microcomputer phenomenon. It is too late, and it has been too late for a long time. To act or react, that is the question. The existence of microcomputers cannot be denied, nor can they be ignored by the data processing department. As microcomputers change the way businesses operate, the people running data processing departments will have to show they can adapt.

Most data processing departments, possibly belatedly, have swung into action. They recognize that user activity on micros can relieve the data processing burden. They also recognize that users need support. The answer for many companies is some version of an information center.

Setting Up the Information Center

Although the impetus often comes from the users themselves, the data processing department usually takes the lead in establishing an information center. If you are involved in such a project, you will want to follow the steps in the systems life cycle. An early task will be analyzing user requirements. This will involve meetings with representatives from potential user groups. Together you will make decisions on hardware, software, training, consulting, security, and other issues.

The information center is most often set up within the data processing department, with the information center manager reporting to the data processing manager. A major issue to be addressed is payment. How will users be charged for services? This will entail a bit of paperwork planning.

An early issue is the physical facility. It is sometimes tempting to pull together a collection of surplused tables and chairs, but not if you want the center to have credibility. Appropriate office furniture is needed. So are telephones, training facilities, and space for a reference library. Perhaps the most important aspect is a central location, one that is convenient for the greatest number of users.

Some sound advice: Make the original information center limited. Do not try to save the world overnight. You can easily get caught in the trap of offering too much to too many too soon. Try the new center out on a pilot group, users who are known to be enthusiastic and/or who already have some computer experience.

BOX 17-2	**Information Center Success Factors**

Although information centers vary from company to company, the ones considered most successful by their users have several factors in common.

- Commitment to the information center from the top
- A pilot group of users with aptitude, peer credibility, and visibility ("I didn't know anything, but look what I did!")
- Agreement on certain software packages, avoiding unlimited proliferation
- Access to appropriate data
- Staff that has been carefully selected (a good technical/business mix with exceptional communications skills)
- Thorough, plentiful training
- Walk-in service whenever possible
- A user group newsletter that is published periodically and includes success stories
- No oversell of the system—"Any fool can learn to press a button" results in built-in failure. ("I guess it's me, I never did get along with machines.")

One more cautionary note. Although this discussion has been slanted toward microcomputer usage, some information centers are mainframe-oriented or may support both ends of the spectrum.

Note the Information Center Success Factors box. In particular, be careful not to produce any work for the users. That is, do not write programs for them, or do any other work that they should be doing. The function of an information center is to provide assistance to users doing their own work, not to provide the service itself. Do not give them fish; teach them how to fish.

◇ INFORMATION CENTER SERVICES

As with any other service, the success of an information center depends on its acceptance by users. An information center must have a strong service component, a true sense of "the customer always comes first." No edict will mandate the use of an information center: It must sell itself to each new user. Although all centers must emphasize service, they differ in most other ways.

Information centers are not standardized. It could be said that no two are alike. Still, there are enough similarities that we can discuss them here in a general way. The information center stands by to assist the up-and-coming user in these ways.

Software Selection

Information center staff can help users to select appropriate software tools, from spreadsheets to database to graphics to word processing. These may be off the shelf, from the variety of software packages available at the center, or from the marketplace. Once the staff person has some idea of what the user needs, a support process begins. Suppose the solution seems to be a software package called X-PAC (a fictitious name). Rather than just handing the user the X-PAC disk, the staff member would follow the approach that includes sentences such as, "We have an X-PAC class on Tuesday" and "Let's sit down and look at X-PAC" and "Here is an X-PAC tutorial for you to try."

Tutorials are disks with online step-by-step lessons on a given topic. They are usually provided free by the software vendor. Tutorials are a great confidence booster and any information center should be well supplied with them.

In addition to assistance with well-known packages like word processors and spreadsheets, users need assistance in selecting and learning basic software tools such as file transfer utilities and editors. And, perhaps most important of all, they need to learn procedures for making backup diskettes.

| BOX 17-3 | **Stockpiling Software** |

Standard software packages that will be available from the information center are spreadsheets, word processing, database management, integrated software, and miscellaneous applications such as tax management, mailing lists, and schedule planning. The hit parade of specific products changes often, but the rules for purchasing software remain the same.

The first rule is to see what is included in the purchase price. Most software companies and their agents include object code and documentation, but not source code. Installation and training are often included as part of the price. You may also want to consider a maintenance contract, which is normally available at 10% of the cost of the purchase price. Another major consideration is the copyright policy. Will the software manufacturer permit extra copies to be made for internal use? Or, will the software manufacturer give discounts for volume purchases?

User Programming

Experts have been saying for some time that programming as we know it today will disappear. There is no need, they say, to learn artificial lingos like COBOL in order to get a computer to do our bidding. Programmers who worry about their jobs do not have to be too nervous just yet, but they would do well to become acquainted with the languages users like, the so-called fourth-generation languages.

Fourth-generation languages (sometimes abbreviated **4GL**) require no compile/link steps and are at least partly nonprocedural. A nonprocedural language is one that tells the computer *what* to do, not *how*. If you compare this to some of the third-generation languages you know—BASIC, COBOL, Pascal, and so on—you realize that they are all procedural, full of instructions on how to take care of things. Fourth-generation languages are relatively easy to use. Training *is* required—don't be oversold—but a user does not need to understand how to plan the logic of the actions needed, only to be able to state what the desired results are in a form that can be accepted by the language.

Most fourth-generation languages have some underlying concepts in common. The first is that, as already noted, a minimum of skill is required to use it. Another is that a minimum amount of work is required to accomplish a task. When a fourth-generation language is suitable, it is estimated its use speeds development time by a factor of ten over a language like COBOL. This is partly because 4GLs avoid strict language syntax and partly

Some popular business software packages are shown on the left. On the right we see an opened package, showing the disks and the documentation for Symphony, a sophisticated Lotus business product.

because there is no requirement to spell out exactly *how* a task is to be done. From these concepts, it follows that projects undertaken in a fourth-generation language will have few errors and be easy to maintain.

Despite the relative ease of fourth-generation languages, however, for the time being they are most suitable for sophisticated users. Fourth-generation languages have been available for the larger machines for some time, especially the ones associated with database packages, but they are now available for many microcomputers. Information center personnel may train users in the use of fourth-generation languages on their own personal computers or on terminals that access the mainframe computer.

Access to Data

Providing access only to authorized people is difficult, as we have already noted in Chapter 16. Who says data access is authorized? For whom? Where is the data, anyway? Shall corporate files be downloaded to the user micros? Can the data be uploaded again after the users have changed it? What safeguards are needed? Shall users have access to corporate data through their terminals? What access, revision, and deletion rights should any given user have? Who decides who has what rights? These may be the hardest issues you face. There are few quick and easy answers. Advanced and ongoing planning is needed.

Hardware Access and Selection

Many information centers provide hardware on the premises exclusively for users. The inventory may include microcomputers, terminals, word processors, printers, plotters, and other selected items. Some information centers also offer advice on the selection of purchased hardware.

An information cen-
ter "classroom,"
where employees can
get hands-on
training.

Education and Training

Many information center personnel consider education and training the top
priority. Initiating new users and keeping sophisticated users interested is
a constant challenge and one the center must be prepared to meet. Classes
should be constantly available. If they are offered on a scheduled basis, the
schedule should be widely disseminated, possibly through a periodic newsletter. Some classes should be offered on an as-needed basis, meaning that
the information center staff must be flexible.

Tutoring may be another function of the information center personnel.
One-on-one training is sometimes set up for busy executives who cannot
find time to teach themselves new technology. This type of training is notably different in two ways. First, the tutor will go to the executive, rather
than the reverse. Second, the tutor will come on a demand basis—sort of
a tutor-on-call. Some companies refer to this process as "coaching." It works
very well in breaking down initial apprehensions.

Technical Assistance

Users should be provided with technical assistance where appropriate. This
can take many forms: planning the input, interpreting the output, figuring
out what went wrong, and using the system to fit user needs.

BOX 17-4	**Midday Matinee**

Bring a brown bag lunch and sit in on a noontime presentation entitled "What Is a Computer?" This and other basic topics are the daily presentations of the information center at the Petro-Lewis Corporation in Denver. One of the responsibilities of information center personnel is to make people aware of available computer resources. This method works very nicely, in a nonthreatening way.

Decision Support Systems

Decision support systems (DSS) have varied hues, depending on the company under discussion or the expert doing the discussing. In some companies, DSS is considered a subset of the information center. We shall list it here as one of its services.

To the casual observer, DSS software output looks very much like spreadsheet data. But there are significant differences between them. Spreadsheets simply process tabular data through prescribed formulas. Their power lies in their ability to make instant bottom-line changes if one or more of the parameters is changed. **Decision support systems** take this several steps further, supplying statistical analysis and data modeling, all tied together with conventional database, graphics, and reports packages. The marketplace has offered several complete DSS packages, most of which can be menu-driven with little training.

◇ INFORMATION CENTER PERSONNEL

An information center needs staff who combine technical expertise and business acumen with strong interpersonal skills. Each staff member will have some technical specialty, in addition to basic familiarity with the tools offered by the center. It also is common for staff members to cross-train, so they can act as backup for another member.

After technical competence, the person working directly with users must genuinely want them to learn so that they can eventually work independently. The staff person must have patience and recognize that users are not going to learn in one hour something that it took him or her years to learn. Since one of the most sought-after services from an information center is education, computer specialists who are former teachers should be good staff candidates.

One more thought: Try some new graduates in the information center. They are often more receptive to new ideas than traditional data processing people.

◇ INFORMATION CENTER COSTS

Users soon discover that computer resources are not free. Even the information center has associated charges. Providing assistance is a major service, with attendant costs.

Some sort of billing system must be put in place. This can be done in a variety of ways. It may be done at management level, where a group of managers agree in advance to split the costs on some sort of equitable basis.

BOX 17-5   Help Is as Close as Your Telephone

"My terminal's blank. What can I do?" the voice on the other end of the phone said. Staff person Cindy Durkee at the Help Line office inquired to make sure that the terminal was plugged in and that the intensity knob was turned up. She then asked the caller if he had access to another workstation. In this case, he did not, so a terminal maintenance man was sent out to replace the broken equipment.

Another day, another call—one of the almost 50,000 that have been received by the computer Help Line at Dennison Manufacturing Company in Framingham, Massachusetts. The Help Line was established as management's response to user grumbling about pass-the-buck answers to their problems. A central point for coordinating and tracking problems was needed.

Four employees answer the Help Line, and they estimate that they can take care of about 75% of the problems themselves. On nights and weekends, a recorded message instructs callers to determine if their problem is critical or not. If critical, users relay the problem after the first beep, otherwise they wait until the third beep. If critical, a page device automatically signals Help Line employees' beepers so they can, from where they are, locate the proper solution.

A daily log, jokingly referred to as "the globe" because of its size, reflects the current status of problems in several categories, such as hardware, software, network, documentation, or report distribution.

Dennison employees today cannot imagine life without the Help Line. It is another form of service to the user.

One basis is planned usage; with this, user groups contribute to the costs based on the expected percentage of services they use. For example, if 40% of the information center personnel services are devoted to the marketing group, then the marketing group pays 40% of the costs of the information center. The advantage of such a system is service without any red tape.

It is more likely, however, especially in the case of an information center that serves many different departments, that charges will have to be tied to actual, as opposed to expected, usage. Services will have to be priced on a time and resources basis, and reported accordingly. These statistics on services received would be gathered by some paperwork or computer-based mechanism.

◇ USER RESPONSIBILITIES

Service to users has been emphasized, but users have responsibilities, too. Users must cooperate with policies and procedures established by the information center. One center, for example, limits users to four hours of assistance for a given new application use. This policy, of course, was instituted to prevent a few users from monopolizing the center.

Users are responsible for attending courses and seminars when appropriate. It would not make sense to offer private tutoring to an individual when a class on the same topic is easily available. A user sets up his or her own application, preparing the data and possibly coding. All of this can be done with assistance. Once established, a user maintains his or her own application.

◇ USERS GET RESPECT

A user panacea? No. But information centers have become very popular. Once the concept gained credibility, centers spread very quickly, even by the fast-changing standards of the computer industry. As a systems person, your principal association with information centers will be twofold: sending your users there when it is appropriate and receiving users who have been sent to you from the center. You, as a representative of the data processing department, and the information center, complement each other.

FROM THE REAL WORLD

Information Center at Exxon

Exxon has established expertise in the use of information centers that can be valuable to other companies. Sixteen information centers were established at Exxon over a 2-year period. The model center is in Manhattan, serving 1,200 professionals, managers, and support personnel. Exxon named the center the Client Support Center (CSC), because it felt that that name accurately reflected its use. The CSC staff consists of a secretary and four computing professionals who complement each other in the breadth of their computing and business skills.

The CSC's singular goal is to support user computing. This means that the CSC is user-oriented and attempts to accommodate its services to its users' business environment. Except for training, all services are offered on call during normal working hours. Even training is offered as soon as a class can be formed rather than on a fixed schedule.

The CSC provides consultation, training, and technical assistance to all levels of the headquarters staff. Consulting includes discussion to determine if an application is suitable for user computing and, if so, how best to do it. Sufficient training is provided to make the user comfortable with the software or procedure. The term *technical assistance* encompasses a broad range of on-call services aimed at keeping the user functioning and productive in computer work.

If the application is complex enough to require conventional development by systems professionals, the user is directed to the appropriate people in other parts of the computing organization.

CSC services include consulting on applications, justification, software selection, security and control, equipment ordering (paid for by the client), equipment setup, training, assistance, and equipment troubleshooting and maintenance. The CSC even provides advice to employees on the purchase of microcomputers for home use.

—From an article in *Datamation*

SUMMARY

- The term information center refers to direct user support. Support through the information center differs from traditional support in that users get direct support, information center staff are devoted exclusively to this service, and user help is immediate.

- The term backlog means an accumulation of work waiting to be done. This term is commonly used to refer to scheduled data processing work waiting for people resources. Invisible backlog refers to user needs that are not articulated because the users know how long the backlog already is.

- Most data processing departments recognize that user activity on microcomputers can relieve the data processing burden. They also recognize that users need support—hence, the information center.

- The head of the information center should report to the data processing manager. When setting up the information center, consider the issues of payment, physical facilities, and limiting the original center.

- Information center services include help with software selection, user programming, access to data, and hardware access and selection; education and training; technical assistance; and help with decision support systems.

- A tutorial is a disk with online step-by-step lessons on a given topic.

- Fourth-generation languages (4GLs) require no compile/link steps and are at least partly nonprocedural. Fourth-generation languages can be used to express a desired result, rather than telling how to do it. Concepts related to the use of fourth-generation languages are minimum skill requirements, minimum amounts of work, few errors, and ease of maintenance.

- Decision support systems (DSSs) further the power of spreadsheets by supplying statistical analysis and data modeling tied to conventional database, graphics, and reports packages.

- Information center personnel should blend a technical/business expertise with strong interpersonal skills.

- The information center must establish a billing system, probably based on percent usage.

KEY TERMS

Backlog
Decision support systems
  (DSSs)
Fourth-generation languages
  (4GLs)

Information center
Invisible backlog
Tutorial

REVIEW QUESTIONS

1. How does information center help differ from the usual help you might give a user?

2. Differentiate between backlog and invisible backlog. Do you think an information center will affect either of these? Both?

3. This chapter makes it plain that help is available to many users. What about data processing personnel? Do you think there may be computer professionals who know little about microcomputers and other information center services? Do you think there are computer professionals who know less than their users on these topics? Do you think the information center should provide this same type of advice for computer professionals who want it?

4. Do you think an information center would change the attitude of some users toward the data processing department? In what ways?

5. The chapter points out that the information center should have a central location. Why is this so important? What are the obstacles to achieving this goal? How might you handle resistance to a particular central site?

6. If you were setting up an information center, which services do you think you probably would want to put in place initially? How would you make final decisions on what those services should be?

7. There are many fourth-generation languages. We have not listed them in the book because products change rapidly. Can you mention some "brand-name" fourth-generation languages that you have heard of or seen advertised?

8. What problems might you anticipate from dissatisfied users?

CASE 17-1:
AND GOOD LUCK TO THE INFORMATION CENTER

Care must be taken to choose suitable employees for the information center. John Bender, as assistant manager of the data processing department at

Waldo Boat Builders, knew that well. He had been assigned the task of finding the very first staff person for the fledgling information center. The person he hired would set the tone for the entire operation.

Waldo had a company policy of hiring from within when possible, so the position description was posted internally and also advertised in the newspapers. After screening and narrowing down the list, John had three contenders for the job. Richard Yates was a pleasant but take-charge type of guy. He had a degree with a dual major in business and data processing, background experience in mainframes, and recent experience on micros and their applications. Virginia Erickson's experience was in producing and marketing microcomputer software. Virginia had been a teacher and had excellent communication skills. Gaston Prentice was a long-time programmer who began his career in assembly language in the fifties. After many years in semiretirement, he had recently been hired by the accounting department at Waldo to help them with their microcomputer applications and BASIC.

All three came with recommendations, but Gaston produced a glowing letter from Derek Mendenhall, the manager of the Waldo accounting department. In fact, Mendenhall, a fellow with clout, paid a personal visit to John, again extolling Gaston's virtues. He mentioned that he hoped John could hire someone who would be able to help out with accounting right away. He also mentioned that he hoped his department would be able to do more than its budget share to support the information center.

John, quite frankly, was puzzled. Gaston seemed like a throwback to the old days, and even spoke with an us-versus-them mentality. (We know this stuff and you don't.) John leaned toward Richard, mostly because of his broad background. John continued to hear from Mendenhall on a daily basis; one day he called twice.

John hired Gaston.

1. How good are you at sniffing out political scenarios? What is going on here?
2. Do you think this case is exaggerated? Why or why not?
3. What chance for success would you give the information center? Explain.

CASE 17-2:
WE AIM TO PLEASE

When Alicia VanBuskirk first took over the information center at the Northgate State Medical Center, she found it in a state of disuse. Although the original planner had given thought to hardware, software, and personnel, the service message had somehow not been received by the user organizations. There were no customers.

Alicia set out to remedy that problem. She decided to make marketing the information center her top priority for the next six months. She targeted potential users and obtained permission to make pitches to them in their weekly staff meetings. She had a professional brochure produced that described the information center services and had it disseminated widely.

When customers began to trickle in, they got the red carpet treatment. Time was not of the essence. Alicia spent many hours with each individual and, in some cases, made substantial contributions to their programs. When things got bogged down, she stayed late and even gave out her home phone number for consultation in off hours. This general picture remained for several months and even escalated somewhat.

By the end of the year Alicia had what she thought she wanted—more business than she could handle. In fact, users were starting to complain that service was too slow. For her part, Alicia would have welcomed two quiet weeks on a remote desert island—she was exhausted.

1. Where did Alicia's enthusiasm go astray?
2. This cannot continue. How can Alicia regain control of the information center?

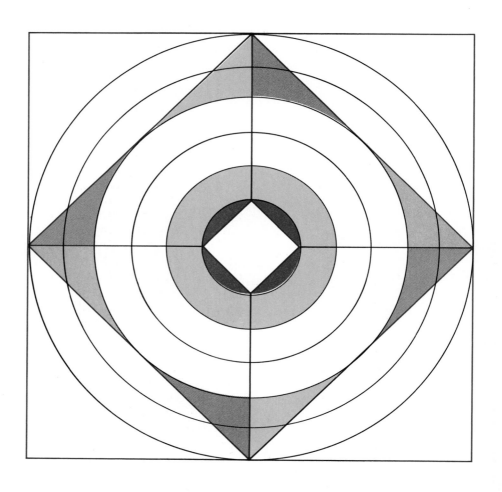

Chapter 18

Office Automation Systems

As editor of the *Journal-Guardian*, a thriving suburban newspaper, Art Turner was familiar with the use of computers in typesetting processes and for word processing in the office. He wanted to take automation a few steps further, but felt resistance from office staff and reporters. He decided to set himself up as an example.

Art purchased a microcomputer with all the trimmings, and set it up in his office at home. He pored over manuals and gradually became knowledgeable in the uses of his machine. Eventually he felt ready for a trial run, and brought the computer into his office at the *Journal-Guardian*.

The prevailing attitude at first was the-boss-has-become-a-computer-nut. The staff began to see the writing on the wall, however, when six similar machines showed up on the desks of Art's chief assistant, the production editor, and four reporters. Furthermore, Art had arranged to have the computers wired so they could send messages to each other. The Computer Six trudged to classes and

even read some manuals, but their real baptism came at a meeting three weeks later, when Art began discussing a subject with which they were not familiar.

They should have known about it, as it turned out, because Art had sent them a report via electronic mail. The information they needed was waiting inside their own computers. That got their attention. Communication was going to be through the computer. With Art leading the way, it was computer-communicate or perish.

There was no more foot-dragging at the *Journal-Guardian*. The office progressed from word processing to electronic mail to news database accessing in just over a year. These and other office computer-related technologies come lumped under the term office automation, often known as simply OA. **Office automation** is the use of technology to help achieve the goals of the office. There are many aspects to office automation: word processing, graphics, electronic mail, spreadsheets, executive calendars, database access, teleconferencing, and more. Few offices have all of these but many have some.

We shall examine all of them in this chapter.

◇ THE SEED: WORD PROCESSING

If you have a typewriter, be sure to save it. It will be a collector's item before long. Those of you who keep the white-out companies in business will never go back once you have tasted the joys of word processing. But aside from hiding your mistakes from the world, word processing offers real benefits to anyone who wishes to communicate on paper. Word processing was described in Chapter 4.

Word processing is so pervasive that few office workers offer serious resistance to it. Besides, a word processor looks and acts like a glorified typewriter. The benefits of word processing are well-established. The principal function of word processing in the context of office automation is as a lead-in to other office automation technology. Many consider word processing a good place to start. Let us now look at the wider ramifications of office automation, beginning with that most important person, the worker.

◇ THE OFFICE WORKER: KEY PLAYER IN THE DRAMA

It is only a matter of time before all office workers have electronic devices of some sort on their desks. They will be able to do a variety of things, from word processing to spreadsheets to graphics to accessing a database.

Workers and Their Needs

And just who is "the worker"? The secretary? Clerical workers? Professionals? Managers? Executives? Yes, all of these people. Office automation

BOX 18-1	Is There Something Electronic on Your Desk?

The ratio used to be 1:3 but now it is approaching 1:1; that is, soon every office desktop will have an electronic device. These devices include electronic typewriters, word processors, and mainframe computer terminals, but the real push will be in microcomputers. More than half of all workstations will be equipped with microcomputers by 1990.

may begin with the secretary, but this is a game that everyone can—and will—play. Office automation technology can be used by employees in the following ways:

- Secretaries and anyone else who produces text documents need word processing.
- Clerical employees need automation of information-handling tasks, such as capture, storage, retrieval, and update of information.
- Professionals need software support tools to help them analyze and process information.
- Executives need software tools to help them in the decision-making process.
- All employees need networking support to communicate more effectively.

Office automation may have begun with support personnel—secretarial and clerical—but it has moved on into the offices of professionals, managers, and even executives. And there also it has found success, although in a qualified way. The impact of office technology on professional workers has been perceived to be overwhelmingly positive. Office automation has raised both the quality and the amount of information professionals receive. Improved availability and the timeliness of vital data contribute to the improved content of reports and analysis. Lawyers, for example, can cross-reference case precedent information to help them plan their own case strategies. Sales representatives can coordinate sales campaign information on clients and sales territories. In addition to improved job performance, professionals report higher job satisfaction directly related to office technology tools.

BOX 18-2	**Where's the Power?**

One of the places you may look for power is the organization chart. That is a good place to start and will tell you something about the formal power, but there is also considerable informal power in any organization. Surely you have encountered the staff assistant or secretary or clerk who has little formal power but is widely acknowledged to be "in the know." How is that? Because that person somehow has access to information.

Now consider the ramifications of giving information to everyone. Or almost everyone. No secrets, no privileges. The balance of power shifts dramatically. If each person has a workstation and can access information for himself or herself, then dependencies diminish and individual power increases.

Vendors have long courted managers and executives, thinking that the way to the office is via the power conduit. Vendor pursuit of executives has been slowed down, however, by the fact that executives do not type, or do not want to type. This barrier—more psychological than real—is gradually being eroded. It is becoming socially acceptable for executives to type on their own office computers. Another device to combat the I-don't-type syndrome is the introduction of **icons**, small pictorial representatives of actions, on screens. An executive can move the cursor, for example, to the little picture of the garbage can, can push the keyboard or mouse button, and the file will be deleted. The use of icons was first mass marketed by Apple in their Macintosh® technology.

When should an executive consider putting office automation technology to work in the inner suite? Part of the hesitation has historically been that the machine would receive relatively low utilization. But this point is not valid when you consider the size of the executive's salary and, consequently, the value of the executive's time. The relatively modest investment for an executive workstation would soon pay for itself in terms of work hours saved. One hundred percent utilization of a workstation is no more valid than insisting that a telephone earn its keep by being used every minute of the day.

There are many reasons an executive might find the price of a workstation justified. These are related to the traditional executive tasks of planning, analyzing, evaluating, and decision-making. An executive may make a move to automation when he or she has a specific business activity to

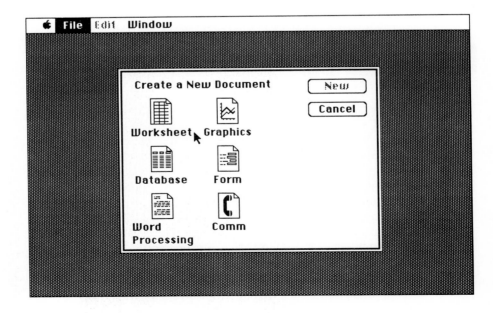

Icons on a screen. The arrow is the cursor.

support—for example, obtaining access to corporate sales information needed for planning an overall marketing policy. Perhaps the most important justification, however, is to become part of a community of office users, sharing data and communicating with each other.

All of the people mentioned in this discussion will interact in an automated office environment. And each, eventually, will have his or her own workstation.

The Workstation: Battle for the Desktop

You may be "keying at your workstation," but no longer will you be "typing at your desk." Note how the terminology is changing. As for what you will be keying on, that opens up a whole topic for discussion that could be titled "the battle for the desktop" (see Figure 18-1).

An astounding number of vendors are plying their wares, seeking to capture the desktop market. Each believes it has the formula that will make its device as common as the telephone. But, at their various workstations, workers are doing things that are related to their jobs: A secretary is preparing a report, a financial analyst is using a spreadsheet, a manager is using database access, and so forth. The various needs of users will inevitably have different features and characteristics, meaning that no vendor is going to be able to offer a single product that is perfect for everyone.

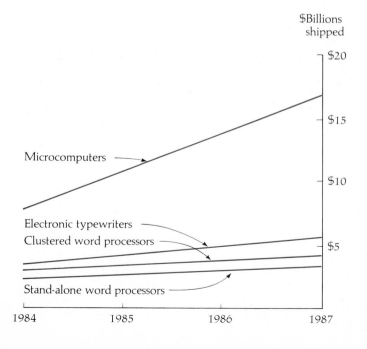

FIGURE 18-1

The electronic desktop. Without including microcomputers for personal home use, this chart shows that micros still outnumber all other types of word processing equipment combined. *Source:* International Data Corporation.

Personal computers
are used by all kinds
of office employees.

There is one characteristic of office work, however, that has influenced a wide range of software products—the idea that office work is done on an interrupted basis. This has lead to the desktop metaphor.

The Desktop Metaphor

Office software, goes the metaphor, should resemble a desktop, in that it should contain many different tools for work tasks, and that these tools will be used by one task interrupting another. There is a tremendous need to work on an interrupt basis. People often work that way. A user gets started on one project, gets interrupted, and then has to spend five minutes on another project.

When interrupted, a user may jump from one tool to another; this is analogous to a user using a desktop tool like a typewriter and then using a calculator and then a notebook. There is separate software for these tasks—

word processors, calculating programs, and file managers. These separate software applications are not so useful on an interrupt basis, however, because it takes time to make them available to the computer—and perhaps even more time if a disk must be located. Worse, using other software when interrupted makes you "lose your place" on the first project.

The basic answer to this problem is to bring together desktop-type tools in a single software package that presents them to users through windows. Windowing software, already discussed in Chapter 8, allows users to see several different outputs on different sections of the same screen. In fact, most windowing software allows a user to "zoom" in on one window and hide the others temporarily. This feature works well with the interrupt mode in which many users work. Most integrated software packages use a window concept.

A type of software package that specializes in an interrupt feature is usually billed under the generic heading "desktop organizer." These packages have also been nicknamed "pops" because the programs needed pop up on the screen. A typical package offers a calculator, notepad, desk calendar, and telephone dialer. In true desktop fashion, any of these tools can be used at any time, and in the middle of another project. At the stroke of a key, you can interrupt a project to use, for example, a calculator, and a window for the calculator will temporarily overlay your work in progress. When you are done with the calculator, another keystroke returns you to

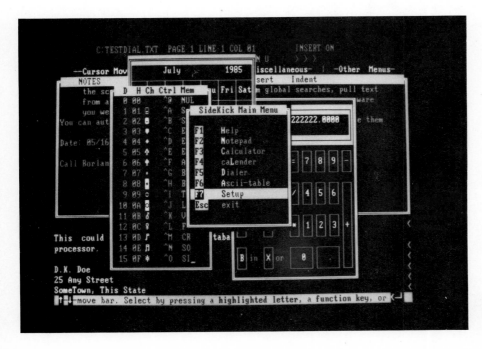

SideKick is a popular "desktop organizer" program.

where you were on the project. The desktop organizer software is always there, "underneath" whatever other software you are using. When you return to the original software, you can continue as if nothing had happened.

◇ OFFICE AUTOMATION IMPLEMENTATION

Implementing an office automation system is not like plugging in an appliance. The hardware and software employed dig right into the firm, shaping it in fundamental ways. Automating an office is a tough job and a lot of work.

The most common approach to office automation has been to begin with stand-alone word processing stations. This probably is not a good approach, however, because most organizations eventually want to establish a network of machines that can communicate with each other. Others are looking cautiously at the overall picture, carefully planning how office automation will be implemented.

Finding the Right Technology

Implementing an office automation system, like any other system, requires a series of steps and a careful planning pattern. Jumping in quickly with the latest technology is an example of what not to do. In fact, there may be nothing more detrimental to the implementation of an automated office than selecting technology before finding a user with a real need for it. Give a kid a hammer and everything's a nail. This solution-in-search-of-a-problem can only lead to disappointment. Consider the following situation.

BOX 18-3	**Steps for Successful Automation Implementation**

Plan strategies carefully.

Obtain the backing of higher management.

Recognize and understand the political issues in your organization.

Identify a need and meet it; do not wait until all needs are known.

Sell office automation to key office personnel.

Develop applications that will help key personnel.

Select a pilot group of enthusiasts.

Jackson Branch, a telecommunications manager, got a message from his company president, who had just toured the experimental video-teleconferencing facilities at an affiliate. Although one would think that electronics was no longer a fad, the president wanted the same "toy." With such support from on high, Jackson should have been delighted with this leading-edge-technology assignment. Unfortunately, Jackson was also accountable for the returns on technological investments. Thus, he was put in the position of being an internal salesperson, promoting this tool to departments, in order to get their financial contribution. So there we have it, a solution ready to go. All we need is the problem.

This example would seem ludicrous if it were not true. And it is but one example of putting the cart before the horse, of designing before analyzing. Life cycle steps apply to office automation systems, too. In any office automation study, user needs assessment must be the dominant theme.

Costs and Benefits of Office Automation

Installing office automation technology implies a much greater expense than just buying workstations. The most frequently used reasoning to justify purchases of office technology goes as follows: Labor costs are high and getting higher, while computer expenses are low and getting lower. So it follows that one should trade an expensive commodity, labor, for an inexpensive commodity, computers. But the economics of office automation is much more complex, partly because its introduction actually increases labor costs for a time. Let us examine the costs more closely.

The *obvious costs* begin with the price of the workstation, which includes the machine itself, and related items such as a printer, cables, hard disk, and modem. Software is the other obvious expense. Costs are also incurred for supplies such as paper, disks, printwheels, and printer ribbons.

Hidden costs begin with related workstation costs that may include a local area network connection, shared long-distance communications expenses, shared database management systems, and shared mainframe access. There also may be a need for changes in the office environment. Such changes might include improving illumination, rewiring the site for additional electric power, increasing the amount of air conditioning, and providing improved acoustics to reduce noise generated by the new equipment.

Startup costs begin with training. Initial training may take several days per person. Follow-up on-the-job training may continue for several weeks. Training costs include the salaries of both the trainers and the trainees, as well as training facilities. Hidden startup costs may include inefficiencies in serving customers by new and unfamiliar means, job interruptions due to unfamiliar procedures, and time spent in peer group meetings to negotiate changes in handling the work.

These costs are balanced by a set of benefits, in addition to the computers-instead-of-labor argument. The justification for office automation usually falls into the following categories.

- Work eliminated
- Improved decision making
- Improved services
- Competitive edge
- Improved quality of work life
- Cost avoidance

Examples abound. Work is eliminated when a secretary can make a small change to a document instead of retyping it. Accountants can improve decision making by using what-if scenarios on a spreadsheet. Customers get better service when the local area network can be used to check stock in the warehouse. The competitive edge can be sharpened when data accessed from the database can be displayed graphically at a sales presentation. Quality of work life is improved for workers who know how to use the computer to reduce or eliminate tedious and redundant tasks. And, as always, all these work advantages help avoid costs.

Office Automation Options

So, having considered technology, costs, and benefits, what will users need? Word processing, of course. Probably a communications network. Possibly microcomputer software packages for spreadsheets, graphics, and database management. Or, shall we up the ante and state their need for access to corporate databases? We can go further and talk about saving travel time and expense with an elaborate teleconferencing system. These options fall into two general categories: microcomputers and networking. We shall begin with microcomputers.

FIGURE 18-2

Acceptance of new technology. As the country's population has increased, so has the speed with which it accepts new technology. Penetration levels it took the telephone 60 years to reach were obtained by the TV in 20; the computer workstation will go the same distance in 10. *Source:* International Data Corporation.

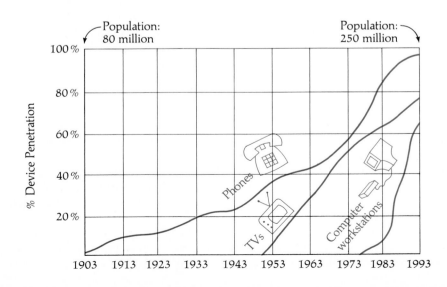

BOX 18-4	**Office Automation at the White House**

Office automation has reached the very innards of the White House, the executive office of the president. The White House staff uses work-stations to schedule Cabinet meetings, analyze the federal budget, and even place a digital version of the president's signature on presidential correspondence.

Office automation began at the White House as it did—and still does—for so many offices, with stand-alone word processors. These were later upgraded to provide graphics, electronic mail, and access to databases. The system now provides access to the White House mainframe computers for spreadsheet data and access to news wire services.

◇ INCORPORATING MICROCOMPUTERS INTO OFFICE AUTOMATION

As we indicated in Chapter 16, microcomputers have had a significant impact on the way people do business. In an office environment, managers know they must control the acquisition and use of micros, but they are not always sure how to do it. The influx of microcomputers in the office took many companies by surprise, with workers initially purchasing small computers and bringing them to the office before any companywide or even officewide policies had been set. Many of these computers are now sitting idle on office desks. Although there are many reasons for this, a primary reason is that the micro may not communicate very effectively with other people or machines in the office as it now exists. That is, compatibility may be a big problem for the office.

The Office Compatibility Issue

In applications like budgeting, microcomputers may be extremely helpful to the individual professionals they serve. Nevertheless, without any companywide coordination, users may have to spend more time thinking about compatibility than about their own business concerns. A user's budget process, for example, may need certain data that is residing in the files of another worker's micro, or perhaps it may need the combined output of these two micros plus the figures produced by the micro of still a third person.

Some corporations have tried dealing with their incompatibility problems with a simplistic hardware approach. Many companies are now recommending—or insisting—that equipment be purchased from a single vendor. This may be a solution in the short run. But, as we noted previously in the battle-for-the-desktop discussion, no vendor is going to have a single solution that works for everyone in the office. Also, locking into a single product or vendor is no guarantee that future office automation techniques will be available or cost-effective.

We have raised many questions here, but provided few answers. The problems continue to rankle. We can say this, however: An office automation strategy should take into account all aspects of office computing, not just the integration of newly arrived microcomputers. Microcomputers should be integrated into existing computing environments as fully as possible, not vice-versa. Part of that strategy, then, will involve the close association of office automation with the established data processing department.

Merging Office Automation and Data Processing Systems

The irony is that, in the beginning, office workers were lulled into believing that the new office technology had nothing to do with "real" computers. Vendors and managers thought it best not to frighten the workers. Patronizing as this attitude seems, now those same benefactors have to convince would-be users that the technology does indeed represent computing power in some way. That is, of course, true.

Each succeeding wave of office automation technology brings OA closer to traditional data processing territory. As the technology gets more complex, more offices turn for help to their systems analysts and other specialists. See Figure 18-3.

The traditionally separate areas of office automation and data processing must become more closely intertwined. A microcomputer, for example, cannot be classified as strictly an office automation tool or strictly a data

FIGURE 18-3

Overlapping technologies. Data processing, data communications, and office automation are intermingled, requiring communication among those with specific technical expertise.

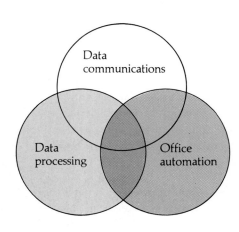

processing tool. Bringing micros into the organization requires that those boundaries be erased. One of the powerful reasons for a cooperative effort is mainframe access from office micros. The micro–mainframe connection, along with several other office features, is possible only with networking.

NETWORKING AND THE OFFICE ENVIRONMENT

Microcomputers allow each worker to do more work in a shorter period of time than ever before. But the work done will be of value only if it is in the context of the office's overall goals, and this often means shared resources. We discussed the impact of local area networks (LANs) in Chapter 16 on microcomputers. LANs are a key factor in sharing resources such as storage, printers, and data files.

But networking can be carried several steps further. Although studies have shown that managers initially approve hardware purchases for spreadsheet and other data-massaging purposes, they soon endorse the concepts of person-to-person communications as well. That next natural step is electronic mail.

Electronic Mail

Anyone who has made endless phone calls in an effort to connect with someone knows the frustrations of telephone-tag. With **electronic mail**, however, workers can use terminals or computers at their workstations to contact other workstations directly. Electronic mail releases workers from the tyranny of the telephone.

It has been hard for some companies to justify the cost of electronic mail systems. But communications experts can now document a startling amount of time wasted in trying to exchange information over the telephone. It takes an average of four tries to complete one phone call. To arrange a meeting among six business professionals can take as many as 100 phone calls! Perhaps there are people in the company who need to work together frequently, people whose communication has high time or money value. Yet these people find communication difficult because they are geographically dispersed or are too active to be reached easily. These people are ideal candidates for electronic mail.

Some people object to the name electronic mail, thinking that it conjures up images of parcel post or even the Pony Express. Nothing could be further from the truth. Perhaps electronic messaging would be a more appropriate name, but electronic mail is the name generally used in the industry.

Whatever it is called, the goal is the same: communicating electronically. A user can send messages to his or her supervisor downstairs, can direct a query across town to that person who is never available for phone calls, even send letters simultaneously to clients in Cleveland, Raleigh, and San Antonio. The beauty of electronic mail, or e-mail, as it is also called, is that a user can send a message to someone and know that the person will receive it. This method of sending mail works, of course, only if the intended receiver has an electronic mail facility to which the sender is connected.

There are several electronic mail options. A user can enlist a third-party service bureau that provides electronic mail service for its customers. Or, another popular option is to use a public data network such as Compu-Serve. Or a user may purchase an electronic mail package for a microcomputer or large computer system. Let us examine each of these.

Electronic mail had its origins in third-party service bureaus. A typical business user who employs such a service bureau has 10 to 20 remote locations that need to communicate with the central office. Sales representatives, for example, may have to keep in touch with headquarters regularly with customer inquiries and orders, or to get financial and inventory information.

The service bureau creates as many electronic "mailboxes"—space on their computer's disk storage—as needed for each user company. (See Figure 18-4.) Users get their mail after going through a typical log-on procedure requiring user identification and password. An interesting feature offered by some services is that users can check to see if their outgoing messages have been "read"—that is, if the recipient has queried the system for mail.

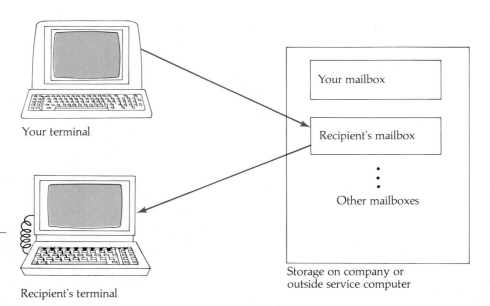

FIGURE 18-4

Electronic mail. One way to send mail from one user to another is through a third-party resource.

Your terminal

Your mailbox

Recipient's mailbox

Other mailboxes

Recipient's terminal

Storage on company or outside service computer

A user may be allowed to cancel any message already sent as long as the designated receiver has not read it. Thus, a brusque message sent in a fit of pique can be retrieved.

Public data services offer their own version of electronic mail. CompuServe users, for example, can send e-mail messages to any other CompuServe subscriber. Users get flashing messages when they log on if mail awaits them. After reading a message, a user can display it again, file it in his or her own storage, or delete it.

Finally, you could consider putting your own custom electronic mail software in place. Major hardware manufacturers are becoming more communications-oriented and are offering software or office systems packages that support e-mail. A key advantage of such packages is that they may be paid for just once; third-party services must be paid for on a continual basis.

Electronic mail users shower it with praise. It is impervious to time zones, can reach many people with the same message, reduces the paper flood, and avoids interruptions to take calls. It is not without limitations, however. The current problem is similar to the problem faced by telephone users a hundred years ago: It is not of much use if you have the only one. As usage of electronic mail escalates, so will its usefulness.

Voice Mail

"Reach out and touch someone" whether the someone is there or not. Here is how a typical **voice mail** system works. A user dials a special number to get on the voice mail system, then tries to complete a call by dialing the desired number in the normal way. If the recipient does not answer, the caller can then dictate his or her message into the system. The voice mail computer system translates the words into digital impulses and stores them in the recipient's "voice mailbox." Later, when the recipient dials his or her mailbox, the system delivers the message in audio form.

Voice mail was originally considered an extension of electronic mail. However, there is one big difference. To use electronic mail you and the mail recipient must have compatible devices with a keyboard, and be able to use them. In contrast, telephones are everywhere and everyone knows how to use them.

Senders can instruct some voice mail systems to redial specific numbers at regular intervals to deliver urgent messages, or simply set one delivery time and date. Another useful feature allows users to circulate messages among associates for comment. This method is far more efficient than circulating the traditional intraoffice memo.

There are some problems, however—not with the technology—but with user acceptance of the technology. Some people do not like talking to a machine. Others will not tell a machine anything substantive. A more serious problem is the lack of editing capability; most users simply cannot organize their thoughts as well verbally as they do in writing. Perhaps this is one of the reasons we are not yet the paperless society the media says we are becoming.

The Myth of the Paperless Society

When networking first began to have an impact on offices, there were predictions that we were moving toward a "paperless society." Why bother with paper, they asked, when most workers have access to terminals that can transfer data very quickly to a preselected audience. These predictions still linger, but the facts have turned out to be quite different. The total use of paper has actually increased.

Paper has decreased, it is true, for some uses. Already it is less expensive to store data on magnetic disk than on paper, and optical disks will be cheaper still. At the same time, networks can carry information at a rate equivalent to transmitting all the words of a very long novel in about a second. Clearly, the use of paper for simply storing and transmitting data is continuing to decline.

But the role of business documents as a management tool continues to gain importance. Paper documents can get attention. They are peppered with attention-getting words like NOTICE or RUSH or IMPORTANT. Papers can be placed in IN boxes or on office wall bulletin boards. Paper can be marked, underlined, scribbled on, and taken home.

The very existence of office automation systems gives more people access to information that can be used to create more activity-producing paper. Office automation, instead of replacing paper as the dominant business medium, has somehow enhanced it.

BOX 18-5	Guidelines for Ergonomic Workstations

Hardware. Provide detachable keyboards, tilt and swivel screens, and adjustable document holders.

Lighting. Avoid glare from windows. Use matte-finished keyboards to reduce glare. Control direct lighting.

Noise. Supply acoustic covers for printers. Place other noisy peripherals in an enclosed area.

Seating. Provide ergonomically designed chairs with a five-legged star base (to avoid tipping) and an adjustable back and seat. Leave plenty of leg room under the desk.

◇ TELECONFERENCING

An office automation wrinkle with great promise is **teleconferencing**, a method of using technology to bring people and ideas together despite geographic barriers. The promise has been there for years, but the acceptance is quite recent. The basic concept of teleconferencing is to let people conduct meetings with others in different geographic locations. Teleconferencing has been described as not just the next best thing to being there—sometimes it's even better.

There are variations on teleconferencing. The simplest is called simply **computer conferencing**, a method of sending, receiving, and storing typed messages as part of a network of users. Computer conferences can be used to coordinate complex projects over great distances and for extended periods of time. Participants can communicate at the same time or in different time frames, at the user's convenience. Conferences can be set up for a limited period of time to discuss a particular problem, as in a traditional office gathering. Or they can be ongoing networks for weeks or months or even years.

A computer conferencing system is a single software package designed to organize text-based communication. The conferencing software runs on a network's host computer, either a mini or mainframe computer. In addition to the host computer and the conferencing software, each participant needs a personal computer or word processor, a telephone and modem, and data communications network software.

In computer conferencing, everyone needs to be able to "talk" to anyone else. It is a many-to-many arrangement. Messages may be sent to a specified individual or set of individuals or "broadcast" to all receivers. Recipients are automatically notified of incoming messages. Some systems allow users to branch off into subconferences, leaving others undisturbed. Researchers working on a joint project, for example, can retreat from a general discussion to plan a report on their area of expertise.

What is a picture worth? Add cameras and you have another form of teleconferencing called **videoconferencing**. The technology varies, but the pieces normally put in place are a large screen (possibly wall size), three-dimensional cameras, and an online system to record communication among participants.

Although this setup is expensive to rent and even more expensive to own, monetary considerations seem trivial when compared to travel expenses for in-person meetings. A travel budget that includes airfare, lodging, and meals for a group of employees is formidable indeed.

Yet, again, there are stumbling blocks that are related more to people than to technology. Some see videoconferencing as another trend that will not happen, mostly because people do not like the look of themselves on camera. We tend to measure our appearance by our expectations from television, and balk when we envision slouching posture, ties askew, and stammering dialogue. There is also fear of the loss of personal contact, a genuine requirement for some business functions, especially those related to sales.

Employees are being trained to overcome their reluctance. Potential users are learning to project their personalities and ideas over the new medium. Some go as far as using professionally produced scripts for presentations such as new-product introductions. Videoconferencing may yet be an idea whose time has come.

◇ OFFICE AUTOMATION AND SECURITY

Office automation is presenting a new variety of problems in the security arena. Office systems and home systems have much in common, so both well-intended users and system abusers find that this compatibility leads to migration of resources to home systems. Another costly problem is that office workers are finding it easy to steal keyboards, modems, software, diskettes, and supplies. Creative fixes are coming to the market. One, for example, sounds a piercing alarm if a component is disconnected. This does not do much, however, about the software-laden floppy disks going out the door in briefcases.

Office systems are particularly vulnerable to security lapses because they use data that is in more finished form. Instead of the masses of detail data being massaged on mainframes, office systems are accumulating correspondence and summary information. Clearly, the need for security attention in office automation systems is paramount.

◇ THE PATH TO IMPROVED PRODUCTIVITY

Predictions for "improved productivity," the phrase that translates directly into cash, vary widely for office automation. Some experts go so far as to mention a possible 500% increase in productivity with the onset of an automated office. Since office automation itself is a generic term for a conglomeration of technologies, we are hard-pressed to say anything definitive about productivity.

Significant productivity savings are apparent nonetheless. The savings on word processing alone are clear to everyone. Office automation had a slower start than originally predicted but, partly because of productivity predictions, it has recently enjoyed an astounding growth pattern of over 30% a year.

◇ OFFICE WORKERS AND TECHNOLOGY:
    BLESSING OR BURDEN?

Few topics in the computer world have stirred as much discussion as office automation. This is because, in addition to the technology, there are social and humanistic issues to be considered in the automated office.

Concerns of office workers are being trumpeted loudly by labor unions, trade organizations, and the national press. Office automation systems have been accused of taking away jobs, reducing job functions, and being the cause of serious health hazards. Others view the automated office as the salvation for organizations mired in paperwork and bureaucratic communications breakdown.

Some workers see their jobs becoming more routinized through office automation, while others welcome the new skills as an opportunity for advancement. Some are afraid of the new technology, but the vast majority—according to repeated polls—report a positive outlook on office automation.

◇ AND SO TO WORK

This is the end of the book but not the end of your association with systems analysis. If you are attracted to the rewards of the field, we wish you well in your career. If you will be pursuing other endeavors, we hope you will feel comfortable in your dealings with the analysis process.

Avon Calling

In 1886, when sales representatives first began going door to door for Avon, the idea was revolutionary. In the first place, the sales product was not snake oil or books or a medical miracle—it was cosmetics. But the most radical departure was the sales force: women selling to their own friends and neighbors.

A hundred years later, Avon has become a fixture in American business. Avon Products, Inc., is now a $3 billion corporation with 30,000 office employees and over a million sales representatives. In order to support this enormous direct selling network, Avon has taken a strong initiative in communications and office automation.

The catalyst was the need for letter-writing. Avon needed frequent and rapid communications with its sales representatives. Avon began with a Wang® system, but the impact of a few operators and slow printers was minimal. The next step was to create an interface with Avon's IBM mainframe. The Wang system was successfully connected to an IBM laser printer. Upon installation of that system, letter output increased from 1,200 a month to 30,000 *per day*.

From that beginning, office automation has spread rapidly through the company. Virtually every department uses OA systems for word processing, file maintenance, tracking, and countless other applications.

The information services department, part of the data processing department, acts in an advisory role to anyone seeking a system. A group of managers meets weekly to discuss the many requests and dispense the work to the proper people.

The user environment has changed dramatically at Avon. Users know exactly what they want and are anxious to participate.

—From an article in *Computerworld Office Automation*

SUMMARY

- Office automation is the use of technology to help achieve the goals of the office. Some aspects of office automation are word processing, graphics, electronic mail, spreadsheets, executive calendars, database access, and teleconferencing.

- The benefits of word processing are well established. The principal function of word processing in the context of office automation is as a lead-in to other office automation technology.

- All office workers—secretaries, clerical employees, professionals, and executives—can use the tools of office automation. Eventually all of them will have workstations on their desks.

- The "battle for the desktop" is unlikely to be won by a single vendor with a single product because the needs of office workers are diverse.

- The desktop metaphor refers to software that allows a user to interrupt a project, use a window software desktop-type tool such as a calculator, and return to the project as if nothing had happened.

- Implementing an office automation system requires careful planning. There are various costs related to office automation. The obvious costs relate to the hardware, software, and supplies. The hidden costs relate to data communications and mainframe access, in addition to possible changes in the physical office environment. Startup costs include training and possible temporary changes in job performance.

- A problem with incorporating microcomputers into office automation is lack of compatibility among brand names, if the micros need to communicate or share data.

- The traditionally separate areas of office automation and data processing are becoming more closely intertwined.

- Using electronic mail, users can use their own terminals or microcomputers to contact other workstations directly. A user can enlist a third-party service bureau that provides electronic mail service for its customers, or use a public data network, or purchase an electronic mail package for a micro or large computer system.

- Using a voice mail system, a user dials a special number to get on the voice mail system and, if the recipient does not answer, dictates a message that is digitized and saved for the recipient, who later receives it in audio form.

- Although the "paperless society" has been predicted, the use of paper has actually increased. Paper use for storing and transmitting data has declined, but paper use as a management tool has increased.

- Teleconferencing uses technology to bring people together despite geographic barriers. Computer conferencing is a method of sending, receiving, and storing typed messages as part of a network of users. Participants may communicate at the same time or at their own convenience. Videoconferencing adds a visual dimension, since group participants are themselves on camera and can see their counterparts in another location.

- Security is an office automation problem in several ways; for example, hardware parts are taken for matching home systems, and diskettes are removed illegally from the premises.

KEY TERMS

Computer conferencing
Electronic mail
Office automation

Teleconferencing
Videoconferencing
Voice mail

REVIEW QUESTIONS

1. What is office automation? What are some of its usual features?

2. What kinds of office workers need office automation? Describe possible uses of office automation by these workers.

3. If you were using a spreadsheet, you might want to call up the calculator from "desktop organizer" software you have available. Can you devise similar scenarios where this type of software might be useful to you on an interrupt basis?

4. List and describe the kinds of costs associated with office automation.

5. Name and describe the different options for electronic mail. Under what circumstances might electronic mail be cost-justified?

6. Describe how voice mail works.

7. The paper debate has not diminished. Do you think the use of paper in the office will decrease as office automation becomes more widespread?

8. Contrast computer conferencing and videoconferencing. Which might you prefer, and why?

9. Describe some potential security problems in an office automation environment. Can you think of other possible problems not mentioned in the book?

10. What are some of the potential people-related problems associated with office automation?

CASE 18-1:
FEAR AND TREMBLING AT PALACE IMPORTS

Not everyone is clamoring for office automation systems. At Palace Imports, a diverse organization with 700 office employees at the headquarters office, operations officer Janet Boggs assigned her staff assistant, Hayes Mulford, to investigate the possibility of office automation. She told him she wanted a thorough study. In particular, she wanted him to check on similar companies, to see if they were using office automation technology effectively.

Hayes was already in favor of word processing for the secretaries, but he was suspicious of the technology that impacted decisions. He worried that increased communications and access to information would dilute control at the top. And, although he did not say so out loud, he saw a possible waning of power for himself—he would lose his special status as conduit to the boss.

Despite these reservations, Hayes decided to try to keep an open mind and pursued the job with vigor. He traveled widely and spoke with several managers about the systems they had in place. He found an amazing diversity. He also found almost universal enthusiasm, but in each case there were problems and fears as well.

Hayes prepared a lengthy report that included reasons for caution. The following themes were mentioned in the report: managers will bicker over OA control; we need to get our data processing systems in shape first; we do not have the technical expertise; our people are not ready for OA; our managers will be doing the work of secretaries; we have been successful without OA so why rock the boat; our people may use these tools for personal tasks; the next generation of OA tools will be better; if we wait OA technology will become less expensive; there always is a compatibility problem; OA technology opens up security problems. In summary, Hayes concluded that office automation was inappropriate for Palace Imports.

Janet Boggs cast a dim eye on the report. She thought—and said—that it looked like a list of alibis. She brought in several office automation vendors for a brief information session. They, of course, were able to counter all the arguments against office automation.

1. Can you suggest what the vendor arguments might have been?

CASE 18-2:
COBWEBS GROW ON THE NEW SYSTEM

A firm called Pension Funds Assistance (PFA) offers evaluation services to help corporations invest employee pension funds wisely. PFA has 175 employees serving 1,300 clients.

PFA management decided to consider an office automation system to improve productivity. In particular, they wanted systems to perform the many calculations done by desktop calculators. Adrienne Wright was assigned the task of planning and implementing such a system.

Adrienne observed the existing office systems briefly and then obtained approval for a system that would include word processing and fund performance calculations systems. Her system design was well planned and she managed the logistics of hardware installation hookups with a minimum of impact on the users. She supplied each user with software disks and the accompanying documentation.

There was a rather serious problem, however, one that did not come to light for a month or so after implementation: Productivity did not improve. After some investigation, it was discovered that the employees simply were not using the new system. The reasons given were varied.

1. Someone is going to have to investigate. What kinds of reasons is an investigator likely to find?
2. Where did Adrienne go wrong?

References

Adams, David R., Michael J. Powers, and V. Arthur Owles. *Computer Information Systems Development: Design and Implementation*. Cincinnati, OH: South-Western, 1985.

Awad, Elias M. *Systems Analysis and Design*. 2d ed. Homewood, IL: Irwin, 1985.

Barcomb, David. *Office Automation: A Survey of Tools and Technology*. Burlington, MA: Digital Press, 1981.

Beizer, Boris. *Software System Testing and Quality Assurance*. New York: Van Nostrand Reinhold, 1984.

Block, Robert. *The Politics of Projects*. New York: Yourdon, 1983.

Boar, Bernard H. *Application Prototyping*. New York: Wiley, 1983.

Brandon, Dick H., and Max Gray. *Project Control Standards*. Melbourne, FL: Krieger, 1980.

Brill, Alan E. *Building Controls into Structured Systems*. New York: Yourdon, 1983.

Brod, Craig. *The Human Cost of the Computer Revolution*. Reading, MA: Addison-Wesley, 1984.

Brooks, Frederick P., Jr. *The Mythical Man-Month*. Reading, MA: Addison-Wesley, 1978.

Buckle, J. K. *Managing Software Products*. Melbourne, FL: Krieger, 1984.

Burch, John G., Jr., Felix R. Strater, and Gary Grudnitski. *Information Systems: Theory and Practice*. New York: Wiley, 1979.

Calmus, Lawrence. *The Business Guide to Small Computers*. New York: Wiley, 1983.

Cameron, John R. *JSP and JSD: The Jackson Approach to Software Development*. Silver Spring, MD: IEEE Computer Society Press, 1983.

Cave, William C., and Gilbert W. Maymon. *Software Lifecycle Management*. New York: Macmillan, 1984.

Cecil, Paula B. *Office Automation*. Menlo Park, CA: Benjamin/Cummings, 1984.

Chow, T. S. *Software Quality Assurance: A Practical Approach*. Silver Spring, MD: IEEE Computer Society Press, 1985.

Clarke, Raymond T., and Charles Prins. *Contemporary Systems Analysis and Design*. Belmont, CA: Wadsworth, 1985.

Cleland, David I. and William R. King. *Project Management Handbook*. New York: Van Nostrand Reinhold, 1983.

Condon, Robert J. *Data Processing Systems Analysis and Design*. 4th ed. Reston, VA: Reston, 1985.

Cougar, Daniel J. *Advanced Systems Development/Feasibility Techniques*. New York: Wiley, 1982.

Curtis, Bill. *Human Factors in Software Development*. 2d ed. Silver Spring, MD: IEEE Computer Society Press, 1985.

Davies, D. W. and W. L. Price. *Security for Computer Networks*. New York: Wiley, 1984.

Davis, William S. *Systems Analysis and Design: A Structured Approach*. Reading, MA: Addison-Wesley, 1983.

Davis, William S. *Tools and Techniques for Structured Systems Analysis and Design*. Reading, MA: Addison-Wesley, 1983.

DeMarco, Tom. *Controlling Software Projects: Management, Measurement, and Estimation*. New York: Yourdon, 1982.

DeMarco, Tom. *Structured Analysis and System Specification*. New York: Yourdon, 1978.

Dickinson, Brian. *Developing Structured Systems: A Methodology Using Structured Techniques*. New York: Yourdon, 1981.

Dolan, Kathleen. *Business Computer Systems Design*. Santa Cruz, CA: Mitchell, 1984.

Dologite, D. G. *Using Small Business Computers*. Englewood Cliffs, NJ: Prentice-Hall, 1984.

Dunn, Robert, and Richard Ullman. *Quality Assurance for Computer Software*. New York: McGraw-Hill, 1982.

Edwards, Perry. *Systems Analysis, Design, and Development*. New York: Holt, Rinehart and Winston, 1985.

Evans, Michael W. *Productive Software Test Management*. New York: Wiley, 1984.

Fairley, Richard E. *Software Engineering Concepts*. New York: McGraw-Hill, 1985.

FitzGerald, Jerry, Ardra F. FitzGerald, and Warren D. Stallings, Jr. *Fundamentals of Systems Analysis*. 2d ed. New York: Wiley, 1981.

Frantzen, Trond, and Kenneth McEvoy. *A Game Plan for Systems Development*. New York: Yourdon, 1985.

Freedman, David M. *Designing Systems with Microcomputers: A Systematic Approach*. Englewood Cliffs, NJ: Prentice-Hall, 1983.

Fritz, James S., Charles F. Kaldenbach, and Louis M. Progar. *Local Area Networks: Selection Guidelines*. Englewood Cliffs, NJ: Prentice-Hall, 1985.

Galitz, Wilbert O. *Human Factors in Office Automation*. Atlanta, GA: LOMA, 1980.

Gall, John. *Systemantics*. New York: Simon and Schuster, 1977.

Gane, Chris, and Trish Sarson. *Structured Systems Analysis: Tools and Techniques*. Englewood Cliffs, NJ: Prentice-Hall, 1979.

Gessford, John Evans. *Modern Information Systems Designed for Decision Support*. Reading, MA: Addison-Wesley, 1980.

Gillett, Will D., and Seymour D. Pollack. *An Introduction to Engineered Software*. New York: Holt, Rinehart and Winston, 1982.

Glass, Robert. *Computing Projects Which Failed*. Seattle, WA: Computing Trends, 1977.

Gore, Marvin, and John Stubbe. *Elements of Systems Analysis*. 3d ed. Dubuque, IA: Wm. C. Brown, 1983.

Hansen, H. D. *Up and Running: Case Study of a Successful Systems Development Project*. New York: Yourdon, 1984.

Harmon, Paul, and David King. *Expert Systems: Artificial Intelligence in Business*. New York: Wiley, 1985.

Hayes-Roth, Frederick, Donald A. Waterman, and Douglas B. Lenat. *Building Expert Systems*. Reading, MA: Addison-Wesley, 1983.

Jackson, Michael A. *Principles of Program Design*. New York: Academic Press, 1975.

Jackson, Michael A. *Systems Development*. Englewood Cliffs, NJ: Prentice-Hall, 1983.

Johnson, Deborah G. *Computer Ethics*. Englewood Cliffs, NJ: Prentice-Hall, 1985.

Kacmar, Charles J. *On-Line Systems Design and Implementation*. Reston, VA: Reston, 1984.

Keller, Robert. *The Practice of Structured Analysis: Exploding Myths*. New York: Yourdon, 1983.

Kindred, Alton R. *Data Systems and Management: An Introduction to Systems Analysis and Design*. 3d ed. Englewood Cliffs, NJ: Prentice-Hall, 1985.

King, David. *Current Practices in Systems Development: Guide to Successful Systems*. New York: Yourdon, 1984.

Landreth, Bill. *Out of the Inner Circle*. Bellevue, WA: Microsoft Press, 1985.

Leeson, Marjorie. *Systems Analysis and Design*. 2d ed. Palo Alto, CA: SRA, 1985.

Licker, Paul S. *The Art of Managing Software Development People*. New York: Wiley, 1985.

Liebowitz, Burt H., and John H. Carson. *Multiple Processor Systems for Real-Time Applications*. Englewoods Cliffs, NJ: Prentice-Hall, 1985.

Lister, Timothy R. *Commonsensible System Implementation*. New York: Yourdon, 1985.

Martin, James. *An Information Systems Manifesto*. Englewood Cliffs, NJ: Prentice-Hall, 1984.

Martin, James. *System Design from Provably Correct Constructs*. Englewood Cliffs, NJ: Prentice-Hall, 1985.

Moder, Joseph J., Cecil R. Phillips, and Edward W. Davis. *Project Management with CPM and PERT*. New York: Van Nostrand Reinhold, 1983.

Murray, Thomas J. *Computer Based Information Systems*. Homewood, IL: Irwin, 1985.

Myers, Glenford J. *The Art of Software Testing*. New York: Wiley, 1979.

Neumann, Ahituv. *Principles of Information Systems for Management*. Dubuque, IA: Wm. C. Brown, 1983.

Newman, William M. *Integrated Office Systems*. New York: McGraw-Hill, 1984.

Nordbotten, Joan C. *The Analysis and Design of Computer-Based Information Systems*. Boston, MA: Houghton Mifflin, 1985.

O'Brien, James A. *Computers in Business Management*. 4th ed. Homewood, IL: Irwin, 1985.

Orr, Kenneth T. *Structured Systems Development*. New York: Yourdon, 1984.

Page-Jones, Meilir. *The Wrong Stuff: DP Management Guidelines*. New York: Yourdon, 1985.

Palmer, John F., and Stephen M. McMenamin. *Essential Systems Analysis*. New York: Yourdon, 1984.

Perry, William E. *The Micro-Mainframe Link*. New York: Wiley, 1985.

Peters, Lawrence J. *Software Design: Methods and Techniques*. New York: Yourdon, 1981.

Powers, Michael J., David R. Adams, and Harlan D. Mills. *Computer Information Systems Development: Analysis and Design*. Cincinnati, OH: South-Western, 1984.

Scott, George M. *Principles of Management Information Systems*. New York: McGraw-Hill, 1985.

Semprevivo, Philip C. *Systems Analysis*. 2d ed. Palo Alto, CA: SRA, 1982.

Semprevivo, Philip C. *Teams in Information Systems Development*. New York: Yourdon, 1980.

Senn, James. *Analysis and Design of Information Systems*. New York: McGraw-Hill, 1984.

Shneiderman, Ben. *Software Psychology: Human Factors in Computer and Information Systems*. Cambridge, MA: Winthrop, 1980.

Sime, M. E., and M. J. Coombs. *Designing for Human–Computer Communication*. New York: Academic Press, 1983.

Squire, Enid. *Introducing Systems Design*. Reading, MA: Addison-Wesley, 1980.

Stallings, William. *Local Network Technology*. 2d ed. Silver Spring, MD: IEEE Computer Society Press, 1985.

Teague, Lavette C., Jr., and Christopher W. Pidgeon. *Structured Analysis Methods for Computer Information Systems*. Palo Alto, CA: SRA, 1985.

Thierauf, Robert J., and George Reynolds. *Systems Analysis and Design: A Case Study Approach*. Columbus, OH: Charles E. Merrill, 1980.

Vick, Charles R., and C. V. Ramamoorthy. *Handbook of Software Engineering*. New York: Van Nostrand Reinhold, 1984.

Walsh, Robert J. *Modern Guide To EDP Design and Analysis Techniques*. Englewood Cliffs, NJ: Prentice-Hall, 1985.

Ward, Paul T. *Systems Development Without Pain: Guide to Modeling Organizational Patterns*. New York: Yourdon, 1984.

Ward, Paul T., and Stephen J. Mellor. *Structured Development for Real-Time Systems*. New York: Yourdon, 1985.

Weinberg, Victor. *An Introduction to General Systems Thinking*. New York: Wiley, 1975.

Weinberg, Victor. *Structured Analysis*. Englewood Cliffs, NJ: Prentice-Hall, 1980.

Wetherbe, James C. *Systems Analysis and Design: Traditional, Structured, and Advanced Concepts and Techniques*. 2d ed. St. Paul, MN: West, 1984.

Yourdon, Edward. *Classics in Software Engineering*. New York: Yourdon, 1979.

Yourdon, Edward. *Managing the System Life Cycle: A Software Development Methodology Overview*. New York: Yourdon, 1982.

Yourdon, Edward, and Larry L. Constantine. *Structured Design*. Englewood Cliffs, NJ: Prentice-Hall, 1979.

Zells, Lois. *Successful Project Management: A Practical Guide*. New York: Yourdon, 1985.

Glossary

Abnormal end debugger   A software tool that supplies information about a problem if the program makes an unscheduled stop.

Acceptance criteria   Set of statements agreed to by the analyst and user on which system acceptance will be based.

Acceptance testing   A subset of system testing used to show that data is behaving correctly in a system.

Access motions   Motions by which disk data is read or written; these include seek time, head switching, rotational delay, and data transfer.

Act fork   A point in a decision tree at which an action may or may not occur.

Activity   On a PERT chart, the use of resources over time.

Analog signal   An electronic signal that represents the form in which data is sent over telephone lines.

Analysis   Phase 2 of the systems life cycle; concerned with understanding the existing system and establishing the system requirements.

Analysis and requirements document   Document describing the existing system and specifying requirements for a new system.

Arbitration   A legally binding procedure, using experts as arbitrators, to settle disputes between two parties out of court.

Attribute   A column of the relation in a relational DBMS model.

Audio output   Computer-created "spoken" output.

Audit trail   A data protection technique by which data can be traced through a system.

Authority for action   One characteristic of an input form.

Backlog   The accumulation of work waiting to be done; commonly refers to scheduled data processing work waiting for people resources.

Backup file   A copy of a file made in case the original is lost or destroyed.

Balance   A concept used to make sure that lower levels of a data flow diagram are set up correctly by ensuring that any process could be replaced by its subdiagram.

Bar-code reader   A device that scans product identification codes and uses the data to store or retrieve data in the related computer system.

Benchmark   A formal performance test for hardware being considered.

Benefits   The tangible and intangible advantages of a new system.

Blank separators   An ergonomic consideration used to distinguish data groups.

Block   A physical record; a group of several logical records.

Blocking   The process of grouping logical records into a block.

Blocking factor   The number of logical records in a physical record.

Bootleg form   An unauthorized form.

Bottom-up coding   Program coding that begins with some complete lower level modules, while higher level modules are merely skeletons.

Breakeven point   The end of the payback period; point at which investment costs have been repaid.

Bus network   A LAN topology in which a portion of network management is assigned to each computer but the system is preserved if one component fails.

Captions   The brief instructions on a form.

Carrier of data   One characteristic of an input form.

Catalyst   A change agent role in which the systems analyst encourages the user to make discoveries and reach conclusions.

Change agent   The role an analyst plays when introducing new or revised systems.

Change theory   The theory that people often find change difficult but usually adapt eventually.

Channel   A set of like positions in a cross section of tape.

Checklist estimating   A method of estimating that uses a detailed breakdown of each task.

Chief programmer team   A systems team model that uses the system concepts of one person, while other team members have supporting roles.

Class check   An edit test that determines whether data in a field is numeric or alphabetic.

Closure   An ergonomic consideration used to give feedback to users at reasonable intervals.

Code generator   A software development tool that turns programmer input parameters into usable code.

Cohesion   The characteristic of a module with a single, well-defined function.

Cold site   An environmentally suitable empty shell in which a company can install its computer in the event of disaster.

Command language tester   A software testing tool that pretests command language instructions before they operate on programs and data.

Communications security   Measures used to protect data communications, including the use of passwords and cryptography.

Comparison check   An edit test that compares related items.

Completeness check   An edit test that determines whether required data items are present.

Computer conferencing   A method of sending, receiving, and storing typed messages as part of a network of users.

Computer system   A system whose components include one or more computers.

Confronter   A change agent role in which the systems analyst forces problem issues in order to bring about change.

Consortium   A joint venture by several organizations to support a complete computer facility in the event of disaster.

Consultant   A title used by a self-employed systems analyst.

Controlling function   A classic management function to monitor progress.

Convelope   Special forms that are mailable envelopes with pressure-sensitive report paper inside, which allows the report to be printed while inside the envelope.

Coordination   One of the three primary functions of the analyst, indicating cooperative work with other people and organizations.

Cost effectiveness   The relationship between development and operational costs and the benefits to be derived from a new system; based on this, the user organization decides whether it can afford a new system.

Coupling   The relationship between modules, which ideally should be weak.

Crash conversion   A system conversion approach in which the old system is switched immediately to the new system.

Critical path   The longest path, timewise, on a PERT chart.

Cross compiler   A software development tool that uses one computer to generate object code that will run on another computer.

Cryptography   The process of encoding data to be sent over lines and then decoding it at the other end.

Custom preprinted forms   Preprinted forms created especially for a particular use.

Cut forms   Output forms that are single sheets of paper.

Cylinder   A set of tracks on a disk accessible by one position of the access arms.

Data analysis   A basic function whereby gathered data is assembled in a meaningful way in order to show how the current system works, provide easily accessible reference material, and set forth the current system as a basis for future comparison with the new system.

Data capture   The procedure of inputting data to the computer for processing.

Data coding   The process of applying a unique identifier code to categories of data to aid in their retrieval.

Data communications   The use of computers to send data over communication channels.

Data content   The input data required.

Data control   Control of data's accuracy, scheduling, security, and auditing.

Data cross-reference chart   A chart on which data elements are cross-referenced against the forms on which they appear.

Data dictionary   A file of data about data, including a list of all data names associated with the system.

Data element listing   A chart made for each form and report, listing all the data elements in each.

Data flow diagram (DFD)   A diagram showing the flow of data through a system.

Data format   A description of the input data.

Data gathering   The process of finding out everything about the existing system, using written materials, interviews, questionnaires, observation, and sampling.

Data security   Measures taken to protect data from damage and unauthorized access.

Data transfer   The fourth of four access motions for reading and writing data directly to disk; actually moving the data to or from disk.

Database A collection of data that is organized to minimize redundancy and maximize access.

Database management system (DBMS) A set of programs that allow users to access a database in a logical way.

Decision support systems (DSSs) Software that furthers the power of spreadsheets by supplying statistical analysis and data modeling to conventional database, graphics, and reports packages.

Decision table A standardized table of the logical decisions to be made regarding the conditions and consequent actions that may occur in a given system.

Decision tree A graphic depiction of combinations of conditions that lead to certain actions.

Deferred online processing Processing in which transactions are accepted in real time but processed later.

Deliverable documents Documents that record the system's progress and mark milestone events.

Demodulation Conversion from analog to digital signals.

Density The number of characters per inch on a magnetic tape.

Design Phase 3 of the systems life cycle; concerned with planning the new system.

Design methodology A set of design techniques based on a concept.

Detail design document Document describing the input-process-output design in detail, as well as covering related topics.

Detail design phase Subphase of the design phase of the systems life cycle, when every facet of the new system is planned in detail.

Detail design review Review of the detail design of the system; the first review in the systems life cycle with a technical orientation.

Detail report A report listing every record of the file in use.

Development Phase 4 of the systems life cycle; the actual work required to bring the system into being, based on specifications prepared during the design phase.

Development costs One-time costs for all resources used to develop the system.

Development document Development phase document, consisting mostly of program listings and test results.

Digital signal The form in which data is used in computing devices.

Digitizer   A graphics input device that converts a picture to digitized data to be processed by the computer.

Direct file   A file whose records are stored so they can be accessed directly, that is, individually.

Directing function   A classic management function that involves leading and instructing subordinates.

Disaster recovery plan   A plan for restoring data processing operations in case of major damage or destruction.

Disk pack   A set of magnetic disks on a spindle.

Distributed data processing   A decentralized data communications system that includes processing in remote locations.

Distribution   The process of moving output from the place where it is produced to the place where it is used.

Dot-matrix printer   Printer that forms characters as a series of dots.

Double envelope method   A method used to improve the return rate of questionnaires where respondents are in-house; the outer envelope gives the employee name (for accountability) but the inner envelope is unmarked (for anonymity).

Downloading   The process of sending data from a minicomputer or a mainframe to a microcomputer.

Economic constraints   Financial limitations of the user organization within which the analyst must work.

Electronic mail   Mail sent by users directly from their own terminals or microcomputers to other workstations.

Equipment conversion   Converting old system equipment to new system equipment.

Ergonomics   The study of human factors in computing; concerned with the people/machine interface.

Event   On a PERT chart, the beginning or end of an activity.

Event fork   A point in a decision tree at which a decision must be made concerning an action that has occurred.

Exception report   A report that lists only out-of-the-ordinary circumstances.

Expert system   A software package that presents the computer as an expert on some topic.

Extended Binary Coded Decimal Interchange Code (EBCDIC)   A common data representation code.

Feasibility The reasonableness of a project considered from technical, operational, and cost aspects.

Feasibility study *See* preliminary investigation.

Federal Privacy Act A law stipulating that there can be no secret personal computer files.

File access One consideration in designing a file, concerned with the way the file will be used.

File activity One consideration in designing a file, concerned with frequency of use.

File characteristics Attributes of a file to be considered in file design.

File conversion Converting old system files to the new system.

File processing The process of planning separate data files to match program logic.

File volatility One consideration in designing a file, concerned with frequency of change.

File volume One consideration in designing a file, concerned with amount of data.

File On a data flow diagram, an element indicating a data repository and represented by an open-ended box.

Fixed length A characteristic of records on a file, in which all have the same length.

Form size One consideration in forms design.

Forms management group A group that monitors forms, assigning form names and numbers and keeping records of forms in use.

Forms specialists Form designers.

Fourth-generation languages (4GLs) Nonprocedural languages used to write programs quickly.

Freedom of Information Act A law allowing citizens to have access to data gathered about them by federal agencies.

Gantt chart A bar chart used to schedule system activities.

Generation basis A characteristic of batch systems in which prior versions of master files are kept for backup.

Goals of systems design The aims of a systems design, including suitability, reliability, ease of use, simplicity, and economy.

Graphics input Input in the form of graphs.

Graphics output   Output in the form of graphs.

Graphics pad   A graphics input device with a pressure-sensitive surface on which a stylus is applied.

Hard-copy graphics   Graphics displayed on a physical medium.

Hard-wired   An online system whose components are physically wired together.

Hardware acquisition   The process of acquiring hardware through purchase or lease.

Hardware requirements   Hardware required by the system.

Hashing   The process of finding a direct location for a record on a disk or locating a record already written directly.

Head switching   The second of four access motions for reading or writing data to disk; chooses the correct track on the cylinder.

HELP functions   Functions that give a user on-screen information related to the current activity.

Hierarchical network   A network that connects a central host to lower level computers that may themselves be hosts for smaller computers.

Hierarchy chart   *See* structure chart.

Hierarchy model   A DBMS model in which data elements are logically associated in a parent–child relationship.

Highlighting   A method for emphasizing material on a screen by making it brighter or having it blink on and off.

HIPO (hierarchy plus input-process-output)   A set of diagrams that describes system functions from a general to a detailed level.

Historical estimating   A method of estimating that uses data extrapolated from historical records.

Hit   An update to a file.

Hit rate   The percentage of records updated in a given time frame.

Host   The central computer in a network.

Hot site   A fully equipped computer center to be used in case of emergency.

Image file   A file of master file records before and after revisions.

Impact printer   A printer that physically strikes the paper with the printing element.

Impetus for change   Stimulus to change an existing system that comes from inside or outside sources.

Implementation   Phase 5 of the systems life cycle; the process of changing over to the new system.

Implementation planning   Planning for the implementation phase of the systems life cycle.

Imposer   A change agent role in which the systems analyst invokes higher authority in order to bring about change.

Indexed file   A file written in sequential order that has an index, thus allowing sequential or direct (through the index) accessing.

Information center   A generic term referring to an organized center for direct user support.

Input devices   Devices used to input data to a computer system.

Input form   A form that carries data to be input to a computer system; it has authority for action and is a visible record of data input and authority given.

Instructions   One consideration in forms design; must be concise and unambiguous.

Intangible benefits   Benefits that are difficult to measure.

Integrated software   Software that combines individual applications packages while coordinating the data among them.

Interblock gap (IBG)   The space between physical records (blocks) on a magnetic tape.

Interviews   A data gathering technique involving person-to-person interaction.

Invalid screen entries   Input errors, instructions, or data.

Invisible backlog   User needs that are not expressed because users know how long the actual backlog already is.

Join relational operation   A relational operation that combines two relations based on common attributes.

Journal file   A file of copies of transactions.

Joystick   A graphics input device that is manipulated by hand to control the cursor on the screen.

Key   A field chosen to identify a record; also, a key is used to order the records in a sequential file.

Key-to-disk device   An input device that keys data to disk for later use or for immediate online updating.

Key-to-tape device   An input device that keys data to tape in low-intensity batch systems.

Labeled vectors   An element in a data flow diagram indicating the flow of data and the type of data being carried; represented by arrows.

Laser-printed forms   Output forms created by laser printers.

Leighton diagram   A hierarchical diagram showing the relationship among processes and inputs, outputs, and files.

Letter-quality printer   Printer that forms solid characters of the same quality as a good typewriter.

Leveling   A system of breaking down data flow diagrams into smaller components that can be represented on separate sheets of paper.

Light pen   A graphics input device for drawing or altering lines on a screen.

Linear probing   During direct access, the process of searching nearby locations on a disk file for a place to put a synonym.

Local area network (LAN)   A network, usually situated in the same building, for sharing hardware, usually microcomputers, and information.

Logic analyzer   A software testing tool that tells the programmer about program logic errors.

Logical Construction of Programs (LCP)   A design methodology, invented by Jean-Dominique Warnier and Ken Orr, based on the concept of data structure.

Logical grouping of data   An ergonomic consideration used to keep similar data together.

Logical record   A record written by an applications program.

Magnetic disk   A secondary storage device that is a platter, often combined with others into a disk pack, covered with ferrous oxide, which can be magnetized to represent data.

Magnetic ink character recognition (MICR)   An input device used by banks to machine-read data from checks.

Magnetic tape   A secondary storage device with an iron oxide coating that can be magnetized.

Mailing labels   Address labels that can be computer printed.

Management authority   The go-ahead needed to begin and to continue a systems project.

Master file   A file that contains data records of a semipermanent nature.

Menu input   A screen design approach that gives users a series of choices.

Micro migration   The conversion of old systems to run on microcomputers.

Microfiche   4-by-6 inch sheets of film with 200 printed page images per sheet.

Microfilm   Special computer output on microfilm.

Milestone chart   A chart that presents tangible activities with a certain due date.

Modem   A device for converting digital signals to analog and vice versa.

Modular strength   Characteristic of a module that is highly functional.

Modulation   Conversion from digital to analog signals.

Module   A set of logically related program statements that perform a specific function.

Mouse   A graphics input device that is rolled by hand on a flat surface, causing corresponding cursor movement on the screen.

Multidrop line configuration   A line configuration that connects several terminals to one line, which is then connected to the computer.

Multiple forms   Forms that provide multiple copies.

Multistar network   A network connecting two or more star networks through their host computers.

Nassi–Shneiderman chart   A drawing that uses rectangles and triangles to depict process logic; also called structured flowchart.

Nature of the problem   The true definition of the problem—what the problem actually is—as opposed to its symptoms.

Net savings   Savings created by a new system minus the costs of the system.

Network model   A DBMS model similar to a hierarchy model, except that data elements can have more than one parent.

Node   Each microcomputer in a LAN.

Nonimpact printer   Printer that forms characters on the printed page without actually touching it.

Nonstructured interview   An interview that can vary considerably from the way it was originally planned.

Objectives   The goals of the system.

Observation   A data gathering technique in which the analyst watches the current system in action; particularly suitable for watching the flow of data.

Office automation   The use of technology to help achieve the goals of the office.

Online debugger   A software testing tool that allows the programmer to stop an online program at preplanned breakpoints to monitor the program's progress.

Online survey   A questionnaire read and answered on a computer screen.

Online systems Systems that interact directly with the computer.

Operational constraints Possible impediments to hardware or software needed to keep the system running.

Operational feasibility The possibility of a plan to meet the stated objectives.

Operations costs Recurring costs of running the completed system.

Operations documentation Documentation describing run times, inputs and outputs, delivery schedules and routing, files and retention, libraries, update schedules, backup and recovery procedures, and security controls.

Optical character recognition (OCR) An input device that scans and recognizes character data.

Optical mark recognition An input device that senses penciled marks on paper.

Optical recognition Input methods that use a light source to scan data and input it to the computer.

Optimizing compiler A software development tool that takes standard code and generates highly efficient object code.

Order and grouping of data One consideration in forms design.

Organization chart A hierarchical diagram of the formal relationships within an organization.

Organizing function A classic management function; assembling the material and personnel resources to carry out the vision of top management.

Packaged software Software that is purchased.

Parallel conversion A system conversion approach in which both old and new systems operate at the same time for a period of time, before changing to the new system.

Participant observation A form of observation in which the analyst temporarily joins the activities of the group.

Passwords A security measure to limit access to software and data to authorized persons.

Payback period The amount of time it takes to pay back the investment funds with net savings from the new system.

Peer group team A systems team model in which members are all equals.

Persuader A change agent role in which the systems analyst gradually convinces the user that the change is in his or her best interests.

PERT (program evaluation review technique) chart A chart showing a network of events connected by activities.

Phase conversion   A system conversion approach in which only part of a new system is introduced at a time.

Physical layout chart   A diagram of the organization work area and the paper flow within it.

Physical record   *See* block.

Physical security   Security of the computer system environment.

Pilot conversion   A system conversion approach in which a chosen group of users is the first to use the new system.

Pixel   Addressable point on the computer screen.

Planning function   A classic management function; envisioning and preparing for future needs.

Point-of-sale (POS) terminal   A cash register with automated features, including the ability to retrieve prices and to capture sales data.

Point-to-point line configuration   A line configuration that connects each terminal directly to the computer.

Political constraints   Potential constraints on the new system based on political power in the organization.

Precompiler   A software development tool that converts shorthand code to a full-blown version of the code in the appropriate language.

Preliminary design document   Document stating the concept for the new design and the rationale for its suitability; it also notes the resources needed to complete the plan.

Preliminary design phase   Subphase of the design phase of the systems life cycle where an overall design is planned that will meet system requirements.

Preliminary design review   Review of the design for management, presented by the systems analyst.

Preliminary investigation   Phase 1 of the systems life cycle; an investigation to determine the nature of the problem, its scope, and the objectives of the system.

Preliminary investigation report   Report discussing the nature and scope of a problem, stating system objectives, and suggesting a possible solution.

Problem definition   Determination of the nature of a problem, its scope, and system objectives.

Procedures conversion   Part of the conversion process in which old procedures are converted to ones that match the new system.

Process design   Design plan of system processes.

Processes   An element of a data flow diagram indicating actions taken on the data; represented by circles.

Progress report   A brief document that accounts for the project since the last report.

Project lead   *See* project manager.

Project management   The process covering system planning, priority setting, effective design, and awareness of project status.

Project management software   Software that allows a manager to monitor a project, including completion dates and resources used.

Project manager   The project leader; has both technical and managerial functions.

Project notebook   Notebook containing all pertinent project-related communications.

Project relational operation   A relational operation that selects certain attributes from a relation and eliminates any resulting duplicate tuples.

Project team   *See* systems team.

Prompt input   A screen design approach that asks users simple questions or prompts with a symbol.

Prototype   A limiting working system that is developed quickly using high-level software tools.

Pseudocode   An English-like way of expressing design that is more precise than English but less precise than a programming language.

Psychological constraints   Potential constraints on the new system based on psychological considerations, such as unwillingness of new users to accept it.

Quality assurance   A discipline that includes a set of tools and techniques to ensure that software is reliable and meets certain standards.

Query   Input that produces replies in a real-time environment.

Query language   An English-like interactive language used to access a database directly.

Questionnaires   A data gathering technique used to obtain information from a large number of people; although inexpensive, the types of questions are limited and the results are subject to misinterpretation.

Randomizing   *See* hashing.

Range check   An edit test that compares data to upper and/or lower limits.

Reasonableness check   An edit test that evaluates data against some reasonable standard.

Recovery file   General name for image and journal files.

Relation   A table in a DBMS relational model.

Relational model   A DBMS model in which data is organized logically in tables.

Relational operations   The processes that manipulate database relations.

Reliability   A characteristic of gathered data that indicates it is dependable enough to use.

Report generator   A software development tool that allows the programmer to write a shorthand code using special report parameters that is processed to produce usable code.

Request For Proposal (RFP)   A document for vendors specifying the computer equipment needs for the system.

Resolution   A screen output consideration based on the number of pixels available; the higher the number of pixels, the greater the screen resolution.

Response time   The time that elapses between user request and system response.

Reverse video   A method for emphasizing material presented on a screen by reversing brightness and darkness.

Ring network   A circle of connected computers with no host computer.

Risk analysis   A method of assigning a dollar value to potential loss and a percentage figure to the probability of that loss and multiplying to get the expected loss.

Risk factor   The likelihood of having a particular problem.

Robotics   The study of the design, construction, and use of robots.

Rotational delay   The third of four access motions for reading or writing data to disk; waiting for the disk to rotate until the correct track area is under the read/write head.

Rules   The lines that divide a form into various sections.

Sampling   A data gathering technique in which only a representative subset of the data is gathered.

Schema   The logical method of data organization in a database.

Scope   The range of a project; what the new or revised system is supposed to do.

Screen design   The format used to request input from users, including menu, prompt, and template.

Screen dump   The process of recreating screen graphics on paper.

Screen ergonomics The subject concerned with the impact of screen design on user comfort, accuracy, and productivity.

Screen output considerations Matters related to screen output including screen size, shape, resolution, and color.

Screen output forms Forms on which to design screen output.

Screen shape An output consideration that can be related to the application.

Screen size An output consideration sometimes related to the application.

Secured waste A data protection technique in which waste is destroyed or protected from theft.

Security A system of safeguards designed to protect computer systems and data from damage or unauthorized access.

Security software package Software designed for data file security.

Seek time The first of four access motions for reading and writing data to disk; positions the access arm over the correct cylinder.

Select relational operation A relational operation that selects certain tuples from a relation.

Sequence check An edit test that checks sequence order.

Sequential file A file whose records are ordered in some sequence by key.

Server The central computer of a star network.

Single entry One of the first goals in constructing a module.

Single exit One of the first goals in constructing a module.

Sinks An element in a data flow diagram indicating the destination for data going outside the organization and represented by a square.

Site preparation Site changes necessary for a major equipment installation.

Software security Protection of software from damage or illicit copying.

Software tool Any software product that can facilitate the programming process and improve productivity.

Solid character printer Printer that produces fully formed characters.

Sources An element in a data flow diagram indicating the origin for data outside the organization and represented by a square.

Spacing One consideration in forms design.

Specialist team A systems team model in which members include constant people plus specialists in particular areas.

Split carbon set A multiple form with partial carbon to duplicate only some data.

Split-screen editor A software development tool that allows a user to view and update two or more files on separate windows of the screen.

Staffing function A classic management function.

Standard formula estimating An estimating method that proposes weights for aspects of a project.

Star network A network that has a host computer connected to several subordinate computers at remote locations.

Structure chart A drawing that shows system design software components and their relationships.

Structured design A design methodology, invented by Larry Constantine, based on the concept of data flow, which takes a top-down modular approach.

Structured English A tool used to describe in detail the processes drawn in a data flow diagram.

Structured interview An interview in which questions are preplanned and no deviations are permitted.

Structured walkthrough A formal review of program design and code by a peer group.

Subschema The logical organization of data in a database for a particular user application.

Subsystems Components of a system that are systems themselves.

Summary report A report that presents an overview of the data.

Surge protector A device that prevents electrical problems from affecting data files.

Synonyms In direct access processing, identical disk addresses for two different records.

System A set of related components that work together to fulfill a purpose.

System candidate A possible alternative system.

System constraints Possible problems that may impact the new system.

System conversion Converting the old system to the new, using one of a variety of available approaches.

System evaluation The process of measuring the existing system against expectations.

System flowchart A drawing that shows the flow of control through a system.

System maintenance The continuing process of correcting, modifying, and improving the system.

System requirements   A precise list of user needs.

System savings   Savings created for the user organization by use of the new system.

System survey   *See* preliminary investigation.

System testing   Testing the flow of data through the entire system.

Systems analyst   Person who understands computer technology and the systems life cycle and who develops new systems.

Systems life cycle   A step-by-step process to develop a new system.

Systems team   The group of people—analysts, users, technical personnel—put together to work on a project; also known as the project team.

Table file   A file that contains reference data that is relatively static.

Tangible benefits   Benefits that are easy to measure.

Technical feasibility   The possibility of a plan to acquire needed hardware and software resources.

Technical writer   Person who writes systems documentation.

Telecommuting   Working at home but being connected to the office using a microcomputer or terminal.

Teleconferencing   Using technology to bring people together, including visual images, despite geographic barriers.

Teleprocessing   Accessing centralized computer power from a remote location.

Template input   Input to a template, which has the appearance of a printed form on the screen.

Template page   A template screen.

Test plan   The testing strategy.

Testing documentation   Documentation useful in planning the testing, measuring testing progress, and documenting the results of testing.

Time-stamped   An online system control in which the transaction generation time is added to each transaction.

Top-down approach   A key concept of structured design in which major software components are broken down into subcomponents, and so on.

Top-down coding   Program coding that begins with coding of the high-level modules and leaves the lower level modules to be filled in later.

Topology   The physical geometric layout of a local area network.

Touch screen   An input device that uses the touch of a finger on the screen to make selections.

Track   A set of like positions in a horizontal cross section of magnetic tape.

Training evaluation   Measuring trainee performance through a test or survey form.

Training tools   Tools needed in training, including microcomputers, terminals, printers, practice data, and so on.

Transaction file   A file that contains records with new data used to update master files.

Transaction processing   The process of accepting individual data transactions to retrieve and update data.

Tuple   A row of a relation in a relational DBMS model.

Tutorial   An online program that teaches a user how to use a software package or system.

Uncrowded screen   An ergonomic consideration used to increase the speed of data use.

Uninterrupted power supply (UPS)   A battery unit that prevents computers from crashing in case of a main power line failure.

Unit set   Multiple forms with tear-off copies.

Unit testing   The initial testing of a program or module, using test data.

Uploading   The process of sending data from a microcomputer to a minicomputer or mainframe.

User documentation   Written and/or online documentation that describes how software and systems work.

User involvement   The theory that users must be involved in a project from its inception; considered the key to a project's success.

User organization   The organization for whom the new or revised system is to be put into effect.

User review   A formal process in which work on the system is subject to user scrutiny.

Users   Employees of the organization needing a new or revised system.

Validity   A characteristic of gathered data that indicates that the questions asked were appropriate and unbiased.

Validity check   An edit test used to establish the validity of data.

Variable length   A characteristic of records on a file, in which the lengths vary.

Videoconferencing   A form of teleconferencing in which cameras are added for a visual dimension.

Vision recognition   Systems that use a camera to take a picture of an object, digitize it, and feed the input to the computer.

Voice mail   A system whereby a number is dialed and, if a recipient does not answer, a message is dictated, digitized, and saved for the recipient, who gets it later in audio form.

Voice recognition   Systems that convert voice input to digital code that can be accepted by the computer.

Volume testing   The process of sending large amounts of real data through a system to test its capacity and its ability to perform under realistic circumstances.

Wand reader   An input device used to scan products being sold and input their identification data.

Weighted criteria method   A formal method for comparing system candidates that weights desirable characteristics and computes a total for each system.

Window   A portion of the screen used to display one type of output while other parts of the screen display a different output or outputs.

Windowing   The ability to display several different outputs in different locations on the screen at the same time.

Word processing   Software that enables a user to easily save, revise, resave, and reprint computer-produced documentation.

Written materials   A data gathering technique in which all written materials relevant to the existing system are gathered together.

Credits and Acknowledgments

FROM THE REAL WORLD

Page 47, excerpt headed "Thinking Big at Ford" from *Computerworld*, March 19, 1985, p. 42. Reprinted with permission.

Page 71, excerpt headed "Stocks Online" from *PC World*, March 1984, p. 232. Reprinted with permission.

Page 92, excerpt headed "Computers on the Campaign Trail" from *PC Magazine*, March 1984, p. 193. Reprinted with permission.

Page 131, excerpt headed "Crime Online" from *Business Computer Systems*, February 1984, p. 108. Reprinted with permission.

Page 168, excerpt headed "Factory Management at Corning Glass" from *Softalk*, February 1984, p. 57. Reprinted with permission.

Page 193, excerpt headed "Computers at the Superbowl" from *Government Computer News*, February 1984, p. 18. Reprinted with permission.

Page 224, excerpt headed "Ergonomists Take on the Traders" from *Fortune*, March 1984, p. 60. Reprinted with permission.

Page 301, excerpt headed "Climbing High with Databases" from *Business Computer Systems*, August 1983, p. 31. Reprinted with permission.

Page 330, excerpt headed "Software Tools as Child's Play" from *Computerworld Special Report*, May 28, 1984, p. 58. Reprinted with permission.

Page 351, excerpt headed "Training First, Micros Second" from *Popular Computing*, February 1984, p. 75. Reprinted with permission.

Page 384, excerpt headed "PC Flies High with Airborne Freight" from *PC Magazine*, February 21, 1984, p. 249. Reprinted with permission.

Page 414, excerpt headed "Fire at the Westin Hotel" from *Computerworld*, January 9, 1984, p. 13. Reprinted with permission.

Page 439, excerpt headed "Computers Move into Real Estate" from *PC Magazine*, November 1983, p. 491. Reprinted with permission.

Page 456, excerpt headed "Information Center at Exxon" from *Datamation*, January 1984, p. 137. Reprinted with permission.

Page 482, excerpt headed "Avon Calling" from *Computerworld Office Automation*. Reprinted with permission.

PHOTOS

Page 28, © 1979 Herman Kokojan/Black Star.

Page 47, © 1983 Michael Hayman/Black Star.

Page 55, © Barbara Alper/Stock, Boston.

Page 71, © Christopher Morrow/Stock, Boston.

Page 92, © 1980 John Chao/Woodfin Camp & Associates.

Page 103, © Ann McQueen/Stock, Boston.

Page 111, UPI/Bettman Newsphotos.

Page 131, © Ellis Herwig/Stock, Boston.

Page 147, Sperry-Univac, a Division of Sperry Corporation.

Page 148, © Courtesy Lotus Development Corporation.

Page 168, Courtesy Corning Glass Works.

Page 181, Courtesy Xerox Corporation.

Page 182, Courtesy Radio Shack, a Division of Tandy Corporation.

Page 184, Courtesy Calcomp, a Sanders Company.

Page 193, © Peter Menzel.

Page 201, National Semiconductor Corporation.

Page 207, left, courtesy Lotus Development Corporation; right, courtesy Apple Computer, Inc.

Page 210, © Gordon Baer/Black Star.

Page 216, Hewlett-Packard Company.

Page 224, © 1984 Peter Menzel.

Page 266, © Ellis Herwig/Stock, Boston.

Page 330, © Peter Menzel.

Page 301, © 1982 Peter Menzel.

Page 313, © Hazel Hankin/Stock, Boston.

Page 341, © 1983 Frank Siteman/Stock, Boston.

Page 351, © Richard Sobol/Stock, Boston.

Page 365, © 1983 Jim Pickerell/Black Star.

Page 367, © Bill Gallery/Stock, Boston.

Page 376, both photos courtesy Microsoft.

Page 384, courtesy Airborne Express.

Page 400, © Dennis Brack/Black Star.

Page 401, courtesy Essex Technologies Corporation.

Page 406, courtesy Radio Shack, a Division of Tandy Corporation.

Page 414, © Susan Ylvisaker/Jeroboam.

Page 431, clockwise from upper left: courtesy Radio Shack, a Division of Tandy Corporation; courtesy Apple Computer; © 1985 Bruce Surber/courtesy Microsoft; courtesy Xerox Corporation.

Page 439, © Peter Menzel.

Page 451, courtesy Lotus Development Corporation.

Page 452, © John Blaustein/Woodfin Camp & Associates.

Page 456, courtesy Exxon Corporation.

Page 466, courtesy Lotus Development Corporation.

Page 468, clockwise from upper left: © 1981 Marc Pokempner/Black Star; © Donald Dietz/Stock, Boston; courtesy Apple Computer; © 1984 Andy Levin/Black Star.

Page 469, courtesy Borland International.

Page 482, courtesy Avon Products, Inc.

TRADEMARKS

Apple is a registered trademark of Apple Computers, Inc.

Campaign Manager is a trademark of Aristotle Industries.

Chryon is a trademark of the Chryon Corporation.

Cyber 176 and Cyber 760 are trademarks of Control Data Corporation.

DEC is a registered trademark of Digital Equipment Corporation.

Huttonline is a trademark of E. F. Hutton.

IBM PC, IBM 360, and IBM 3083 are registered trademarks of International Business Machines, Inc.

Job Control Language (JCL) is a registered trademark of International Business Machines, Inc.

KayPro is a registered trademark of Kaypro Corporation.

Macintosh is a registered trademark of Apple Computers, Inc.

Materials Requirement Planning (MRP) is a trademark of Black and Decker, Inc.

OCH is a registered trademark of International Business Machines, Inc.

RAMAC is a registered trademark of International Business Machines, Inc.

Solon is a trademark of Q Systems Research Corporation.

System-80 is a trademark of Phoenix Systems, Inc.

VAX is a registered trademark of Digital Equipment Corporation.

Wang is a registered trademark of Wang Laboratories.

Index